Expander Families and Cayley Graphs

Consider the two graphs here. Each has 46 vertices, and each has 3 edges per vertex. Which one would make a better communications network, and why?

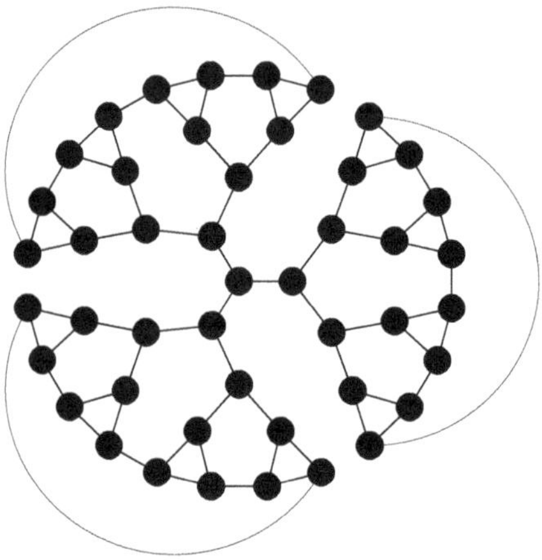

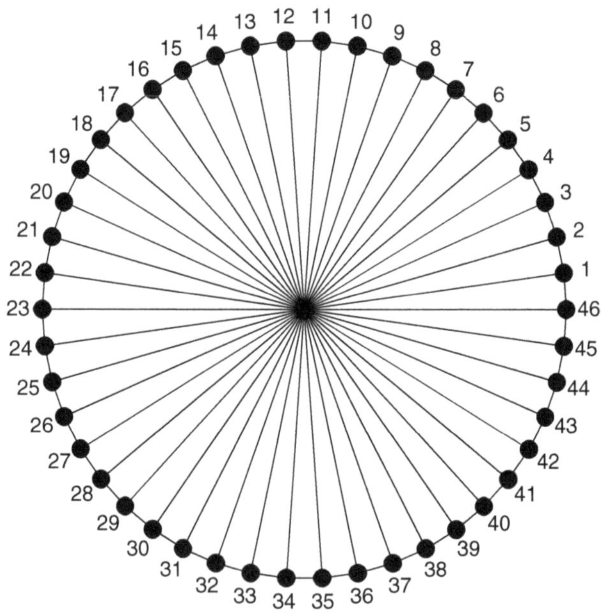

See the Introduction for more details and a discussion of how this question leads us naturally into the fascinating theory of *expander families*.

Expander Families and Cayley Graphs

A Beginner's Guide

MIKE KREBS

AND

ANTHONY SHAHEEN

OXFORD
UNIVERSITY PRESS

OXFORD
UNIVERSITY PRESS

Oxford University Press, Inc., publishes works that further
Oxford University's objective of excellence
in research, scholarship, and education.

Oxford New York
Auckland Cape Town Dar es Salaam Hong Kong Karachi
Kuala Lumpur Madrid Melbourne Mexico City Nairobi
New Delhi Shanghai Taipei Toronto

With offices in
Argentina Austria Brazil Chile Czech Republic France Greece
Guatemala Hungary Italy Japan Poland Portugal Singapore
South Korea Switzerland Thailand Turkey Ukraine Vietnam

Published by Oxford University Press, Inc.
198 Madison Avenue, New York, New York 10016
www.oup.com

CIP to come

ISBN-13: 978-0-19-976711-3

CONTENTS

PART TWO Combinatorial Techniques

PART THREE Representation-Theoretic Techniques

PREFACE

ABOUT THIS BOOK

This book provides an introduction to the mathematical theory of expander families. It is intended for advanced undergraduates, graduate students, and faculty.

The prerequisites for this book are as follows. No graph theory is assumed; we develop it all from scratch. One course on introductory undergraduate group theory is assumed. One course on linear algebra is assumed. We provide Appendix A as a refresher on linear algebra for those who need it. In particular, those who are reading the book should either know the spectral theorem (Theorem A.53) or be willing to take it for granted. With these parameters, the book is self-contained. Analysis is helpful but not necessary; we occasionally encounter statements of the form "For all $\epsilon > 0$, there exists N such that if $n \geq N$, then ... " Appendix B gives a review of big oh notation and the limit inferior of a sequence. This appendix uses the definition of the limit of a sequence. At one point in Chapter 8, we use the fact that a continuous function obtains its maximum on a compact subset of $\mathbb{C}^n$. In Section 6 of Chapter 7, we use the fact that the multiplicative subgroup of the set of integers modulo a prime is cyclic.

Figure P.1 gives a flow chart for the book. For example, Chapter 1 is a prerequisite for every other chapter. To read Chapter 7 one must first read Chapters 1 and 6.

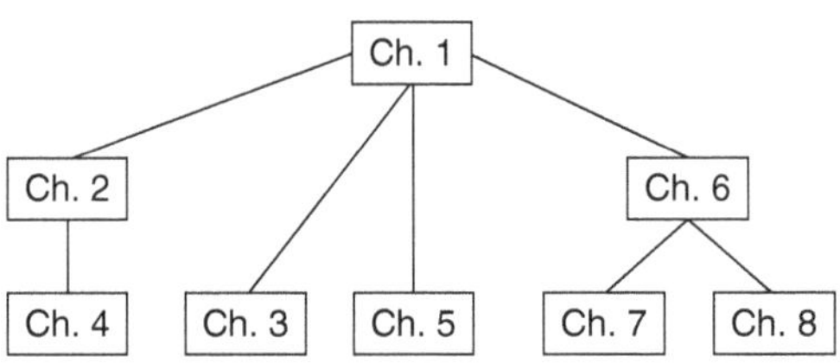

Figure P.1

We make several points with regard to Figure P.1.

1. Two concepts briefly introduced in Chapter 3 are the definition of Ramanujan graph and the notation $\lambda(X)$, both of which are used often in later chapters. We have not indicated this dependence in Figure P.1.

When readers encounter these objects, they can quickly read the definition in Chapter 3—there is no need to read all of Chapter 3.

2. The sections on subgroups and quotients in Chapters 4 and 8 make use of concepts and notations from Chapter 2 (especially the section on Schreier generators).
3. Chapter 5 may be read directly after reading Chapter 1. However, if one wants to read Section 5 of Chapter 5, one must learn the definition of the semidirect product. This definition is given in Section 6 of Chapter 4.
4. Chapter 7 uses only Section 1–4 of Chapter 6. Section 6 of Chapter 6 is used only in Section 4 of Chapter 8.

Each chapter of the book contains a Notes section, where we provide additional information on expander families and related material. We also give references to the literature. The Notes sections are extensive, and we encourage the reader to browse through them.

Many chapters end with a list of student research problems. These problems are intended for independent studies, Research Experience for Undergraduates programs (REUs), and research projects for theses or capstone projects. The problems can be attacked by advanced undergraduates, graduate students, and faculty interested in the area.

The authors have used this book for several courses. In a 10-week course for advanced undergraduates and master's students we covered the following: all of Chapter 1; Section 1 and Section 2 of Chapter 3; Section 1, Section 2, and Section 3 of Chapter 4; and all of Chapter 5. Shaheen used the textbook in a 10-week undergraduate capstone course. In that course, he covered some of Chapter 1. Shaheen used the course for several undergraduate and independent studies courses. In one course he covered Chapter 1 and Chapter 3; in another he covered Chapter 6; in another he covered Chapter 6 and Chapter 8.

ACKNOWLEDGMENTS

We thank Franque Bains, Mai Barker, Anthony Caldera, Dwane Christensen, Marco Cuellar, Keith Dsouza, Harald Flecke, Jennifer Fong, Salvador Guerrero, Dan Guo, Carrie Huang, Marcia Japutra, Karoon John, Keith Lui, Sui Man, Maritza Noemi Mejia, Seth Miller, Gomini Mistry, Alessandro Moreno, Erik Pachas, Tuyetdong Phan-Yamada, Novita Phua, Odilon Ramirez, Sergio Rivas, Andrea Williams, Lulu Yamashita, and May Zhang for many helpful and insightful comments, as well as for suffering through 10 weeks of lectures that revealed the glaring need for this assistance. We acknowledge David Beydler for pointing out the clever trick in Exercise 18. Shaheen thanks his Math 490 class (Hakob Antonyan, Marcia Japutra, Andrea Kulier, Todd Matsuzaki, Cruz Osornio, Jorge Rodriguez, Manuel Segura, Matt Stevenson, Quang Tran, and Antonio Vizcaino) for reading through the first draft of the algebraic graph theory chapter; his independent studies class (Jeff Derbidge, Novita Phua, Udani Ranasinghe, and Sergio Rivas) for reading through the first draft of the Kazhdan constant chapter; and his student Isaac Langarica for reading an early draft of several chapters. We thank Avi Wigderson for generously sharing some LaTeX code.

NOTATIONS AND CONVENTIONS

If V is a set, we write $S \subset V$ to indicate that S is a (not necessarily proper) subset of V.

If S and V are multisets, then the notation $S \subset V$ will indicate that the multiplicity of any element of S is less than or equal to its multiplicity as an element of V.

Let A and B be sets. We write $A \setminus B$ for set difference; that is $A \setminus B = \{x \in A \mid x \notin B\}$.

If X, Y, and Z are sets with $X \subset Y$ and $f : Y \to Z$, then we use the notation $f|_X$ to denote the restriction of f to X.

If G is a group, we write $H < G$ to indicate that H is a subgroup of G and $H \lhd G$ to indicate that H is a normal subgroup of G.

We write $G = 1$ to indicate that G is the trivial group.

We denote the group of integers modulo n under addition by $\mathbb{Z}_n$.

We denote the sets of integers, real numbers, and complex numbers, respectively, by $\mathbb{Z}$, $\mathbb{R}$, and $\mathbb{C}$.

Let x be a real number. We write $\lfloor x \rfloor$ for the greatest integer less than or equal to x. For example, $\lfloor 1.64 \rfloor = 1$.

We sometimes write $\exp(z)$ for e^z.

If $\sigma, \tau \in S_n$, then we multiply as follows: $\sigma\tau = \sigma \circ \tau$. We write elements of S_n in cycle notation, using commas to seperate the entries. For example, $(1, 2)(1, 3) = (1, 3, 2)$.

The dihedral group is given by

$$D_n = \{1, r, r^2, \dots, r^{n-1}, s, sr, \dots, sr^{n-1}\}$$

where $r^n = 1$, $s^2 = 1$, and $rs = sr^{-1}$.

We use ^{t} to denote the transpose of a matrix. To save space, we write column vectors as the transpose of a row vector; that is,

$$(x_1, x_2, \dots, x_n)^t = \begin{pmatrix} x_1 \\ x_2 \\ \vdots \\ x_n \end{pmatrix}.$$

INTRODUCTION

1. WHAT IS AN EXPANDER FAMILY?

A graph is a collection of dots (*vertices*, singular *vertex*) and connections (*edges*) between them. (We make this definition precise in Chapter 1; for now, our discussion will be informal and intuitive.) Figure I.1 shows a graph Y, and Figure I.2 shows a graph Z.

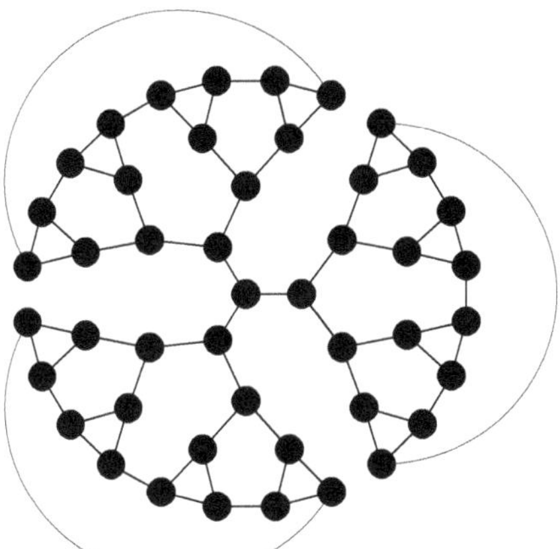

Figure I.1 A 3-regular graph Y with 46 vertices

We introduce a way of thinking about graphs that will help us develop many of the mathematical concepts in this book. Regard a graph as a communications network. The vertices represent entities we wish to have communicate with one another, such as computers, telephones, or small children holding tin cans. The edges represent connections between them, such as fiber optic cables, telephone lines, or pieces of string tied to the cans. Two vertices can communicate directly with one another iff there is an edge that runs between them (i.e., if the two vertices are *adjacent*). Communication is instantaneous across edges, but there may be delays at the vertices (because that's where the humans are). The distance

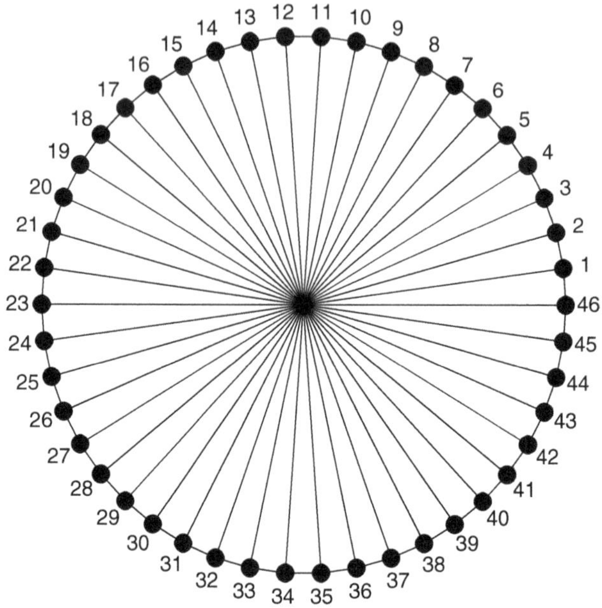

Figure I.2 A 3-regular graph Z with 46 vertices

between two vertices is irrelevant. Likewise, whether two edges appear to cross is irrelevant. (We can always run one wire above another, putting some space between them.) We would like to have a large number of vertices, so that many people can communicate with one another. However, cables are expensive, so we want to get by with as few edges as possible.

Suppose we are told to design a communications network with 46 vertices such that each vertex is adjacent to exactly 3 other vertices (in which case, we say that the graph is 3-*regular*). One way to do this would be as in the graph Y in Figure I.1. An alternative is to use the graph Z in Figure I.2. (Note that in the graph Z, there is no vertex in the center; this is simply an illusion caused by all the edges that appear to cross there. Each vertex in Z is adjacent to the two next to it as well as the one directly across from it. Vertex 3, for example, is adjacent to vertices 2, 4, and 26.)

Of the graphs Y and Z, which one is a better communications network, and why?

(We realize that this question is not precise, for we have not told you what "better" means. Indeed, the point here is for *you* to think about what constitutes a "better" communications network. Enjoy this opportunity to take a stab at a math question that truly has no right or wrong answer!)

We could make a strong case for Y by arguing that it is a *faster* network than Z. In Y, we can get from any vertex to any other vertex in no more than eight steps, by first going to the center of the graph and then back outward. To get from vertex 1 to vertex 12 in Z, however, requires at least 11 steps.

On the other hand, one could argue equally well that Z is the better communications network, for it is *more reliable* than Y. If we cut just one of the edges in Y attached to the center vertex, then the graph becomes disconnected: 15 people will have no way to communicate with the other 31. In contrast, you can cut any edge in Z, and the graph will remain connected, that is, any two vertices can still send messages to one another. (In fact, cutting any *two* edges in Z still leaves behind a connected graph.)

Perhaps to the reader's dismay, we leave unresolved the question of whether Y or Z is the superior communications network. Discuss the matter at your own peril; we take no responsibility for any barroom brawls that may ensue. Our intention in asking this question was merely to find some natural ways of measuring a graph's quality as a network, and now we have two such ways, namely, its speed (as measured by the minimum number of edges required, even in the worst case, to get from one vertex to another, i.e., the graph's *diameter*) and its reliability (as measured by the minimum number of edge cuts needed to disconnect the graph, i.e., its *edge connectivity*).

As it turns out, a single quantity gives us information about both the speed and the reliability of a communications network. Let's probe the graphs Y and Z a bit more deeply to see a sense in which we can view diameter and edge connectivity as two sides of the same coin.

Consider the set V_n of vertices we can get to in no more than n steps from a fixed base vertex. If a communications network is to be fast—that is, if the graph is to have small diameter—then we expect to have many edges from V_n to its complement. In that way, the sets V_n grow rapidly as n increases; in other words, we can reach many points in relatively few steps. (One technicality: we must restrict our attention to sets V_n which contain no more than half of the graph's vertices, or else we will start running out of vertices in the complement.)

Let's see what goes wrong in our "slow" graph Z. Consider the set S of vertices we can reach in no more than four steps, starting at vertex 1. Then $S = \{1, 2, 3, 4, 5, 21, 22, 23, 24, 25, 26, 27, 43, 44, 45, 46\}$, as shown by the white vertices in Figure I.3. Note that while S contains sixteen vertices, there are only six edges going from S to its complement. (These are the six dashed edges in Figure I.3. They connect 5 and 28, 20 and 43, 5 and 6, 20 to 21, 27 to 28, and 42 and 43.) The graph Z would make a slow network because it contains this set S with many vertices but relatively few edges from S to its complement.

Now, what makes Y unreliable? Let S be the set of 15 white vertices in Figure I.4. Then there is only one edge going from S to its complement. (It's the dotted edge in Figure I.4.) Cutting this edge disconnects those 15 vertices from the rest of the graph. The graph Y would make an unreliable network because it contains this set S with many vertices but relatively few edges from S to its complement.

Our discussion suggests a graph invariant that we may wish to measure. For the reason cited earlier, we restrict ourselves to those sets containing no more than half of the graph's vertices. For all such sets S, we take the minimum ratio of the number of edges between S and its complement to the size of the set S. This invariant is called the *isoperimetric constant* of the graph X and is denoted $h(X)$.

Let's now consider an infinite family of 2-regular graphs: the cycle graphs C_n, some of which are depicted in Figure I.5.

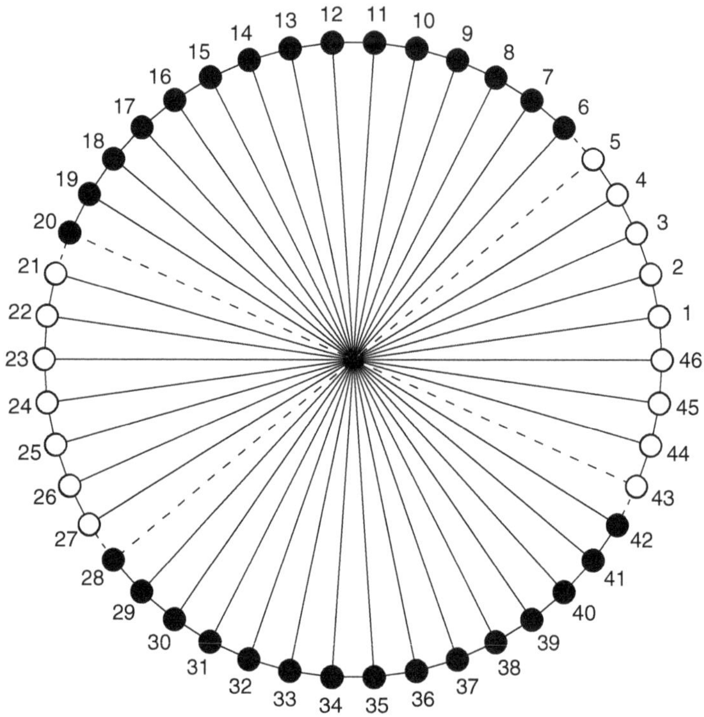

Figure I.3 Graph *Z* and set *S* with dotted lines for boundary

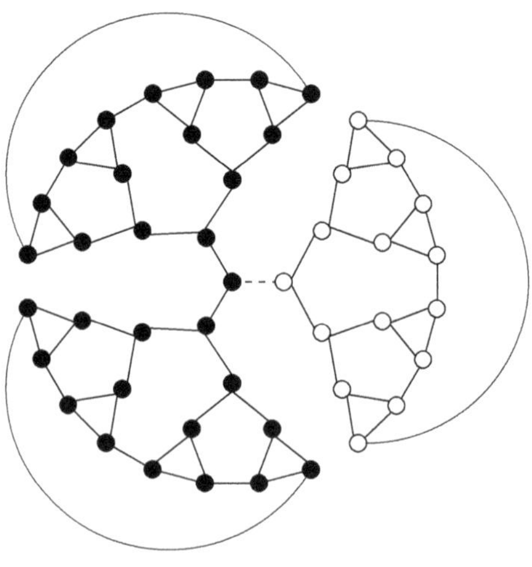

Figure I.4 Graph *Y* and set *S* with dotted lines for boundary

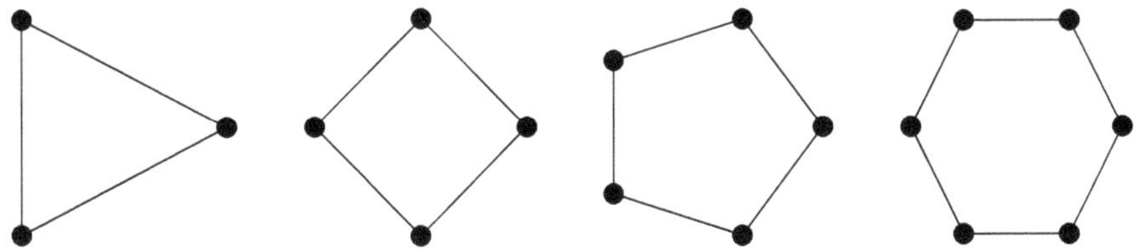

Figure I.5 The cycle graphs C_3, C_4, C_5, and C_6

Take the set S to be the "bottom half" of vertices in C_n. For example, the white vertices in Figure I.6 give S for C_8. (We assume for now that n is even.) Then S contains $n/2$ vertices, and there are two edges going from S to its complement. So we get the ratio $2/(n/2) = 4/n$. This ratio may or may not be the minimum, so $h(C_n) \leq 4/n$ for n even. A similar argument shows that $h(C_n) \leq 4/(n-1)$ for n odd. Hence we see that as $n \to \infty$, the isoperimetric constant of the cycle graph C_n goes to 0. This fact matches well with our intuition that the quality of the cycle graphs as communications networks becomes poorer as they become larger.

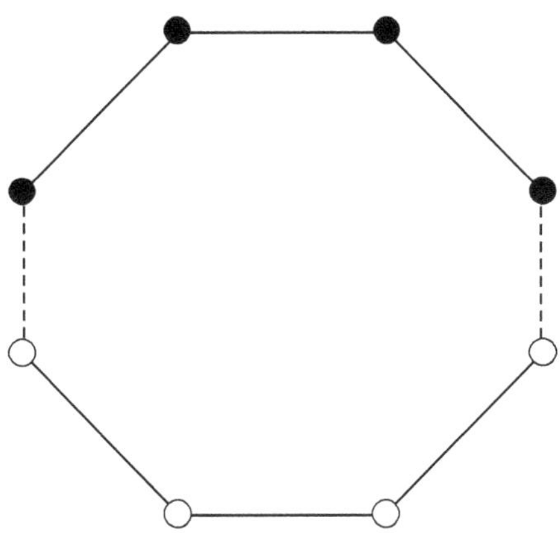

Figure I.6 C_8 with bottom half chosen and boundary dotted

We may be tempted to conjecture, based on this example, that if d is a fixed natural number and (X_k) is a sequence of d-regular finite graphs, where the number of vertices goes to infinity, then $h(X_k) \to 0$. Surprisingly, this statement is false. In 1973, Pinsker [108] used a probabalistic argument to demonstrate that for any integer $d \geq 3$, there exists $\epsilon > 0$ and an unbounded sequence (X_k) of d-regular graphs such that $h(X_k) \geq \epsilon$ for all k. Such a sequence (X_k) is called an *expander family*.

Pinsker's existence proof was non-constructive. In 1973, Margulis [92] used high-powered algebraic techniques to give the first explicit construction of an expander family. In the early twenty-first century, mathematicians began employing more elementary combinatorial methods to explicitly construct expander families (we discuss one of these methods in Chapter 5). The purpose of this book is to give a brief and accessible introduction to the theory of expander families, particularly in light of recent developments in combinatorial techniques.

It turns out that if one chooses a sequence of d-regular graphs "at random," it is almost certain to be an expander family. (See the Notes sections at the end of Chapters 1 and 3 for some precise statements in this vein.) Nevertheless, explicitly constructing an expander family is nontrivial. The situation is not unlike that of transcendental numbers. If one chooses a real number "at random," it is almost certain to be transcendental. However, it is by no means easy to prove that any particular number is transcendental.

2. WHAT IS A CAYLEY GRAPH?

Before attempting to construct expander families, we first tackle a much more basic question: how does one produce an unbounded sequence of d-regular graphs at all?

We have seen one such construction already, namely, the cycle graphs C_n. Label the vertices of C_n with the elements of the group $\mathbb{Z}_n$ of integers modulo n, as in Figure I.7. The vertex a is adjacent to the vertices $a+1$ and $a-1$. We might say that two ingredients give us this graph, namely, the group $\mathbb{Z}_n$ and the set of "generators" 1 and -1. Moreover, the graph is 2-regular because we have chosen two generators.

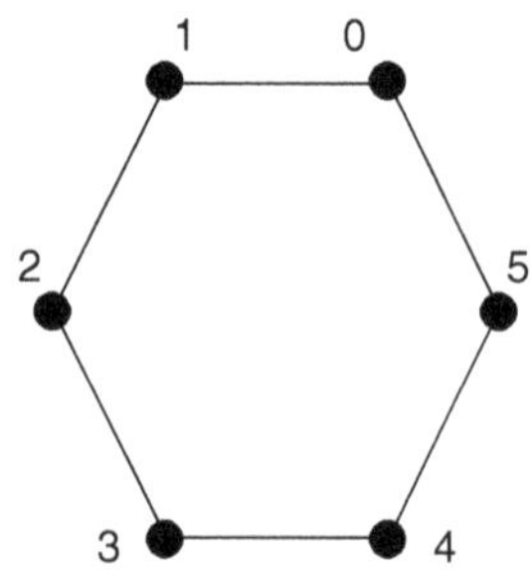

Figure I.7 Cycle graph C_6

The same two ingredients produce the graph Z in Figure I.2. Here, the group is $\mathbb{Z}_{46}$ and the generators are 1, -1, and 23. There are three generators, hence Z is 3-regular.

The general recipe is the classical Cayley graph construction. We start by taking a group G and some subset Γ of G. The elements of G become the vertices of the graph. We draw an edge from x to $x\gamma$ for every $\gamma \in \Gamma$. (Some care must be taken to ensure that the graph is undirected—see Section 1.2 for all the gory details.)

Armed with the Cayley graph construction, we can produce a vast wonderland of regular graphs, for the theory of finite groups provides us with a wide variety of possible starting points: $\mathbb{Z}_n$, dihedral groups, symmetric groups, alternating groups, matrix groups, direct products, semidirect products, wreath products, and so on. Moreover, the richness of group theory means that we can take advantage of the algebraic structure underlying a Cayley graph to extract information about the graph. This theme permeates throughout this book, especially Chapter 2, Chapter 4, and Part III.

We remark that not every regular graph is a Cayley graph. The graph Y in Figure I.1, for example, is not a Cayley graph—see Exercise 10 of Chapter 2.

3. A TALE OF FOUR INVARIANTS

Computing the isoperimetric constant $h(X)$ of a graph X with n vertices requires minimizing over *all* sets with $\leq n/2$ vertices. The number of such sets grows exponentially as a function of n. Hence, generally speaking it is not feasible to determine $h(X)$ when n is large. So in our search for expander families (X_k), we seek indirect methods for showing that $h(X_k)$ is uniformly bounded away from 0. Typically, one looks for graph invariants that are related to the isoperimetric constant but more tractable to work with. We briefly describe two such invariants; the precise details of what they are and why and how they are related to the questions at hand occupies the remainder of this book.

Consider the complex vector space of functions from V to $\mathbb{C}$, where V is the vertex set of a graph X. Define a linear operator A from this vector space to itself, where the value of Af at a vertex v equals the sum of the values of f at all vertices adjacent to v. All eigenvalues of A are real, as we state in Def. 1.38. Denote by λ_1 the second largest eigenvalue of A. We shall see that λ_1 and $h(X)$ are intimately related.

In the special case where X is a Cayley graph, we obtain another important invariant by considering the underlying group G. We can decompose the operator A in terms of irreducible representations (irreps) of G. We define the *Kazhdan constant* κ relative to the group G and a subset Γ of G to be the minimum value ϵ such that for any nontrivial irrep ρ of G and any vector $\mathbf{v}$ in the representation space, some element $\gamma \in \Gamma$ moves $\mathbf{v}$, via ρ, a distance of at least ϵ.

Four invariants—the isoperimetric constant, the second largest eigenvalue, the diameter, and the Kazhdan constant—recur throughout this book. Each measures in some way the expansion quality of a Cayley graph. Table I.1 lists the four invariants, the notation used to denote them, the place they're defined, and some minor variations on them.

Table I.1 FOUR INVARIANTS

Invariant	**Notation**	**Defined in**	**Variation**
Isoperimetric constant	h	Def. 1.63	—
Second largest eigenvalue	λ_1	Def. 1.38	λ
Diameter	diam	Def. 1.18	—
Kazhdan constant	κ	Def. 8.5	$\hat{\kappa}$

We repeatedly find occasion to ask three questions pertaining to these four invariants.

1. How do they relate to one another?
2. How do they behave with respect to subgroups and quotients?
3. For large regular graphs with fixed degree, what are their best-case scenarios?

We provide the following answers.

(1) Table I.2 shows inequalities that relate each invariant to the others for a finite d-regular graph X.

(2) For quotients, we have the following. Let G be a finite group; $H \lhd G$; Γ a symmetric subset of G; $X = \text{Cay}(G, \Gamma)$; and $Y = \text{Cay}(G/H, \overline{\Gamma})$, where $\overline{\Gamma}$ is the image of Γ under the canonical homomorphism. Then:

$$h(X) \leq h(Y) \qquad \text{(Lemma 2.17)}$$

$$\lambda_1(X) \geq \lambda_1(Y) \qquad \text{(Prop. 2.26)}$$

$$\text{diam}(X) \geq \text{diam}(Y) \qquad \text{(Prop. 4.35)}$$

$$\kappa(G, \Gamma) \leq \kappa(G/H, \overline{\Gamma}) \qquad \text{(Prop. 8.28)}$$

For subgroups, we have the following. Let G be a finite group; $H < G$; Γ a symmetric subset of G; $d = |\Gamma|$; $\hat{\Gamma}$ a set of Schreier generators (see Definition 2.30); $X = \text{Cay}(G, \Gamma)$; and $Z = \text{Cay}(H, \hat{\Gamma})$. Then:

$$h(X) \leq \frac{h(Z)}{[G : H]} \qquad \text{(Lemma 2.41)}$$

$$\lambda_1(X) \geq \frac{\lambda_1(Z)}{[G : H]} \qquad \text{(Lemma 2.47)}$$

$$\text{diam}(X) \geq \text{diam}(Z) \qquad \text{(Prop. 4.35)}$$

$$\kappa(G, \Gamma) \leq d^{1/2}\kappa(H, \hat{\Gamma}) \qquad \text{(Prop. 8.30)}$$

Roughly speaking, the moral of the story in each case is that a group can do no better than its subgroups and quotients.

(3) As previously noted, d-regular expander families exist for all $d \geq 3$. In other words, the best-case scenario for h is better than $o(1)$. For a sequence of d-regular Cayley graphs, the three quantities h, $d - \lambda_1$, and κ have the property that if one of them goes to 0, then so do the other two. (This fact follows from the inequalities in Table I.2.) So the best-case growth rates for $d - \lambda_1$ and κ are also better than $o(1)$.

For λ, we can be more precise. The Alon-Boppona theorem asserts that for fixed d and arbitrary $\epsilon > 0$, if X is a sufficiently large d-regular graph, then $\lambda \geq 2\sqrt{d-1} - \epsilon$. Chapter 3 discusses this theorem in detail. *Ramanujan graphs* achieve the asymptotically optimal bound $\lambda \leq 2\sqrt{d-1}$. Section 6 and Note 4 of Chapter 7 deal with two interesting families of Ramanujan graphs.

The best-case growth rate for diameters is logarithmic as a function of the number of vertices of the graph (see Prop. 4.6). Expander families achieve this optimal growth rate. Chapter 4 deals with this fact and the severe restrictions it places

Table I.2 How the Four Invariants Relate to One Another

	$\boldsymbol{\lambda_1}$ **or** $\boldsymbol{\lambda}$	**diam**	$\boldsymbol{\kappa}$ **or** $\boldsymbol{\hat{\kappa}}$
$\boldsymbol{h}$	$\frac{d-\lambda_1(X)}{2} \leq h(X) \leq \sqrt{2d(d-\lambda_1(X))}$ (Prop. 1.84)	$\mathrm{diam}(X) \leq \frac{2}{\log\left(1+\frac{h(X)}{d}\right)} \log(\lvert X\rvert)$ (Prop. 4.6)	$h \geq \frac{\kappa^2}{4d}$ (Prop. 8.17)
$\boldsymbol{\lambda_1}$ **or** $\boldsymbol{\lambda}$	—	$\mathrm{diam}(X) \leq \lceil \log(n-1)/\log(d/\lvert\lambda_1(X)\rvert)\rceil$ (Note 6 from Chapter 4)	$\kappa \geq \sqrt{\frac{2(d-\lambda_1)}{d}}$ (Prop. 8.18)
diam	—	—	$\kappa(G,\Gamma) > \frac{\sqrt{2}}{\mathrm{diam}(\mathrm{Cay}(G,\Gamma))}$ (Exercise 6 of Chapter 8)

on the types of sequences of groups that can possibly yield expander families via the Cayley graph construction.

4. APPLICATIONS OF EXPANDER FAMILIES

The explosion of interest in expander families in recent decades stems in no small part because of the wealth of applications they enjoy, particularly in computer science. In this section we give a brief overview of some of these applications. This book deals only with the mathematical theory underlying expander families, but we want to briefly mention some applications so that the reader is aware they exist. Discussing them in detail is beyond the scope of this book. Instead, we point the reader to as many resources as we can.

4.1 Computer Science

There are many objects of interest to computer scientists that depend on expander families for their construction. We begin by quoting a paragraph from M. Klawe's article [79]. We have changed the citation numbers to match the ones in the back of this book.

> The study of the complexity of graphs with special connectivity properties originated in switching theory, motivated by problems of designing networks able to connect many disjoint sets of users, while only using a small number of switches. An example of this type of graph is a superconcentrator, which is an acyclic directed graph with n inputs and n outputs such that given any pair of subsets A and B of the same size, of inputs and outputs respectively, there exists a set of disjoint paths joining the inputs in A to the outputs in B. Some other examples are concentrators, nonblocking connectors and generalized connectors (see [38], [110]). There is a large body of work searching for optimal constructions of these graphs ([108], [17], [34], [105], [95], [110], [109]). So far all optimal explicit constructions depend on expanding graphs of some sort.

Along those lines, Lubotzky [87, p. 4] presents a result of Gabber and Galil [63] that shows how to construct superconcentrators using expanders. The survey article by Hoory, Linial, and Wigderson [70] discusses the construction of superconcentrators via expanders. Chung [38] discusses nonblocking networks and superconcentrators, along with constructions of these objects using expander families and Ramanujan graphs. She cites [38], [93], [108], and [109] for a history of expanders and how they apply to communication networks.

The introduction in Klawe [79] gives many applications of expanders to computer science. These include the following two results. Ajtai, Komlos, and Szemeredi [2] use expanders to describe a sorting network of size $O(n \log n)$ and depth $O(\log n)$. Their construction solved a problem on sorting networks that had been open for 20 years [27]. Erdös, Graham, and Szemeredi [54] use expanding graphs to construct sparse graphs with dense long paths. These types of graphs occur in the study of Boolean functions and fault-tolerant microelectronic chips. The survey article [70] gives other applications of expanders to computer science, including applications to complexity theory.

Alon [5] uses finite geometries to construct expander families. These graphs enable him to improve previous results on a parallel sorting problem that arises in structural modeling. He also gives applications to Ramsey theory. Pippenger [111] applies expanders to sorting and selecting in rounds. Bellar, Goldreich, and Goldwasser [20] use expanders to study quantitative aspects of randomness in interactive proofs.

The publications here—especially [70], [79], and [87]—contain many more applications of expanders in computer science and elsewhere.

4.2 Random Walks

Let X be a regular graph. Suppose one begins at a vertex of X and at each step chooses a neighboring vertex uniformly at random (i.e., of equal probability) to move to next. Let $\pi_i(x)$ be the probability that one is at vertex x after i steps of this random process. If X is nonbipartite (see Def. 1.16) one can show that $\lim_{i\to\infty} \pi_i(x) = 1/n$, where n is the number of vertices in X. That is, after many steps the distribution is approximately uniform. Moreover, one can show that the rate at which this distribution tends to the uniform distribution is controlled by $\lambda_1(X)$. The convergence is rapid in expander families—in other words, one gets lost quickly when walking randomly in expanders. This leads to a number of applications that use walks in expander families. For more information on random walks in graphs see Diaconis's book [49], the survey article [70], and Terras's introductory book [129].

Charles, Lauter, and Goren [36] construct provably collision resistant hash functions that are used in cryptography. They construct these hash functions using walks on expander families in which finding cycles is hard. They use two different families of expander graphs: the family constructed by Lubotzky, Phillips, and Sarnak [89] and Pizer graphs [112]. Random walks on expander families can be used to reduce the error in probabilistic algorithms while trying to save on random bits used [70].

4.3 Error-Correcting Codes

We outline an application of expander families to error-correcting codes. See the survey article by Hoory, Linial, and Wigderson [70] for more information.

Suppose that two parties wish to communicate over a noisy channel. Let n be fixed and $C \subset \{0, 1\}^n$ be a code. Define the Hamming distance between x and y in C, denoted by $d_H(x, y)$, as the number of bits that need to be flipped to get from x to y. Define $\text{dist}(C) = \min_{\substack{x,y\in C \\ x\neq y}} d_H(x, y)$. One scheme to communicate over the noisy channel is to do the following. Transmit information using codewords from C. When a codeword is received, use the Hamming distance to see which codeword in C is closest to it. This will be taken as the codeword that was originally sent. One can show that if the number of bit-flips introduced in transmission is bounded by the greatest integer less than or equal to $(\text{dist}(C) - 1)/2$, then the above scheme is guaranteed to work.

Let $\text{rate}(C) = \log(|C|)/n$. A family of codes $C_n \subset \{0, 1\}^n$ is asymptotically good if there are some fixed constants $r > 0$ and $\delta > 0$ such that $\text{dist}(C_n) > \delta n$ and $\text{rate}(C_n) > r$ for all n. Note that an asymptotically good family of codes gives a way to create codes that simultaneously ensure a lower bound on both distance and rate. Probabilistic arguments are able to establish the fact that such families exist. Amazingly, explicit constructions of expander graphs give explicit constructions of such codes. For more details see [70].

PART ONE

Basics

1

Graph Eigenvalues and the Isoperimetric Constant

In this chapter, we discuss the two graph invariants that will occupy the bulk of our attention throughout this book: the isoperimetric constant, and the second-largest eigenvalue. Because we assume no prior knowledge of graph theory, we begin by giving several basic definitions and facts from that field.

If we know that the isoperimetric constant of a finite graph X is at least a, then it follows that for any set S containing no more than half the vertices of X, there are at least $a|S|$ edges that connect a vertex in S to a vertex not in S. Roughly speaking, the larger the isoperimetric constant is, the faster and more reliable the graph is as a communications network.

The eigenvalues of a finite graph are defined to be the eigenvalues of its adjacency operator. The isoperimetric constant of a finite regular graph X is closely related to its second largest eigenvalue $\lambda_1(X)$. In Section 7, we state the fundamental inequality (Prop. 1.84) that relates these two invariants.

In this chapter, we provide some techniques to estimate $\lambda_1(X)$. The most frequently used technique is a version of the Rayleigh-Ritz theorem from linear algebra. To make this book self-contained, we have included Appendix A. This appendix contains information on most of the topics from linear algebra that we will need: eigenvalues, eigenvectors, symmetric matrices, and so on. For further information on linear algebra, we refer the reader to the textbook by Friedberg, Insel, and Spence [56].

1. BASIC DEFINITIONS FROM GRAPH THEORY

Definition 1.1 A *multiset* is a collection of objects where objects may appear in the collection more than once. The number of times that a certain object appears in a multiset is called the *multiplicity* of that object. If S is a multiset, then $|S|$ is the number of elements in S.

Example 1.2
Consider the multiset $S = \{a, a, 4, a, -1, -1, x, 15\}$. The multiplicity of the element a is three, the multiplicity of the element -1 is two, and the elements 4, x, and 15 each have multiplicity one. The size of S is $|S| = 8$.

Remark 1.3
The notation for sets carries over for multisets. For example, we write $a \in S$ where S is from Example 1.2.

One can a multiset more rigorously as a function from a set U to $\{n \in \mathbb{Z} \mid n \geq 0\}$. For example, the multiset in Example 1.2 can be defined as the function f where $f(a) = 3, f(4) = 1, f(-1) = 2, f(x) = 1$, and $f(15) = 1$.

Definition 1.4 A *graph* is composed of a *vertex* set V and an *edge* multiset E. The vertex set V can be any set. The edge multiset E is a multiset whose elements are sets of the form $\{v, w\}$ or $\{v\}$ where v and w are distinct vertices. An edge of the form $\{v\}$ is called a *loop*. We sometimes denote V by V_X.

If $\{v, w\} \in E$, then we say that v and w are *adjacent* or *neighbors*. We also say that the edge $\{v, w\}$ is *incident* to v and w. If $\{v\}$ is in E, then we say that v is *adjacent* to itself. We also say that the loop $\{v\}$ is *incident* to v.

If $v \in V$, then the *degree* of v is the number of edges $e \in E$ such that v is incident to e. We write $\deg(v)$ to denote the degree of a vertex.

The *order* of a graph X, denoted by $|X|$, is the number of vertices in the graph.

Remark 1.5
Throughout this book we mainly deal with graphs that have finitely many vertices and edges. In fact, the only infinite graph in this book is the universal cover of a regular graph given in Def. 3.25.

We draw graphs as follows. For each vertex we draw a dot. If two vertices are adjacent, then we connect the corresponding vertices with a line (or curve) in our picture. If a vertex is adjacent to itself, then we draw a loop at that vertex. When drawing a graph, the apparent distance between two vertices does not matter. Also, it doesn't matter if edges appear to cross each other.

Example 1.6
Consider the graph in Figure 1.1. The vertex set is given by

$$V = \{v_1, v_2, v_3, v_4\}$$

and the edge set is given by

$$E = \{\{v_1\}, \{v_1, v_2\}, \{v_2, v_3\}, \{v_2, v_3\}, \{v_3, v_4\}\}.$$

The degrees of the vertices are

$$\deg(v_1) = 2, \quad \deg(v_2) = 3, \quad \deg(v_3) = 3, \quad \text{and } \deg(v_4) = 1.$$

The order of the graph is 4.

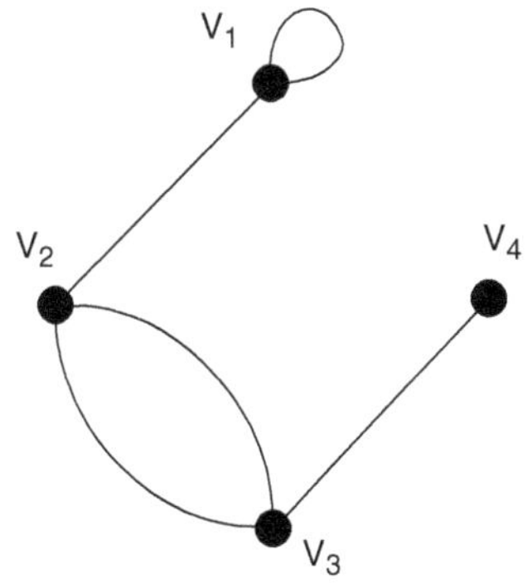

Figure 1.1

Remark 1.7
Consider Example 1.6. Some other books say that $\deg(v_1) = 3$, because they count each loop as contributing 2 to the degree, not 1. Our definition will be more natural for our purposes. See Note 3 and Remark 1.30.

We say that a graph has *multiple edges* if there are two distinct edges connecting the same pair of vertices, that is, if some edge has multiplicity greater than 1. For instance, in Example 1.6, there are multiple edges between v_2 and v_3. Note that the definition of a graph allows loops and multiple edges in our graphs. In many texts, this type of graph is called a multigraph. The reason we allow loops and multiple edges is that many of the constructions in this book (e.g., Schreier generators, zig-zag products) force us to consider them.

In the definition of a graph, the edges are defined as sets, hence the edges are given no direction. In Section 8 of this chapter, we consider directed graphs.

Definition 1.8 We say that a graph is *d-regular* if every vertex has degree *d*.

Example 1.9
The graph in Figure 1.2 is 3-regular. The graph in Figure 1.1 is not regular because the degrees of the vertices are not all equal.

Definition 1.10 A *walk* in a graph X with vertex set V and edge multiset E is a finite sequence of the form

$$w = (v_0, e_0, v_1, e_1, \ldots, v_{n-1}, e_{n-1}, v_n), \tag{1}$$

where $v_i \in V$ and $e_i \in E$, v_i is adjacent to v_{i+1} for $i = 0, \ldots, n-1$, and e_i is an edge that is incident to v_i and v_{i+1} for $i = 0, \ldots, n-1$. For the walk w in (1), we say that w is a walk of *length* n from v_0 to v_n.

Remark 1.11
If there are no multiple edges in the graph X, then we omit the edges when listing a walk. See Example 1.12. If there are multiple edges, we may regard distinct copies of an edge as distinct edges in the walk. See Example 1.13.

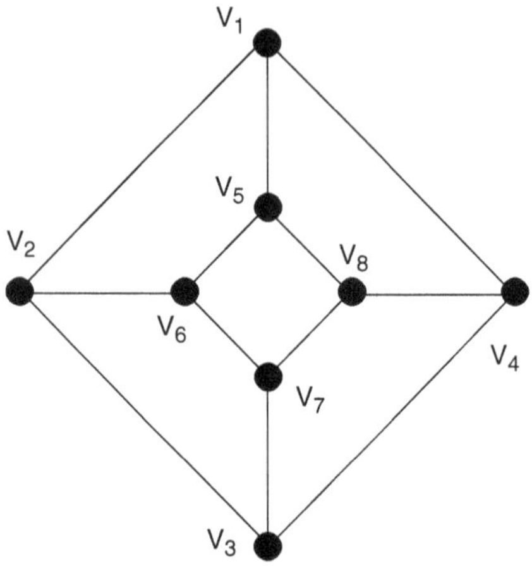

Figure 1.2

Example 1.12
Consider the graph in Figure 1.2. The walk

$$w_1 = (v_1, \{v_1, v_2\}, v_2, \{v_2, v_6\}, v_6, \{v_6, v_5\}, v_5)$$

has length 3. Because the graph has no multiple edges, we can also write the walk w_1 as (v_1, v_2, v_6, v_5). Think of this as traveling from v_1 to v_5 by taking three steps. A shorter walk between v_1 and v_5 is $w_2 = (v_1, v_5)$, which has length 1.

Example 1.13
In the graph in Figure 1.1, the edge $\{v_2, v_3\}$ has multiplicity 2. Regard this as two distinct edges $\{v_2, v_3\}_1$ and $\{v_2, v_3\}_2$. Hence $(v_1, \{v_1, v_2\}, v_2, \{v_2, v_3\}_1, v_3)$ and $(v_1, \{v_1, v_2\}, v_2, \{v_2, v_3\}_2, v_3)$ are two distinct walks of length 2 from v_1 to v_3.

Definition 1.14 A graph X with vertex set V is *connected* if for every $x, y \in V$ there exists a walk from x to y. Otherwise, we say that the graph is *disconnected.*

Example 1.15
The graph in Figure 1.2 is connected. The graph in Figure 1.3 is disconnected, since there is no walk from a to b.

Definition 1.16 A graph X with vertex set V is *bipartite* if there exist $V_1, V_2 \subset V$ such that

1. $V = V_1 \cup V_2$,
2. $V_1 \cap V_2 = \emptyset$,
3. every edge of X is incident to a vertex in V_1 and a vertex in V_2.

In this case, we call (V_1, V_2) a *bipartition* of V.

Equivalently, a bipartite graph is one where the vertices can be colored with two colors so that no two adjacent vertices have the same color.

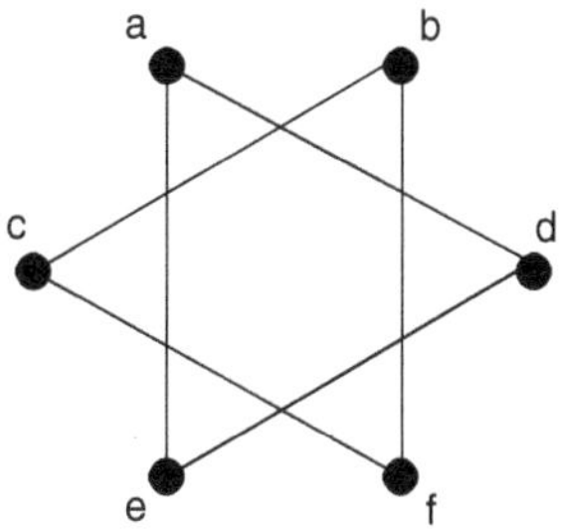

Figure 1.3

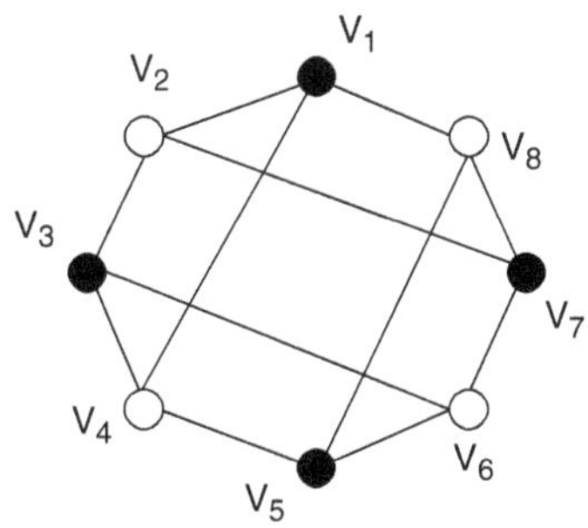

Figure 1.4

Example 1.17
The graph in Figure 1.4 is bipartite. Consider the following bipartition: $V_1 = \{v_1, v_3, v_5, v_7\}$ and $V_2 = \{v_2, v_4, v_6, v_8\}$. In the figure, the vertices from V_1 are colored black, and the vertices from V_2 are colored white. Note that no two vertices of the same color are adjacent to one another.

Definition 1.18 Let X be a graph with vertex set V. Given $x, y \in V$, the *distance* between x and y, denoted by $\text{dist}(x, y)$, is the minimal length of any walk between x and y. The *diameter* of X is given by $\text{diam}(X) = \max_{x,y\in V} \text{dist}(x, y)$.

Remark 1.19
If X is a disconnected graph and $x, y \in X$, then there may not be a walk from x to y. In that case we say that $\text{dist}(x, y)$ is infinite and $\text{diam}(X)$ is infinite.

Remark 1.20
The function dist defines a metric on V. Hence the vertex set of a graph is a metric space.

Example 1.21
Let X be the graph in Figure 1.4. There are several walks of length three from v_1 to v_6. For example, one of them is (v_1, v_8, v_5, v_6). There are no walks of length less than 3 from v_1 to v_6. Thus, $\text{dist}(v_1, v_6) = 3$.

Similarly, $\text{dist}(v_3, v_8) = 3$, $\text{dist}(v_2, v_6) = 2$, and $\text{dist}(v_4, v_5) = 1$. In fact, $\text{dist}(v, w) \leq 3$ for all vertices v, w. Hence, $\text{diam}(X) = 3$.

2. CAYLEY GRAPHS

Given a group G and a certain kind of multi-subset Γ of G (to be defined shortly), one can construct a graph called a Cayley graph. These graphs are highly symmetric. We can derive properties of the graph from properties of the group, thereby giving us a bridge between graph theory and group theory.

Definition 1.22 Let G be a group and Γ be a multi-subset of G. We say that Γ is *symmetric* if whenever γ is an element of Γ with multiplicity n, then γ^{-1} is an element of Γ of multiplicity n. If Γ is a symmetric multi-subset of G, then we write $\Gamma \circledS G$.

Example 1.23
Consider the group $\mathbb{Z}_6$. The multi-subset $\Gamma_1 = \{1, 1, 2, 4, 5, 5\}$ is symmetric. The multi-subset $\Gamma_2 = \{1, 5, 5\}$ is not symmetric because 5 occurs with multiplicity two, while its inverse 1 occurs with multiplicity one. The multi-subset $\Gamma_3 = \{1, 2, 4\}$ is not symmetric as the inverse of 1 does not even appear.

Definition 1.24 Let G be a group and $\Gamma \circledS G$. The Cayley graph of G with respect to Γ, denoted by $\mathrm{Cay}(G, \Gamma)$, is defined as follows. The vertices of $\mathrm{Cay}(G, \Gamma)$ are the elements of G. Two vertices $x, y \in G$ are adjacent if and only if there exists $\gamma \in \Gamma$ such that $x = y\gamma$. (In other words, $y^{-1}x \in \Gamma$.) The multiplicity of the edge $\{x, y\}$ in the edge multiset E equals the multiplicity of $y^{-1}x$ in Γ.

Example 1.25
The Cayley graph $\mathrm{Cay}(\mathbb{Z}_4, \{1, 1, 3, 3\})$ is shown in Figure 1.5. There are two edges between 1 and 2, for example, because $-1 + 2 = 1$ has multiplicity 2 in $\{1, 1, 3, 3\}$.

Remark 1.26
Why do we want Γ to be symmetric in Def. 1.24? Suppose that $x = y\gamma$ where $\gamma \in \Gamma$. Then x and y are adjacent. But to make the definition well defined there should exist a $\gamma' \in \Gamma$ such that $y = x\gamma'$. This would imply that $\gamma' = \gamma^{-1}$.

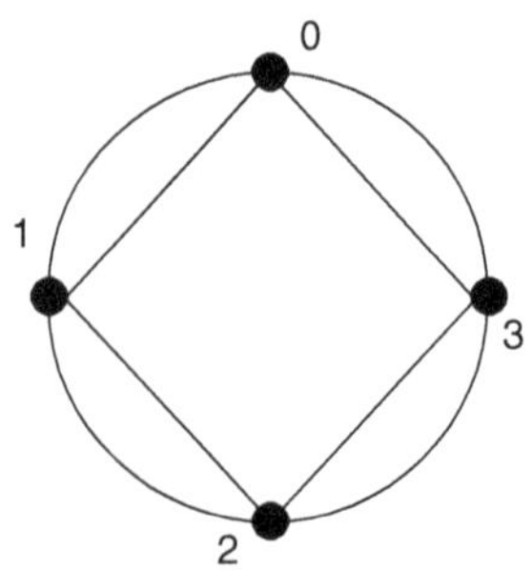

Figure 1.5 $\mathrm{Cay}(\mathbb{Z}_4, \{1, 1, 3, 3\})$

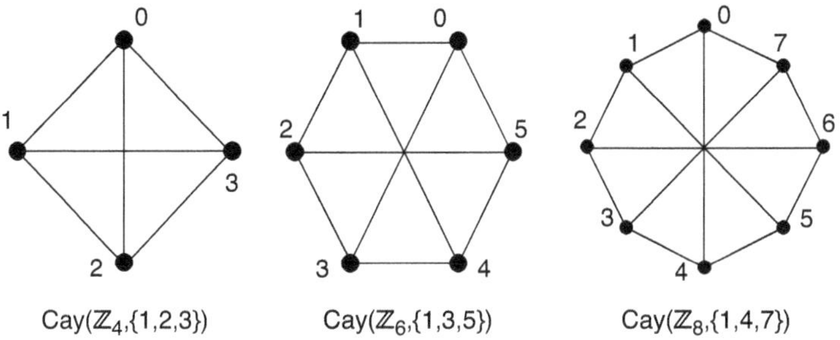

Figure 1.6 Three Cayley graphs

If you relax this condition and let Γ be any multi-subset of G, you end up with a directed Cayley graph. See Def. 1.101.

Example 1.27
Figure 1.6 shows the first few graphs in the sequence $(\mathrm{Cay}(\mathbb{Z}_{2n}, \{1, n, 2n-1\}))$.

Example 1.28
Recall our conventions regarding the symmetric group S_n. Figure 7.1 shows the Cayley graph $\mathrm{Cay}(S_3, \{(1,2), (2,3), (1,2,3), (1,3,2)\})$. You will come to recognize this graph as our "end-of-proof" symbol.

Proposition 1.29
Let G be a group and $\Gamma \Subset G$. Then the following are true:

1. $\mathrm{Cay}(G, \Gamma)$ is $|\Gamma|$-regular.
2. $\mathrm{Cay}(G, \Gamma)$ is connected if and only if Γ generates G as a group.

Proof

1. Suppose that $g \in G$ is a vertex of $\mathrm{Cay}(G, \Gamma)$ and that $\Gamma = \{\gamma_1, \ldots, \gamma_d\}$. Then the neighbors of g are the vertices $g\gamma_1, g\gamma_2, \ldots, g\gamma_d$ (counted with multiplicity). Hence the degree of the vertex g is $d = |\Gamma|$.
2. Let 1_G be the identity element of G. Then Γ generates G as a group if and only if for every $g \in G$ there exists $\gamma_1, \ldots, \gamma_k \in \Gamma$ such that $g = \gamma_1 \cdots \gamma_k = 1_G\gamma_1 \cdots \gamma_k$. This is equivalent to saying that for every element $g \in G$, there is a walk in the graph X from 1_G to g. (The equation $g = \gamma_1 \cdots \gamma_k = 1_G\gamma_1 \cdots \gamma_k$ gives the walk $(1_G, 1_G\gamma_1, 1_G\gamma_1\gamma_2, \cdots, 1_G\gamma_1\gamma_2 \cdots \gamma_k)$.) This is equivalent to the fact that X is a connected graph. (For if $g, h \in G$, reverse the walk from g to 1_G, then walk from 1_G to h.)

Remark 1.30
Note that if we had counted a loop as contributing 2 to the degree and $1_G \in \Gamma$, then Prop. 1.29(1) would fail.

3. THE ADJACENCY OPERATOR

Definition 1.31 Let S be a finite set. We define the complex vector space $L^2(S)$ by

$$L^2(S) = \{f : S \to \mathbb{C}\}.$$

Let $f, g \in L^2(S)$ and $\alpha \in \mathbb{C}$. The vector space sum in $L^2(S)$ is given by $(f+g)(x) = f(x) + g(x)$. Scalar multiplication is given by $(\alpha f)(x) = \alpha f(x)$. The standard inner product and norm are given by

$$\langle f, g \rangle_2 = \sum_{x \in S} f(x)\overline{g(x)} \quad \text{and} \quad \|f\|_2 = \sqrt{\langle f, f \rangle_2} = \sqrt{\sum_{x \in S} |f(x)|^2}.$$

We drop the subscript and just write $\langle \cdot, \cdot \rangle$ or $\| \cdot \|$ when the inner product, or norm, is understood to be the standard one. See Appendix A for a refresher on inner products.

Usually in this book, we deal with the space $L^2(V)$ where V is the vertex set of some graph X. To simplify matters we define $L^2(X) = L^2(V)$. When doing this, we may think of an element f of $L^2(X)$ in several different ways: (a) as a function; (b) as a picture, where we draw the graph X and label each vertex $v \in V$ with the value of f at the vertex v, that is, with $f(v)$; and (c) as a vector, where we give the vertices of V a particular ordering $v_1, \ldots, v_n$ and then think of f as the vector $(f(v_1), \ldots, f(v_n))^t$. We frequently use $L^2(E)$ where E is the edge multiset of X. Here we think of $f \in L^2(E)$ as a function on the edge multiset of X.

Example 1.32
Consider the graph C_3 given in Figure 1.7 with vertex set V ordered as v_1, v_2, v_3. Define the functions $f, g \in L^2(C_3)$ by

$$f(v) = \begin{cases} 10 & \text{if } v = v_1 \\ i & \text{if } v = v_2 \\ -2+i & \text{if } v = v_3 \end{cases} \quad \text{and} \quad g(v) = \begin{cases} 1-4i & \text{if } v = v_1 \\ 0 & \text{if } v = v_2 \\ -1 & \text{if } v = v_3 \end{cases}.$$

Figure 1.7 shows how we think of g as a picture. We may also think of g as the vector $(1-4i, 0, -1)^t$. Note that we needed to order the vertices of C_3 to think

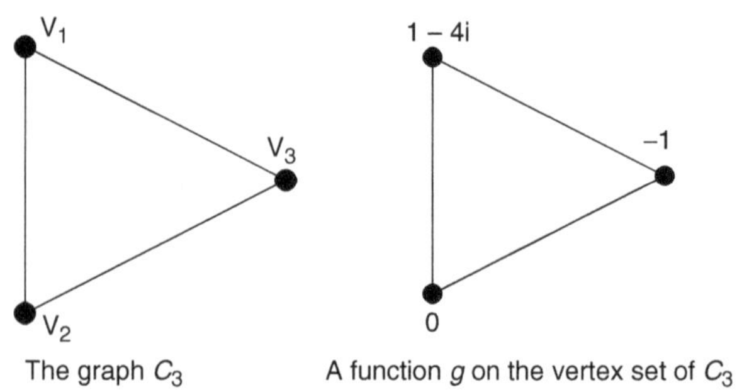

Figure 1.7

of g as a vector; a different ordering of the vertices would have resulted in a different vector for g.

The inner product of f and g is

$$\begin{aligned}\langle f, g\rangle_2 &= 10\overline{(1-4i)} + i\,\overline{0} + (-2+i)\,\overline{(-1)} \\ &= 10(1+4i) + (-2+i)(-1) \\ &= 12 + 39i.\end{aligned}$$

Note that if we think of f and g as vectors, then the inner product is the same as the complex inner product of the vectors. The norm of f is

$$\|f\|_2 = \sqrt{10\,\overline{10} + i\,\overline{i} + (-2+i)\,\overline{(-2+i)}} = \sqrt{106}.$$

Remark 1.33

Consider a set $S = \{x_1, x_2, \ldots, x_n\}$. Let $\beta = \{\delta_{x_1}, \delta_{x_2}, \ldots, \delta_{x_n}\} \subset L^2(S)$, where $\delta_{x_i}(x_j) = 1$ if $i = j$, and $\delta_{x_i}(x_j) = 0$ if $i \neq j$.

If $f \in L^2(S)$, then

$$f(x) = f(x_1)\delta_{x_1}(x) + \cdots + f(x_n)\delta_{x_n}(x).$$

Hence β spans the vector space $L^2(S)$. It is easy to see that the functions δ_{x_i} are mutually orthogonal. By Proposition A.16 this implies that $\delta_{x_1}, \ldots, \delta_{x_n}$ are linearly independent. Therefore β is a basis for $L^2(S)$ called the *standard basis* for $L^2(S)$. This implies that $L^2(S)$ has dimension n as a vector space over $\mathbb{C}$.

Definition 1.34 Let X be a graph with an ordering of its vertices given by $v_1, v_2, \ldots, v_n$. Then the *adjacency matrix* for X is the matrix A, where $A_{i,j}$ is the number of edges that are incident to both v_i and v_j.

If x and y are vertices of X, then we sometimes write $A_{x,y}$ for the number of edges that are incident to x and y. This way we may refer to an "entry" of A without having to order the vertices.

Note that in the definition, a loop is counted only once, while it is counted twice in some graph theory texts.

Example 1.35

Consider the graph in Figure 1.1 with vertices ordered as v_1, v_2, v_3, v_4. Then the adjacency matrix of the graph is

$$A = \begin{pmatrix} 1 & 1 & 0 & 0 \\ 1 & 0 & 2 & 0 \\ 0 & 2 & 0 & 1 \\ 0 & 0 & 1 & 0 \end{pmatrix}.$$

Remark 1.36

If X is a graph with adjacency matrix A, and v and w are vertices of X, then $A_{v,w} = A_{w,v}$. Thus, A is a symmetric matrix.

Remark 1.37
Suppose that X is a graph and A_1 and A_2 are two adjacency matrices of X using different orderings of the vertices of X. Then, by Exercise 1, the matrices A_1 and A_2 have the same eigenvalues.

Definition 1.38 Let A be an adjacency matrix for a graph X with n vertices. By Remark 1.37, the eigenvalues of A do not depend on the choice of ordering of the vertices of X. Because A is symmetric, by Theorem A.53 the eigenvalues of A are real. We order them as follows:

$$\lambda_{n-1}(X) \leq \lambda_{n-2}(X) \leq \ldots \leq \lambda_1(X) \leq \lambda_0(X)$$

We call the multiset of eigenvalues of X the *spectrum of* X. If the spectrum consists of the distinct eigenvalues $\mu_1, \mu_2, \ldots, \mu_r$ with multiplicities $m_1, m_2, \ldots, m_r$, respectively, then we sometimes write

$$\mathrm{Spec}(X) = \begin{pmatrix} \mu_1 & \mu_2 & \cdots & \mu_r \\ m_1 & m_2 & \cdots & m_r \end{pmatrix}.$$

Example 1.39
Consider the graph C_3 given in Figure 1.8. The adjacency matrix for C_3 is given by

$$A = \begin{pmatrix} 0 & 1 & 1 \\ 1 & 0 & 1 \\ 1 & 1 & 0 \end{pmatrix}.$$

The eigenvalues of C_3 are the roots of the characteristic polynomial

$$p_A(x) = \det(A - xI) = \begin{pmatrix} -x & 1 & 1 \\ 1 & -x & 1 \\ 1 & 1 & -x \end{pmatrix} = -(x-2)(x+1)^2.$$

Thus, the spectrum of $\mathrm{Spec}(C_3) = \begin{pmatrix} -1 & 2 \\ 2 & 1 \end{pmatrix}$. So $\lambda_0(C_3) = 2$ and $\lambda_1(C_3) = \lambda_2(C_3) = -1$.

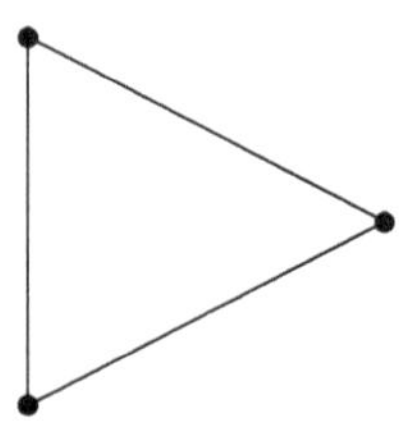

Figure 1.8 C_3

Example 1.40
Consider the 3-regular graph X given in Figure 1.4. The graph is bipartite; a bipartition is given by the sets of black and white vertices. The adjacency matrix for the graph, using the ordering $v_1, \ldots, v_8$ of the vertices, is

$$A = \begin{pmatrix} 0 & 1 & 0 & 1 & 0 & 0 & 0 & 1 \\ 1 & 0 & 1 & 0 & 0 & 0 & 1 & 0 \\ 0 & 1 & 0 & 1 & 0 & 1 & 0 & 0 \\ 1 & 0 & 1 & 0 & 1 & 0 & 0 & 0 \\ 0 & 0 & 0 & 1 & 0 & 1 & 0 & 1 \\ 0 & 0 & 1 & 0 & 1 & 0 & 1 & 0 \\ 0 & 1 & 0 & 0 & 0 & 1 & 0 & 1 \\ 1 & 0 & 0 & 0 & 1 & 0 & 1 & 0 \end{pmatrix}.$$

Using software (such as Mathematica), one can compute that $\operatorname{Spec}(X) = \begin{pmatrix} -3 & -1 & 1 & 3 \\ 1 & 3 & 3 & 1 \end{pmatrix}$. So $\lambda_0(X) = 3, \lambda_1(X) = \lambda_2(X) = \lambda_3(X) = 1, \lambda_4(X) = \lambda_5(X) = \lambda_6(X) = -1$, and $\lambda_7(X) = -3$.

Notice that the graph is 3-regular and that 3 shows up as an eigenvalue. Also notice that -3 shows up as an eigenvalue (this is because the graph is bipartite). These facts will be proven in general in Proposition 1.48.

Remark 1.41
We will *not* be spending our time explicitly computing spectra of graphs—for large graphs that would be utterly impractical. We see later that $\lambda_1(X)$ is the eigenvalue we care about most. Our main task henceforth will be developing techniques to get good upper and lower bounds for λ_1.

Example 1.42
Consider the graph C_4 shown in Figure 1.9. Define the function

$$f(v) = \begin{cases} 1 & \text{if } v = v_1 \text{ or } v = v_3 \\ -1 & \text{if } v = v_2 \text{ or } v = v_4 \end{cases}$$

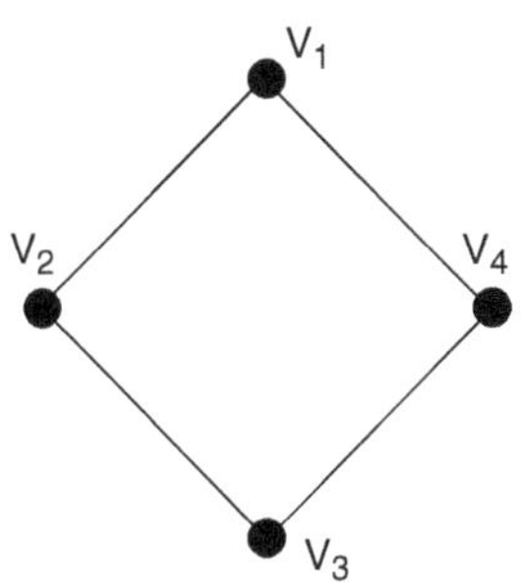

Figure 1.9 The graph C_4

on the vertices of C_4. If we order the vertices of C_4 as v_1, v_2, v_3, v_4, then we may think of f as the vector $(1, -1, 1, -1)^t$. If A is the adjacency matrix of C_4 using the same ordering, then

$$Af = \begin{pmatrix} 0 & 1 & 0 & 1 \\ 1 & 0 & 1 & 0 \\ 0 & 1 & 0 & 1 \\ 1 & 0 & 1 & 0 \end{pmatrix} \begin{pmatrix} 1 \\ -1 \\ 1 \\ -1 \end{pmatrix} = \begin{pmatrix} -1-1 \\ 1+1 \\ -1-1 \\ 1+1 \end{pmatrix} = -2f.$$

Thus, f is an eigenfunction of A associated with the eigenvalue -2.

Notice what A does to f. As an example, note that the value of the vector Af at the vertex v_1 is the sum of the values of f at the neighboring vertices v_2 and v_4. In general, how is the value of $(Af)(v)$ related to the values of f at the neighbors of v? (Hint: The answer is given in the next remark.)

Remark 1.43
Suppose that X is a graph with vertex set V ordered as $v_1, v_2, \ldots, v_n$, and let A be the adjacency matrix of X using this ordering. Given $f \in L^2(X)$ we may think of f as a vector in $\mathbb{C}^n$. Then

$$Af = \begin{pmatrix} A_{1,1} & A_{1,2} & \cdots & A_{1,n} \\ A_{2,1} & A_{2,2} & \cdots & A_{2,n} \\ \vdots & \vdots & \ddots & \vdots \\ A_{n,1} & A_{n,2} & \cdots & A_{n,n} \end{pmatrix} \begin{pmatrix} f(v_1) \\ f(v_2) \\ \vdots \\ f(v_n) \end{pmatrix} = \begin{pmatrix} \sum_{j=1}^n A_{1,j} f(v_j) \\ \sum_{j=1}^n A_{2,j} f(v_j) \\ \vdots \\ \sum_{j=1}^n A_{n,j} f(v_j) \end{pmatrix}.$$

Therefore, we may think of A as a linear transformation from $L^2(X)$ to $L^2(X)$ given by the formula

$$(Af)(v) = \sum_{w \in V} A_{v,w} f(w). \tag{2}$$

Definition 1.44 The linear operator A defined by equation (2) is called the *adjacency operator* of X.

Remark 1.45
Up until now, we had been thinking of A as a matrix that acts on vectors by matrix multiplication. When we think of A in terms of Equation (2) we are thinking of f as an element of $L^2(X)$ and A as a function that maps $L^2(X)$ to itself. Note that if we order the vertices, then the adjacency matrix is the matrix associated to the adjacency operator, with respect to the standard basis. (See Remark 1.33 and Def. A.38.)

Remark 1.46
If $X = \text{Cay}(G, \Gamma)$ is a Cayley graph, then we get a nice formula for the action of the adjacency operator A on a function $f \in L^2(G)$. This action is given by

$$(Af)(g) = \sum_{\gamma \in \Gamma} f(g\gamma)$$

for all $g \in G$.

For example, consider the Cayley graph $X = \text{Cay}(\mathbb{Z}_4, \{1, 1, 3, 3\})$ that is shown in Figure 1.5. Let A be the adjacency operator of X and let $f \in L^2(X)$ be given by $f(0) = -1, f(1) = 5, f(2) = 1$, and $f(3) = 0$. Then $(Af)(0) = f(0+1) + f(0+1) + f(0+3) + f(0+3) = 10$.

4. EIGENVALUES OF REGULAR GRAPHS

Definition 1.47 Suppose that X is a d-regular graph. We say that the spectrum of X is *symmetric about* 0 if whenever λ is an eigenvalue of X of multiplicity k, then $-\lambda$ is also an eigenvalue of X with multiplicity k.

The following proposition relates some of the structural properties of a regular graph X to its eigenvalues. Exercise 8 generalizes part (3) of the proposition.

Proposition 1.48
If X is a d-regular graph with n vertices, then

1. d is an eigenvalue of X.
2. $|\lambda_i(X)| \leq d$ for $i = 0, \ldots, n-1$.
3. $\lambda_1(X) < \lambda_0(X)$ if and only if X is a connected graph.
4. If X is bipartite, then the spectrum of X is symmetric about 0.
5. If $-d$ is an eigenvalue of X, then X is bipartite.

Proof
Throughout this proof let V denote the vertex set of X and A denote the adjacency operator of X.

1. We show that d is an eigenvalue of A. The trick is that a constant function on the vertices is an eigenfunction associated with d. Let $f_0 \in L^2(V)$ be defined as $f_0(x) = 1$ for all $x \in V$. Then

$$(Af_0)(x) = \sum_{y\in V} A_{x,y} f_0(y) = \sum_{y \in V} A_{x,y} = d = d \cdot f_0(x).$$

   Thus, d is an eigenvalue of A.
2. Let λ be an eigenvalue of A and f be a real-valued eigenfunction of A associated with λ. (By Theorem A.53 we know such an f exists.) Pick an $x \in V$ such that $|f(x)| = \max_{y \in V} |f(y)|$. Note that $f(x) \neq 0$ since f is an eigenfunction of A. By the definition of x we see that

$$|\lambda| \, |f(x)| = |(Af)(x)| = \left| \sum_{y \in V} A_{x,y} f(y) \right| \leq \sum_{y\in V} |A_{x,y}| \, |f(y)|$$
$$\leq |f(x)| \sum_{y \in V} |A_{x,y}| = d\, |f(x)| .$$

   Thus, $|\lambda| \leq d$.

3. Since A is a symmetric matrix, by Theorem A.53, the multiplicity of d as an eigenvalue of A is equal to $\dim(E_d(A))$, where

$$E_d(A) = \{f \in L^2(V) \mid Af = d \cdot f\}$$

is the eigenspace of A associated with the eigenvalue d. We will show that $\dim(E_d(A)) = 1$ if and only if X is connected.

Suppose that X is connected and f is a real-valued eigenfunction associated with d. We will show that f is a constant function on V, from which it follows that $\dim(E_d(A)) = 1$. Pick $x \in V$ such that $|f(x)| = \max_{y \in V} |f(y)|$. We may assume that $f(x) > 0$, since $-f$ is also a real-valued eigenfunction of A associated with d. We see that

$$f(x) = \frac{(Af)(x)}{d} = \sum_{y \in V} \frac{A_{x,y}}{d} f(y).$$

Suppose $f(y_0) < f(x)$ for some y_0 adjacent to x. Then, because $f(y) \leq f(x)$ for all $y \in V$,

$$f(x) = \sum_{y \in V} \frac{A_{x,y}}{d} f(y) < \sum_{y \in V} \frac{A_{x,y}}{d} f(x) = f(x).$$

This would be a contradiction. Hence, $f(y) = f(x)$ for each y that is adjacent to x. Now repeat this process for each y that is adjacent to x. Continuing in this fashion, since X is connected, we eventually reach every vertex of X. (To make this precise, use induction on the distance from v to x.) Hence f is constant on X.

Now suppose that X is disconnected. Let v be a vertex of X. Let V_1 be the set of all vertices w such that there exists a walk in X from v to w. Let $V_2 = V \setminus V_1$. Note that if $u \in V$ is adjacent to a vertex in V_1, then $u \in V_1$. Ditto for V_2. So X "splits" into two d-regular graphs, with vertex sets V_1, V_2, respectively. Define the functions

$$f_1(x) = \begin{cases} 1 & \text{if } x \in V_1 \\ 0 & \text{if } x \in V_2 \end{cases} \quad \text{and} \quad f_2(x) = \begin{cases} 0 & \text{if } x \in V_1. \\ 1 & \text{if } x \in V_2. \end{cases}$$

Then f_1 and f_2 are linearly independent eigenfunctions of A associated with the eigenvalue d. Hence, $\dim(E_d(A)) > 1$.

4. Suppose that X is a bipartite graph and $V = V_1 \cup V_2$ is a bipartition of V. Let λ be an eigenvalue of A with multiplicity k. By Theorem A.53 there exist linearly independent real-valued eigenfunctions $f_1, \ldots, f_k$ of A associated with λ. Define the functions

$$g_i(x) = \begin{cases} f_i(x) & x \in V_1 \\ -f_i(x) & x \in V_2 \end{cases}$$

for $i = 1, \ldots, k$.

We will now show that each g_i is an eigenfunction of A associated with $-\lambda$. Suppose that $x \in V_1$. Then, since every y adjacent to x is in V_2, we see that

$$(Ag_i)(x) = \sum_{y \in V_2} A_{x,y} g_i(y) = -\sum_{y \in V} A_{x,y} f_i(y) = -(Af_i)(x)$$
$$= -\lambda f_i(x) = -\lambda g_i(x).$$

Similarly, if $x \in V_2$, then $(Ag_i)(x) = -\lambda g_i(x)$. One can check that $g_1, \ldots, g_k$ form a linearly independent set. Hence $-\lambda$ is an eigenvalue of A with multiplicity $m \geq k$. The same argument, reversing the roles of λ and $-\lambda$ shows that $k \geq m$.

5. First assume that X is connected. Now suppose that $-d$ is an eigenvalue of A. Let f be a real-valued eigenfunction of A associated with $-d$. Pick an $x \in V$ such that $|f(x)| = \max_{y \in V} |f(y)|$. We may assume that $f(x) > 0$ because $-f$ is also a real-valued eigenfunction of A associated with $-d$. We have that

$$f(x) = \frac{(Af)(x)}{-d} = \sum_{y \in V} \frac{A_{x,y}}{d}(-f(y)).$$

By the same argument as in the first direction of part (3) of this proposition, since $-f(y) \leq f(x)$ for each y adjacent to x, we must have that $f(y) = -f(x)$ for all y adjacent to x. Since X is connected, using this same logic to the vertices that are distance two from x, and then distance 3 from x and so on, we eventually reach every vertex of the graph and get that

$$f(y) = \begin{cases} f(x) & \text{if dist}(x, y) \text{ is even} \\ -f(x) & \text{if dist}(x, y) \text{ is odd} \end{cases}.$$

This gives a bipartition of V where $V_1 = \{y \in V \mid f(y) = f(x)\}$ and $V_2 = \{y \in V \mid f(y) = -f(x)\}$.

If X is disconnected use Exercise 8 and apply the foregoing argument to each connected component of X to complete the proof. ◈

Prop. 1.48 is absolutely fundamental to everything that follows in this book. We urge the reader to immediately do Exercise 17, which has been designed specifically to gain familiarity with Prop. 1.48.

Example 1.49
The 4-regular graph X given in Figure 1.10 is called the Chvatal graph. Using software, one can compute the spectrum of X.

$$\mathrm{Spec}(X) = \begin{pmatrix} -3 & \frac{-1-\sqrt{17}}{2} & -1 & 0 & 1 & \frac{-1+\sqrt{17}}{2} & 4 \\ 2 & 1 & 1 & 2 & 4 & 1 & 1 \end{pmatrix}$$

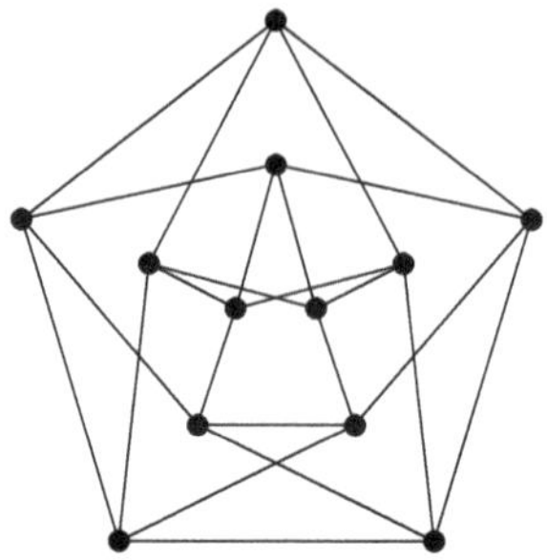

Figure 1.10 Chvatal graph

In accordance with Proposition 1.48 all of the eigenvalues of X (including $\frac{-1-\sqrt{17}}{2} \approx -2.56155$ and $\frac{-1+\sqrt{17}}{2} \approx 1.56155$) lie in the interval $[-4, 4]$. The fact that -4 is not an eigenvalue implies that X is not bipartite.

Example 1.50
The 2-regular graph given in Figure 1.3 is not connected. Its spectrum is $\{-1, -1, -1, -1, 2, 2\}$. Note that the spectrum is the union of the spectra of two C_3 graphs (see Example 1.39 for the spectrum of C_3).

The following lemma will be used frequently throughout this book.

Lemma 1.51
Let $n \geq 2$ and a be integers. Let $\xi = \exp(2\pi i/n)$. Then

$$\sum_{j=0}^{n-1} (\xi^a)^j = \begin{cases} 0 & \text{if } n \text{ does not divide } a \\ n & \text{otherwise} \end{cases}.$$

Proof
If n divides a, then $\xi^a = 1$. Hence $\sum_{j=0}^{n-1} (\xi^a)^j = n$. If n does not divide a, then $\xi^a \neq 1$. Recall that $1 + z + z^2 + \cdots + z^m = \frac{z^{m+1}-1}{z-1}$ if $z \neq 1$. Hence

$$\sum_{j=0}^{n-1} (\xi^a)^j = \frac{(\xi^a)^n - 1}{\xi^a - 1} = \frac{1-1}{\xi^a - 1} = 0.$$

Example 1.52
The *complete graph* K_n is the graph with n vertices where vertices v and w are adjacent, via an edge of multiplicity 1, if and only if $v \neq w$. Some examples are given in Figure 1.11. Note that $K_n = \text{Cay}(\mathbb{Z}_n, \{1, 2, \ldots, n-1\})$.

The adjacency matrix for K_n is given by

$$\begin{pmatrix} 0 & 1 & 1 & \ldots & 1 \\ 1 & 0 & 1 & \ldots & 1 \\ 1 & 1 & 0 & \ldots & 1 \\ \vdots & \vdots & \vdots & \ddots & \vdots \\ 1 & 1 & 1 & \ldots & 0 \end{pmatrix}.$$

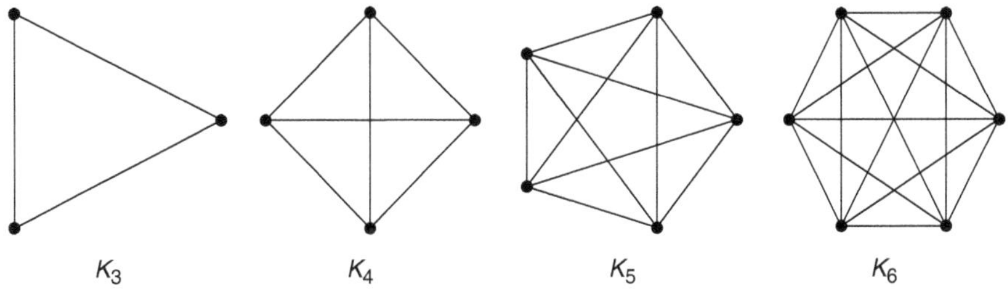

Figure 1.11 Complete graphs

This is a circulant matrix (see Definition A.63). Therefore, by Proposition A.64, the eigenvalues of K_n are given by $\chi_a = \sum_{j=1}^{n-1} \xi^{aj}$, where $\xi = \exp(2\pi i/n)$ and a is an integer with $0 \leq a \leq n-1$. If $a = 0$, then $\chi_0 = n-1$. If $0 < a \leq n-1$, by Lemma 1.51,

$$\chi_a = \sum_{j=1}^{n-1} (\xi^a)^j = \sum_{j=0}^{n-1} (\xi^a)^j - 1 = -1.$$

Therefore,

$$\mathrm{Spec}(K_n) = \begin{pmatrix} -1 & n-1 \\ n-1 & 1 \end{pmatrix}.$$

Note that if $n \geq 3$ then K_n is not bipartite, because $-(n-1)$ is not an eigenvalue of K_n. (We can also quickly use the definition to check that K_n is nonbipartite.)

Example 1.53
The Cayley graph $\mathrm{Cay}(\mathbb{Z}_n, \{1, -1\})$ is called the *cycle graph* on n vertices and is denoted by C_n. The first few cycle graphs are shown in Figure I.5.

The adjacency matrix for C_n is the circulant matrix

$$\begin{pmatrix} 0 & 1 & 0 & 0 & 0 & \dots & 0 & 1 \\ 1 & 0 & 1 & 0 & 0 & \dots & 0 & 0 \\ 0 & 1 & 0 & 1 & 0 & \dots & 0 & 0 \\ \vdots & \vdots & \vdots & \vdots & \vdots & \ddots & \vdots & \vdots \\ 1 & 0 & 0 & 0 & 0 & \dots & 1 & 0 \end{pmatrix}.$$

Let $\xi = \exp(2\pi i/n)$. By Proposition A.64 the eigenvalues of C_n are

$$\begin{aligned}\chi_a &= \xi^a + \xi^{a(n-1)} \\ &= \cos\left(\tfrac{2\pi a}{n}\right) + i\sin\left(\tfrac{2\pi a}{n}\right) + \cos\left(\tfrac{2\pi a(n-1)}{n}\right) + i\sin\left(\tfrac{2\pi a(n-1)}{n}\right),\end{aligned}$$

where $a = 0, 1, 2, \ldots, n-1$. Note that $\cos(2\pi a(n-1)/n) = \cos(-2\pi a/n) = \cos(2\pi a/n)$. Similarly, $\sin(2\pi a(n-1)/n) = -\sin(2\pi a/n)$. Therefore

$$\chi_a = 2\cos\left(\frac{2\pi a}{n}\right).$$

If n is even, then

$$\mathrm{Spec}(C_n) = \begin{pmatrix} -2 & 2\cos\left(\frac{2\pi(n/2-1)}{n}\right) & \cdots & 2\cos\left(\frac{4\pi}{n}\right) & 2\cos\left(\frac{2\pi}{n}\right) & 2 \\ 1 & 2 & \cdots & 2 & 2 & 1 \end{pmatrix}.$$

If n is odd, then

$$\mathrm{Spec}(C_n) = \begin{pmatrix} 2\cos\left(\frac{2\pi(n/2-1)}{n}\right) & \cdots & 2\cos\left(\frac{4\pi}{n}\right) & 2\cos\left(\frac{2\pi}{n}\right) & 2 \\ 2 & \cdots & 2 & 2 & 1 \end{pmatrix}.$$

Later in this chapter, we will see that the second largest eigenvalue is of particular importance for our purposes. So take note of the fact that $\lambda_1(C_n) = 2\cos(2\pi/n)$.

Remark 1.54

You may wonder why we defined $L^2(X)$ as the set of functions from X into the complex numbers, as opposed to functions from X into the real numbers. After all, we are mainly interested in how the adjacency operator (which has real entries when realized as a matrix) interacts with these functions. Indeed, in many of the proofs in this section, we restricted ourselves to real-valued eigenfunctions of the adjacency matrix. However, when we bring representation theory into the picture, we need the complex structure of $L^2(X)$.

5. THE LAPLACIAN

The adjacency operator is one of the many linear operators associated to a graph. In this section, we discuss another one, called the Laplacian. Recall from multivariable calculus that the ordinary Laplacian is defined by $\Delta(f) = \mathrm{div}(\mathrm{grad}(f))$ and that it provides a measure of a function's rate of change. The Laplacian on a graph is a discrete analogue. There are several reasons for introducing this new operator. One is that the results in this section simplify the proofs of many of our main theorems. Another reason is that many facts from differential geometry about the Laplacian on manifolds extend, with appropriate modifications to graphs. Such results usually hold for arbitrary graphs, unlike the results in this book, where we deal almost exclusively with regular graphs. We do not adopt this geometric viewpoint here; consult [40] for a thorough treatment.

Let X be a graph with vertex set V and edge set E. Give the multiset of edges of X an arbitrary orientation. That is, for each edge $e \in E$, label one endpoint e^+ and the other endpoint e^-. We call e^- the *origin* of e, and e^+ the *extremity* of e. See Figure 1.12. Here we treat multiple edges as distinct edges. If e is a loop, then $e^+ = e^-$ is the vertex that is incident to e.

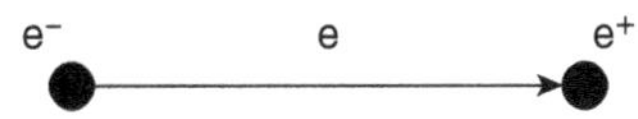

Figure 1.12

We first define a finite analogue of the gradient operator. Let $d : L^2(V) \to L^2(E)$ be defined for each $f \in L^2(V)$ as

$$(df)(e) = f(e^+) - f(e^-).$$

That is, $(df)(e)$ measures the change of f along the edge e of the graph.

We now define a finite analogue of the divergence operator. Let $d^* : L^2(E) \to L^2(V)$ be defined for each $f \in L^2(E)$ as

$$(d^*f)(v) = \sum_{\substack{e\in E \\ v=e^+}} f(e) - \sum_{\substack{e\in E \\ v=e^-}} f(e).$$

That is, if we think of the function f as a flow on the edges of the graph X, then $(d^*f)(v)$ measures the total inward flow at the vertex v.

Remark 1.55
In algebraic topology, d is called the simplicial coboundary operator and d^* is called the simplicial boundary operator.

Example 1.56
Figure 1.13 shows a graph X with vertex set V and edge multiset E and an orientation on E, as well as a function $f \in L^2(E)$ and a function $g \in L^2(V)$. Then we have that

$$\begin{aligned} (d^*f)(v) &= 5 + 4 - (2 + 1 + 3) = 3, \\ (d^*f)(w) &= 2 - (2 + 0 + 5) = -5, \\ (dg)(e_1) &= 3 - 0 = 3, \\ \text{and } (dg)(e_2) &= 2 - 2 = 0. \end{aligned}$$

In $(d^*f)(w)$, note that the loop at w contributes 2 to both the inflow and outflow, thus having no net effect. Note that $d^*dg \in L^2(V)$ and that

$$(d^*dg)(w) = \sum_{\substack{e\in E \\ w=e^+}} dg(e) - \sum_{\substack{e\in E \\ w=e^-}} dg(e) = [2-2] - [(3-2) + (0-2)] = 1.$$

Definition 1.57 Suppose X is a graph with vertex set V and edge set E. Given an orientation on the edges, define the Laplacian operator $\Delta : L^2(V) \to L^2(V)$ to be $\Delta = d^*d$.

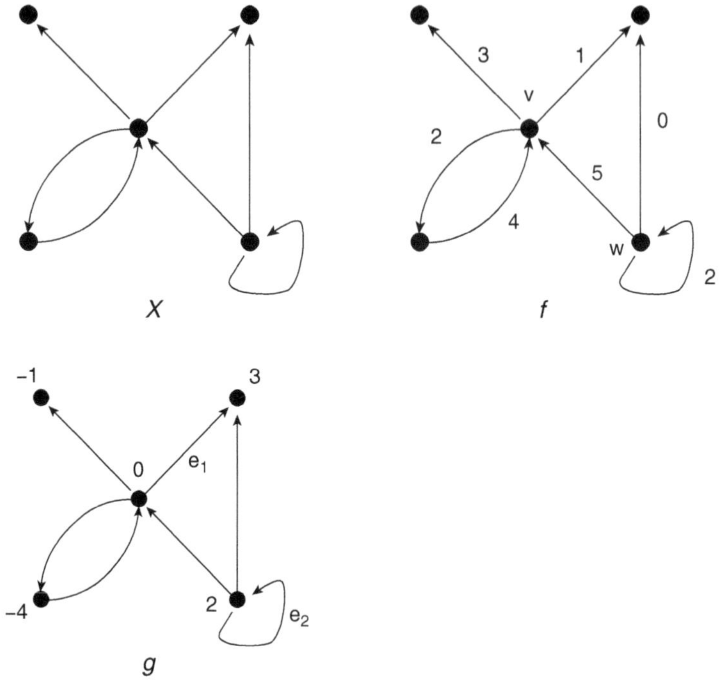

Figure 1.13

The maps d and d^* depend on the choice of orientation. The Laplacian, however, does not. The next lemma demonstrates this fact for regular graphs, in which case the Laplacian has a particularly nice relationship to the adjacency operator.

Lemma 1.58
If X is a k-regular graph with vertex set V, edge multiset E, and adjacency operator A, then $\Delta = kI - A$.

Proof
Let $f \in L^2(V)$ and $x \in V$. Then

$$
\begin{aligned}
(\Delta f)(x) &= (d^*(df))(x) \\
&= \sum_{\substack{e\in E \\ x=e^+}} (df)(e) - \sum_{\substack{e\in E \\ x=e^-}} (df)(e) \\
&= \left(\sum_{\substack{e\in E \\ x=e^+}} f(x) - \sum_{\substack{e\in E \\ x=e^+ \text{ and } y=e^-}} f(y) \right) && (3) \\
&\quad - \left(\sum_{\substack{e\in E \\ x=e^- \text{ and } y=e^+}} f(y) - \sum_{\substack{e\in E \\ x=e^-}} f(x) \right) && (4)
\end{aligned}
$$

$$= kf(x) - \sum_{y \in V} A_{x,y} f(y)$$
$$= ((kI - A)f)(x).$$

The argument is correct, but there is one subtlety that can be confusing. Suppose that e has is a loop incident to x. Then, $x = e^+$ and $x = e^-$. So the portion of the expression that e contributes to in lines (3) and (4) is

$$f(x) - f(x) - (f(x) - f(x)) = 2f(x) - 2f(x) = f(x) - A_{x,x} f(x),$$

which is consistent with the derivation given. Here we are using the fact that a loop at a vertex is counted once in the adjacency matrix A. ⊛

Remark 1.59
Since $\Delta = kI - A$, Δ is a linear transformation from $L^2(X)$ to $L^2(X)$. In particular, $\Delta(\alpha f) = \alpha \Delta f$ where $\alpha \in \mathbb{C}$.

The equation $\Delta = kI - A$ from Lemma 1.58 immediately allows us to express the eigenvalues of Δ in terms of the eigenvalues of A. Recall Prop. 1.48. Later in this book, we will see that the eigenvalues of a graph X are related to the inner product $\langle \Delta f, f \rangle_2$. Our next proposition gives us a useful form for this expression.

Proposition 1.60
Suppose X is a k-regular graph with vertex set V and edge multiset E. Let $n = |V|$. Orient the edges of X.

1. The eigenvalues of Δ are given by

$$0 = k - \lambda_0(X) \leq k - \lambda_1(X) \leq \cdots \leq k - \lambda_{n-1}(X).$$

In particular, the eigenvalues of Δ lie in the interval $[0, 2k]$.

2. Let $f \in L^2(V)$, and $g \in L^2(E)$. Then $\langle df, g \rangle_2 = \langle f, d^* g \rangle_2$ and

$$\langle \Delta f, f \rangle_2 = \sum_{e \in E} \left| f(e^+) - f(e^-) \right|^2.$$

Proof

1. Let $f \in L^2(V)$. Note that $Af = \lambda f$ if and only if $(kI - A)f = (k - \lambda)f$. The result follows from Lemma 1.58 and Prop. 1.48.
2. Note that

$$\langle df, g \rangle_2 = \sum_{e \in E} (df)(e)\overline{g(e)} = \sum_{e \in E} \left[f(e^+) - f(e^-) \right] \overline{g(e)}$$
$$= \sum_{e \in E} f(e^+)\overline{g(e)} - \sum_{e \in E} f(e^-)\overline{g(e)}$$
$$= \sum_{v \in V} f(v) \sum_{\substack{e \in E \\ v = e^+}} \overline{g(e)} - \sum_{v \in V} f(v) \sum_{\substack{e \in E \\ v = e^-}} \overline{g(e)}$$

$$= \sum_{v \in V} f(v)\overline{(d^*g)(v)}$$

$$= \langle f, d^*g \rangle_2 .$$

Thus,

$$\langle \Delta f, f \rangle_2 = \langle d^* df, f \rangle_2 = \overline{\langle f, d^* df \rangle_2} = \overline{\langle df, df \rangle_2} = \langle df, df \rangle_2 = \|df\|_2^2 ,$$

and

$$\langle df, df \rangle_2 = \sum_{e \in E} (f(e^+) - f(e^-))\overline{(f(e^+) - f(e^-))} = \sum_{e \in E} |f(e^+) - f(e^-)|^2 .$$

6. THE ISOPERIMETRIC CONSTANT

Think of a graph X as a communications network where the vertices represent entities (e.g., people, computers) that want to communicate with one another. Two vertices are connected by an edge iff they can communicate directly with one another.

> The purpose of the network is, of course, to transmit information quickly. Suppose a few individuals know a piece of information. Let us think of time in discrete units and suppose that in one unit of time the information can be carried to all the nearest neighbors of these individuals. In the next unit of time, the information is carried to the neighbors of the neighbors and so forth. How long does it take before all the individuals receive this information? To minimize the transmission time, what is clearly needed is that every subset of vertices has a lot of distinct neighbors. . . . On the other hand, the total length of cables needed to wire a network is also a quantity we would like to minimize for several reasons. . . . To simplify the model, suppose all pieces of cables have the same length. An efficient communications network is thus represented by a graph with a small number of edges and such that every subset of vertices has many distinct neighbors. [22]

We are interested in explicitly constructing very large graphs with that have good "expansion" properties but do not use a lot of edges. To do so, we limit ourselves to regular graphs. In a large regular graph with low degree, edges are very sparse.

In Def. 1.63 we define the isoperimetric constant of a graph. This quantity measures how quickly information can flow through the graph. Expander families are certain sequences of regular graphs so that the isoperimetric constant is uniformly bounded away from 0—see Def. 1.74. In light of the foregoing discussion, we can view expander families as good communication networks.

Definition 1.61 Let X be a graph with vertex set V. Let $F \subset V$. The *boundary* of F, denoted by ∂F, is defined to be the set of edges with one endpoint in F and one endpoint in $V \setminus F$. That is, ∂F is the set of edges connecting F to $V \setminus F$.

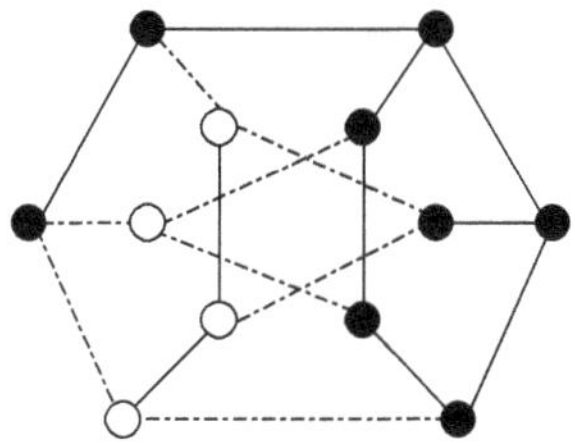

Figure 1.14

Example 1.62
In the graph in Figure 1.14, the set F consists of the white vertices. The boundary of F consists of the dashed lines. Note that F has four vertices and ∂F has eight edges, that is, $|F| = 4$ and $|\partial F| = 8$. Note that reversing the roles of the black dots and the white dots does not change the boundary—in other words, $\partial F = \partial(V \setminus F)$.

Definition 1.63 The *isoperimetric constant* of a graph X with vertex set V is defined as

$$h(X) = \min \left\{ \frac{|\partial F|}{|F|} \;\middle|\; F \subset V \text{ and } |F| \leq \frac{|V|}{2} \right\}.$$

Example 1.64
Let's compute the isoperimetric constant of the graph C_4 given in Figure 1.15.

Because of the symmetry of the graph, we need only perform the three computations illustrated in Figure 1.16. In each picture, the set F is given by the

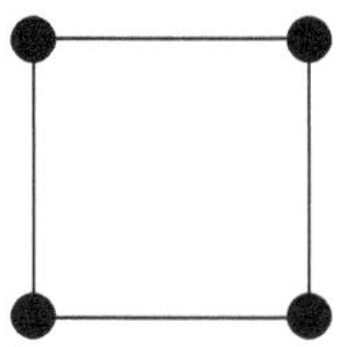

Figure 1.15 C_4

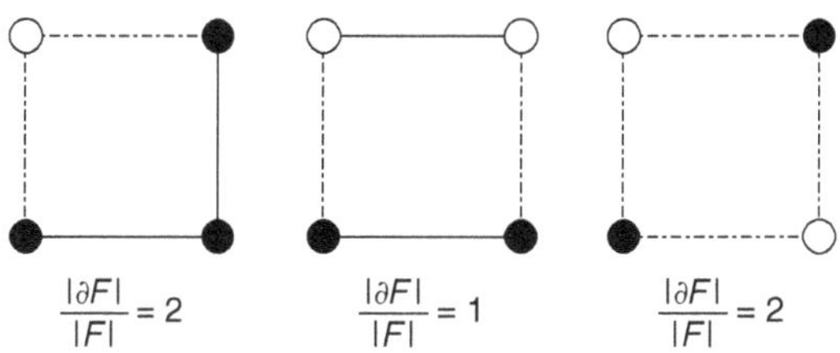

Figure 1.16

white vertices, and the boundary of F consists of the dashed lines. Taking the minimum of the ratios $|\partial F| / |F|$ yields $h(C_4) = 1$.

Remark 1.65
Let X be a graph with vertex set V. Let $n = |V|$. If $F \subset V$ with $|F| \leq n/2$, then $|\partial F| \geq h(X)\,|F|$. That is, the size of the boundary of F is at least $h(X)$ times the size of F. When $h(X)$ is large, every set of no more than half the vertices F will have a lot of distinct neighbors relative to its size.

Remark 1.66
Let X be a graph with vertex set V. Let $n = |V|$. Let us explain why the definition of $h(X)$ only includes subsets of V of size less than or equal to $n/2$. Suppose that $F \subset V$ and $|F| > n/2$. It doesn't make sense to include the ratio $|\partial F| / |F|$ in the calculation of $h(X)$ because F is too large and this ratio measures how much information flows from F into $V \setminus F$ relative to the size of the larger set F. Note that $\partial F = \partial(V \setminus F)$ and $|V \setminus F| \leq n/2$. Hence, the ratio $\left|\partial(V \setminus F)\right| / |V \setminus F|$ is included in the calculation of $h(X)$ and it is more appropriate: it measures how much information flows from $V \setminus F$ into F relative to the size of the smaller set $V \setminus F$. A symmetrical definition of $h(X)$, which is equivalent to Def. 1.63, is as follows.

$$h(X) = \min \left\{ \frac{|\partial F|}{\min\{|F|\,,\,|V \setminus F|\}} \;\middle|\; F \subset V \right\}.$$

Remark 1.67
See Exercises 3 and 4 for some basic facts about $h(X)$. See the notes—especially Note 8—for references to publications that discuss $h(X)$.

Remark 1.68
The isoperimetric constant goes by many other names. It is sometimes called the expansion constant, the edge expansion constant, the conductance, or the Cheeger constant (the latter particularly when emphasizing the geometric connections).

Remark 1.69
Let X be a graph with vertex set V. By the definition of the isoperimetric constant of X, there exists at least one subset F_0 of vertices such that $h(X) = \left|\partial F_0\right| / \left|F_0\right|$. In a sense, this subset measures the worst-case senario for X. That is, every other subset of vertices F of X has a larger boundary, relative to its size, than F_0 does.

Example 1.70
Consider the graph in Figure 1.17. Let F_0 be the set of white vertices. Then $1/7 = \left|\partial F_0\right| / \left|F_0\right| = h(X)$. No other F has a smaller boundary, as a fraction of its size. Moreover, notice that deleting ∂F_0 cuts F_0 off from the rest of the graph. The single edge creates a bottleneck. If the edge from ∂F_0 was removed, the graph would become disconnected. The isoperimetric constant provides some measure of connectivity in a graph.

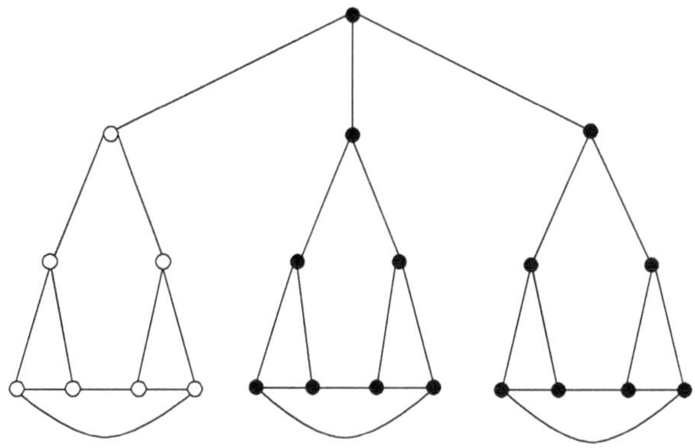

Figure 1.17

Example 1.71
Let's compute the isoperimetric constant of the complete graph K_n from Example 1.52. Let V be the vertex set of K_n. If $F \subset V$, then

$$\frac{|\partial F|}{|F|} = \frac{(|V| - |F|)\,|F|}{|F|} = |V| - |F| = n - |F|\,.$$

Therefore,

$$h(K_n) = \begin{cases} \frac{n}{2} & \text{if } n \text{ is even.} \\ \frac{n+1}{2} & \text{if } n \text{ is odd.} \end{cases}$$

Notice that $h(K_n)$ grows as K_n grows in size. This makes sense: K_n is a very good communications network. (Every vertex is adjacent to every other vertex!) However, K_n is very "expensive." That is, K_n contains many edges.

Definition 1.72 Let (a_n) be a sequence of nonzero real numbers. We say that (a_n) is *bounded away from zero* if there exists a real number $\epsilon > 0$ such that $a_n \geq \epsilon$ for all n.

Example 1.73
Notice that $(\frac{1}{n})$ and $(1 + (-1)^n)$ are not bounded away from zero, while $(\frac{n}{3n+2})$ is.

Our goal is to construct arbitrarily large graphs with large isoperimetric constants, while at the same time controlling the number of edges in our graph. To ensure that we do not use too many edges, we require our graphs to be regular of fixed degree. The following definition encapsulates the kinds of graphs we're looking for.

Definition 1.74 Let d be a positive integer. Let (X_n) be a sequence of d-regular graphs such that $|X_n| \to \infty$ as $n \to \infty$. We say that (X_n) is an *expander family* if the sequence $(h(X_n))$ is bounded away from zero.

Remark 1.75
It might be more accurate to call expander families "expander sequences," with the term "expander family" reserved for arbitrary families $\{X_\alpha\}$ of graphs satisfying $h(X_\alpha) \geq \epsilon$ for all α. The nomenclature, however, has stuck. Even worse, the literature is filled with references to "expander graphs," even though the objects of study are sequences of graphs, not individual graphs. *C'est la mathematique.*

Remark 1.76
There are several other ways to define an expander family. One of these approaches is given in Exercises 19–21.

Example 1.77
We now show that there do not exist expander families of degree 2. Recall the cycle graphs from Example 1.53. Note that (C_n) is a sequence of 2-regular connected graphs. In fact, any connected 2-regular graph must be isomorphic to C_n for some n.

Let V be the vertex set of C_n. If k is a fixed integer such that $1 \leq k \leq n/2$, then

$$\min\left\{\frac{|\partial F|}{|F|} \ : \ |F| = k, F \subset V\right\} = \frac{2}{k}.$$

To see this, note that among the subsets $F \subset V$ of a fixed size, the minimum of $\frac{|\partial F|}{|F|}$ occurs when the vertices of F are bunched up together (i.e., there are no vertices of $V \setminus F$ "between" any vertices of F). See Figure I.6. In this case, $|\partial F| = 2$ and $|F|$ is $n/2$ or $(n-1)/2$, depending on whether n is even or odd. Therefore,

$$h(C_n) = \min_{\substack{0<|F|\leq n/2 \\ F\subset V}} \left\{\frac{|\partial F|}{|F|}\right\} = \begin{cases} \frac{4}{n} & \text{if } n \text{ is even.} \\ \frac{4}{n-1} & \text{if } n \text{ is odd.} \end{cases}$$

Notice that $h(C_n) \to 0$ as $n \to \infty$. Therefore, (C_n) is not an expander family.

Intuitively, this fact makes sense, because for large n, the graphs C_n would be a lousy communications network. Two edges per vertex, then, are simply not enough to design a large, fast, and reliable network.

The next example gives a 3-regular family of Cayley graphs constructed from symmetric groups that do not form an expander family. Later we give several other proofs of this result using different techniques. See Examples 1.91, 4.21, and 8.21. Note that there do, in fact, exist 3-regular expander families—for example, see Note 6.

Example 1.78
Consider the symmetric group S_n. Assume that $n \geq 4$. Let $\Gamma = \{\sigma, \sigma^{-1}, \tau\} \subset S_n$ where $\sigma = (1, 2, \ldots, n)$ and $\tau = (1, 2)$. Let $X_n = \text{Cay}(S_n, \Gamma)$. The graph X_n is called a *bubble-sort graph*, because of its connection to the bubble-sort algorithm from computer science. We now show that (X_n) is not an expander family.

Recall the notation for the floor function $\lfloor \cdot \rfloor$ given in the Introduction. Consider the set $B_m = \{\beta \in S_n | 1 \leq \beta^{-1}(1) \leq m\}$ where $m = \lfloor \frac{n}{2} \rfloor$. Then, $|B_m| = m(n-1)!$. Note that $\alpha \notin B_m$ and $\beta \in B_m$ are adjacent if and only if $\beta = \alpha\gamma = \alpha \circ \gamma$ for some $\gamma \in \Gamma$. We have three cases to consider. Throughout the cases, assume that $\alpha \notin B_m$, $\beta \in B_m$, and $\beta = \alpha\gamma$.

1. Suppose $\gamma = \sigma$. Since $\alpha \notin B_m$, the permutation α cannot map any element between 1 and m to 1. To ensure that $\beta \in B_m$, we must have that $\alpha(m+1) = 1$. This implies that $\beta(m) = 1$. There are $(n-1)!$ such β. We get one edge from σ that leaves B_m for each such β.
2. If $\gamma = \sigma^{-1}$, then we must have that $\alpha(n) = 1$ and $\beta(1) = 1$. Again, there are $(n-1)!$ such β, yielding $(n-1)!$ edges with one extremity in B_m and one extremity in $X_n \setminus B_m$.
3. The case of $\gamma = \tau$ cannot happen.

Therefore,

$$h(X_n) \leq \frac{|\partial B_m|}{|B_m|} = \frac{2(n-1)!}{m(n-1)!} = \frac{2}{m} \to 0$$

as $n \to \infty$. Hence the sequence (X_n) of bubble-sort graphs does not form an expander family.

Remark 1.79
In Chapter 8, we will construct an actual expander family.

7. THE RAYLEIGH-RITZ THEOREM

Recall that $\lambda_1(X)$ is the second-largest eigenvalue of the graph X. The main result of this section states that a sequence (X_n) of d-regular graphs forms an expander family iff the sequence $(d - \lambda_1(X_n))$ is bounded away from zero. To prove this theorem, we employ the Rayleigh-Ritz theorem, which provides a useful method for determining λ_1, or at least getting good bounds for it. The Rayleigh-Ritz theorem plays a role in nearly every major result in this book; we see it again in the proofs of the Alon-Boppana theorem, eigenvalue bounds for zig-zag products, and the relationship between eigenvalues and Kazhdan constants.

Definition 1.80 Let X be a finite set and f_0 be the function that is equal to 1 on all of X. Define

$$L^2(X, \mathbb{R}) = \{f : X \to \mathbb{R}\}$$

and

$$\begin{aligned} L_0^2(X, \mathbb{R}) &= \{f \in L^2(X, \mathbb{R}) \mid \langle f, f_0 \rangle_2 = 0\} \\ &= \{f \in L^2(X, \mathbb{R}) | \sum_{x \in X} f(x) = 0\}. \end{aligned}$$

Note that $L^2(X, \mathbb{R}) \subset L^2(X)$.

Remark 1.81
The inner product of two functions in $L^2(X, \mathbb{R})$ simplifies, since the conjugate of a real number is itself. If $f, g \in L^2(X, \mathbb{R})$, then

$$\langle f, g \rangle_2 = \sum_{x \in X} f(x) g(x) \quad \text{and} \quad \|f\|_2 = \sqrt{\sum_{x \in X} f(x)^2}.$$

Proposition 1.82 (Rayleigh-Ritz)
Let X be a d-regular graph with vertex set V. Then

$$\lambda_1(X) = \max_{f \in L_0^2(V, \mathbb{R})} \frac{\langle Af, f \rangle_2}{\|f\|_2^2} = \max_{\substack{f \in L_0^2(V, \mathbb{R}) \\ \|f\|_2 = 1}} \langle Af, f \rangle_2 .$$

Equivalently, we have that

$$d - \lambda_1(X) = \min_{f \in L_0^2(V, \mathbb{R})} \frac{\langle \Delta f, f \rangle_2}{\|f\|_2^2} = \min_{\substack{f \in L_0^2(V, \mathbb{R}) \\ \|f\|_2 = 1}} \langle \Delta f, f \rangle_2 .$$

Proof
By Theorem A.53, there exists an orthonormal basis $\{f_0, f_1, f_2, \ldots, f_{n-1}\}$ for $L^2(V, \mathbb{R})$, such that each f_i is a real-valued eigenfunction of A associated with the eigenvalue $\lambda_i = \lambda_i(X)$. Recall that f_0 is constant on V. (See the proof of Prop. 1.48 (1).)

Let $f \in L_0^2(V, \mathbb{R})$ with $\|f\|_2 = 1$. Then, $f = c_0 f_0 + c_1 f_1 + \cdots + c_{n-1} f_{n-1}$ for some scalars $c_i \in \mathbb{R}$. Note that

$$0 = \langle f, f_0 \rangle_2 = c_0 \langle f_0, f_0 \rangle_2 + c_1 \langle f_1, f_0 \rangle_2 + \cdots + c_{n-1} \langle f_{n-1}, f_0 \rangle_2 = c_0.$$

So, $f = c_1 f_1 + \cdots + c_{n-1} f_{n-1}$. We also have that

$$\begin{aligned} \langle Af, f \rangle_2 &= \left\langle A \sum_{i=1}^{n-1} c_i f_i, \sum_{j=1}^{n-1} c_j f_j \right\rangle_2 = \left\langle \sum_{i=1}^{n-1} c_i \lambda_i f_i, \sum_{j=1}^{n-1} c_j f_j \right\rangle_2 \\ &= \sum_{i=1}^{n-1} \sum_{j=1}^{n-1} c_i c_j \lambda_i \langle f_i, f_j \rangle_2 = \sum_{i=1}^{n-1} c_i^2 \lambda_i \\ &\leq \lambda_1 \sum_{i=1}^{n-1} c_i^2 = \lambda_1 \|f\|_2^2 = \lambda_1. \end{aligned}$$

Hence

$$\lambda_1(X) \geq \max_{\substack{f \in L_0^2(V, \mathbb{R}) \\ \|f\|_2 = 1}} \langle Af, f \rangle_2 .$$

Note that $f_1 \in L^2_0(V,\mathbb{R})$, $\|f_1\|_2 = 1$, and

$$\langle Af_1, f_1\rangle_2 = \langle \lambda_1 f_1, f_1\rangle_2 = \lambda_1.$$

Thus

$$\lambda_1(X) = \max_{\substack{f\in L^2_0(V,\mathbb{R})\\ \|f\|_2=1}} \langle Af, f\rangle_2 .$$

The second statement, about Δ, follows from the fact that $\Delta = dI - A$ (see Lemma 1.58).

Remark 1.83
Let X be a graph with vertex set V. One application of Proposition 1.82 is finding lower bounds of $\lambda_1(X)$. For if $f \in L^2_0(V,\mathbb{R})$, then

$$\lambda_1(X) \geq \frac{\langle Af, f\rangle_2}{\|f\|_2^2}.$$

By cleverly choosing f one can get good lower bounds on $\lambda_1(X)$. We use this method several times throughout this book.

Proposition 1.84
Let X be a d-regular graph with vertex set V and edge multiset E. Then

$$\frac{d-\lambda_1(X)}{2} \leq h(X) \leq \sqrt{2d(d-\lambda_1(X))}.$$

Proof
We prove the lower bound for $h(X)$. The proof for the upper bound is fairly involved, so we defer it until Section 9.

By the definition of $h(X)$, we may pick a subset $F \subset V$ such that $|F| \leq |V|/2$ and $h(X) = |\partial F| / |F|$. Let $a = |V \setminus F|$ and $b = |F|$.

Define the functions

$$g(x) = \begin{cases} a & \text{if } x \in F \\ -b & \text{if } x \in V \setminus F \end{cases}$$

and $f = g/\|g\|_2$. Because

$$\sum_{v\in V} g(v) = \sum_{v\in F} a - \sum_{v\in V\setminus F} b = ba - ab = 0,$$

we see that $f, g \in L^2_0(V,\mathbb{R})$.

Orient the edges of the graph X. By Proposition 1.60 we have that

$$\langle \Delta g, g \rangle_2 = \sum_{e \in E} |g(e^+) - g(e^-)|^2 = \sum_{e \in \partial F} (b+a)^2 = |\partial F| (b+a)^2 .$$

Also

$$\|g\|_2^2 = \langle g, g \rangle_2 = \sum_{x \in F} a^2 + \sum_{x \in V \setminus F} b^2 = a^2 b + b^2 a = ab(a+b).$$

The fact that $b \leq |V|/2$ implies that $a \geq b$. Recall Remark 1.59. We see that

$$\langle \Delta f, f \rangle_2 = \frac{1}{\|g\|_2^2} \langle \Delta g, g \rangle_2 = \frac{|\partial F| (b+a)}{ba} = \left(1 + \frac{b}{a}\right) h(X) \leq 2h(X).$$

From Prop. 1.82 we know that $d - \lambda_1(X) \leq \langle \Delta f, f \rangle_2$, which completes the proof of the lower bound on $h(X)$. ⊛

Remark 1.85

Note that the expression $d - \lambda_1(X)$ appears on both sides of the double inequality in Proposition 1.84. Roughly speaking, the smaller $\lambda_1(X)$ is, the larger $h(X)$ is. Hence, the smaller $\lambda_1(X)$ is, the better the graph X is as a communications network.

Definition 1.86 If X is a connected d-regular graph, then $d - \lambda_1(X)$ is called the *spectral gap* of X.

The following result follows immediately from Prop. 1.84.

Corollary 1.87

Let (X_n) be a sequence of d-regular graphs with $|X_n| \to \infty$ as $n \to \infty$. Then (X_n) is a family of expanders if and only if the sequence $(d - \lambda_1(X_n))$ is bounded away from zero.

Corollary 1.87 could make a strong claim for the title of "Fundamental Theorem of Expander Families." Isoperimetric constants are difficult to deal with directly; eigenvalues provide more traction. Because of Corollary 1.87, the search for expander families focuses almost exclusively on graph spectra, with particular attention on the second-largest eigenvalue.

Example 1.88

In Example 1.53, we saw that $\lambda_1(C_n) = 2\cos\left(\frac{2\pi}{n}\right)$. Hence, the spectral gap of C_n satisfies $2 - 2\cos\left(\frac{2\pi}{n}\right) \to 0$ as $n \to \infty$. Therefore, $h(C_n) \to 0$ as $n \to \infty$. This gives us another proof that (C_n) is not a family of expanders.

Sometimes we need to consider the set of complex-valued functions that sum to 0 on the vertex set of a graph. This is the complex analogue of $L_0^2(V, \mathbb{R})$. We give it in the following definition.

Definition 1.89 Let X be a finite set and f_0 be the function that is equal to 1 on all of X. Define

$$L_0^2(X) = \{f \in L^2(X) \mid \langle f, f_0 \rangle_2 = 0\}$$
$$= \{f \in L^2(X) \mid \sum_{x \in X} f(x) = 0\}.$$

Remark 1.90
Let X be a graph with vertex set V. Let A be the adjacency matrix of X. Suppose that $f \in L_0^2(V)$. By Lemma A.31 we have that $\overline{\langle Af, f \rangle_2} = \overline{\langle f, Af \rangle_2} = \langle Af, f \rangle_2$. Hence $\langle Af, f \rangle_2$ is real. Using the same logic as in the proof of Prop. 1.82 one can show that

$$\lambda_1(X) = \max_{g \in L_0^2(X)} \frac{\langle Ag, g \rangle_2}{\|g\|_2^2}.$$

For a given $f \in L_0^2(V)$ we have that $\lambda_1(X) \geq \langle Af, f \rangle_2 / \langle f, f \rangle_2$.

We return to the bubble-sort graphs. We saw this family previously in Example 1.78, and we will see it again in Examples 4.21 and 8.21, each time to illustrate a different technique.

Example 1.91
Let $\Gamma = \{\sigma, \sigma^{-1}, \tau\} \subset S_n$ where $\sigma = (1, 2, \ldots, n)$ and $\tau = (1, 2)$. We will prove that $(X_n) = (\text{Cay}(S_n, \Gamma))$ is not an expander family. Note that each X_n is 3-regular. By Corollary 1.87 it is sufficient to show that the spectral gap $3 - \lambda_1(X_n) \to 0$ as $n \to \infty$. We use the Raleigh-Ritz theorem (Prop. 1.82) to do this.

Define the function $f : S_n \to \mathbb{C}$ by $f(\alpha) = \exp\left(\frac{2\pi i}{n} \cdot \alpha^{-1}(1)\right)$. We first show that $f \in L_0^2(S_n)$. By Remark 1.90, this implies that $\lambda_1(X_n) \geq \langle Af, f \rangle_2 / \langle f, f \rangle_2$.

For each a satisfying $1 \leq a \leq n$ there are $(n-1)!$ permutations α of S_n such that $\alpha^{-1}(1) = a$. By Lemma 1.51, we have that

$$\sum_{\alpha \in S_n} f(\alpha) = (n-1)! \sum_{a=1}^{n} \exp\left(\frac{2\pi ia}{n}\right) = 0.$$

So $f \in L_0^2(S_n)$.

Next we calculate $\langle Af, f \rangle_2$. By Remark 1.46, we have that

$$\langle Af, f \rangle_2 = \sum_{\beta \in S_n} f(\beta\sigma)\overline{f(\beta)} + \sum_{\beta \in S_n} f(\beta\sigma^{-1})\overline{f(\beta)} + \sum_{\beta \in S_n} f(\beta\tau)\overline{f(\beta)}.$$

If $\beta^{-1}(1) = a$, then $(\beta\sigma)^{-1}(1) = (\sigma^{-1} \circ \beta^{-1})(1) = \sigma^{-1}(a)$. Therefore,

$$\sum_{\beta \in S_n} f(\beta\sigma)\overline{f(\beta)} = \sum_{\beta^{-1}(1)=1} f(\beta\sigma)\overline{f(\beta)} + \cdots + \sum_{\beta^{-1}(1)=n} f(\beta\sigma)\overline{f(\beta)}$$
$$= (n-1)! \left[e^{\frac{2\pi i}{n} \cdot n} e^{-\frac{2\pi i}{n}} + e^{\frac{2\pi i}{n} \cdot 1} e^{-\frac{2\pi i}{n} \cdot 2} + \cdots + e^{\frac{2\pi i}{n} \cdot (n-1)} e^{-\frac{2\pi i}{n} \cdot n} \right]$$
$$= (n-1)! \left[e^{\frac{2\pi i}{n} \cdot (n-1)} + (n-1) e^{-\frac{2\pi i}{n}} \right]$$
$$= n! \left[e^{-\frac{2\pi i}{n}} \right]$$

(because $\exp(2\pi i(n-1)/n) = \exp(-2\pi i/n)$). Similarly,

$$\sum_{\beta \in S_n} f(\beta\sigma^{-1})\overline{f(\beta)} = n! \left[e^{\frac{2\pi i}{n}} \right]$$

and

$$\sum_{\beta \in S_n} f(\beta\tau)\overline{f(\beta)} = (n-1)! \left[e^{\frac{2\pi i}{n}} + e^{-\frac{2\pi i}{n}} + (n-2) \right].$$

Hence

$$\langle Af, f \rangle_2 = (n-1)! \left[(n+1) e^{\frac{-2\pi i}{n}} + (n+1) e^{\frac{2\pi i}{n}} + (n-2) \right]$$
$$= (n-1)! \left[2(n+1)\mathrm{Re}(e^{\frac{2\pi i}{n}}) + (n-2) \right].$$

Calculating $\langle f, f \rangle_2$ gives

$$\langle f, f \rangle_2 = \sum_{\alpha \in S_n} \exp\left(\frac{2\pi i}{n} \cdot \alpha^{-1}(n) \right) \exp\left(-\frac{2\pi i}{n} \cdot \alpha^{-1}(n) \right) = \sum_{\alpha \in S_n} 1 = n!.$$

By Remark 1.90, we have that

$$\lambda_1(X_n) \geq \frac{2(n+1)\mathrm{Re}(e^{\frac{2\pi i}{n}}) + (n-2)}{n} \to 3$$

as n goes to infinity. So by Corollary 1.87, (X_n) is not an expander family.

8. POWERS AND PRODUCTS OF ADJACENCY MATRICES

Definition 1.92 We define a *directed graph* X to be a set V (the set of *vertices*) along with a multiset E (the set of *directed edges*), where every element of E is an ordered pair (v, w) of elements of V. We visualize the ordered pair (v, w) as an arrow pointing from v to w. We call (v, w) *the directed edge from v to w*. We call v the *initial point* and w the *terminal point* of the directed edge (v, w). A directed edge of the form (v, v) is called a *loop at v*.

Let X be a finite directed graph with vertex set V. Choose an ordering $(v_1, \ldots, v_n)$ of V. Define the adjacency matrix of X with respect to this ordering to be the $n \times n$ matrix whose ij entry equals the number of directed edges in X from v_i to v_j.

Example 1.93
Figure 1.18 depicts a directed graph X with vertex set $\{a, b, c\}$ and edge multiset $\{(b, a), (b, a), (b, c), (c, b), (c, c)\}$. The adjacency matrix of X, with respect to the ordering (a, b, c), is $A_X = \begin{pmatrix} 0 & 0 & 0 \\ 2 & 0 & 1 \\ 0 & 1 & 1 \end{pmatrix}$.

Remark 1.94
Every graph can be viewed as a directed graph, by replacing each nonloop edge $\{v, w\}$ with a directed edge from v to w and a directed edge from w to v. A directed graph X is produced from a graph in this manner iff the multiplicity of the directed edge from v to w equals the multiplicity of the directed edge from w to v for all vertices v, w iff the adjacency matrix of X is symmetric.

Definition 1.95 Let X_1 and X_2 be two finite graphs, each with the same vertex set V. Define $X_1 \cdot X_2$ to be the directed graph with vertex set V, where the multiplicity of the directed edge from v_1 to v_2 equals the number of pairs (e_1, e_2) such that e_1 is a directed edge starting at v_1, and e_2 is a directed edge from the terminal point of e_1 to v_2.

Example 1.96
Figure 1.19 shows a directed graph Y with adjacency matrix $A_Y = \begin{pmatrix} 0 & 1 & 1 \\ 0 & 0 & 0 \\ 0 & 0 & 1 \end{pmatrix}$ with respect to the ordering (a, b, c). Let X be as in Example 1.93. Directed edges in $X \cdot Y$ are obtained by taking a "step" along a directed edge in X, followed by another "step" along a directed edge in Y. Hence there are three directed edges in $X \cdot Y$ from b to c. One comes from following a directed edge

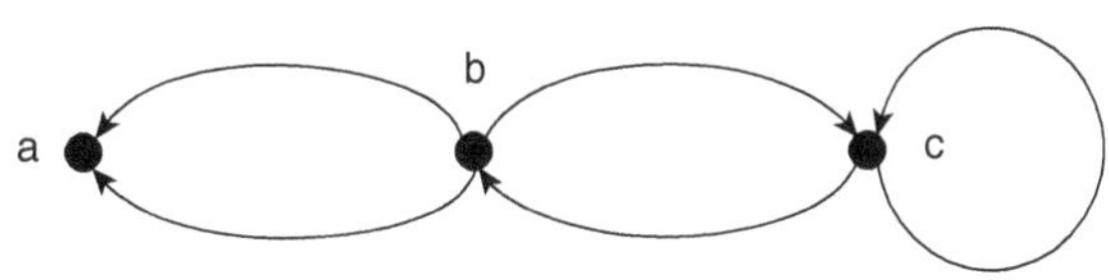

Figure 1.18 A directed graph X

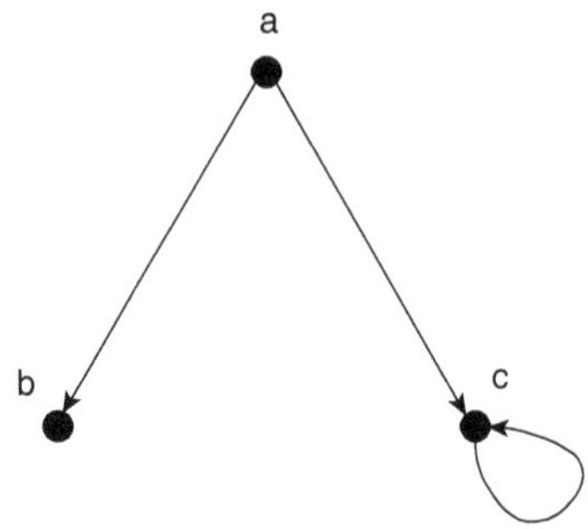

Figure 1.19 A directed graph Y

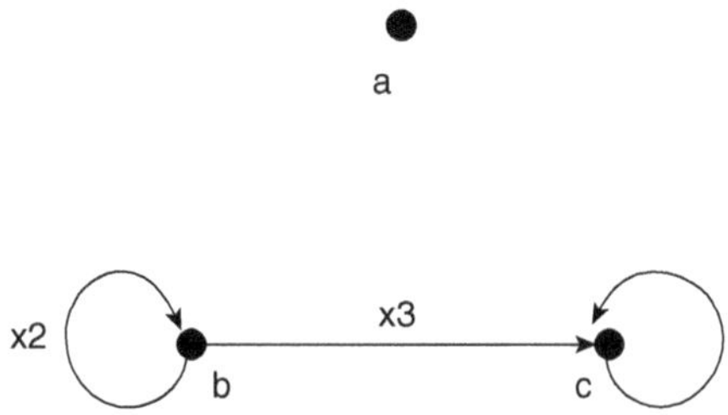

Figure 1.20 The graph $X \cdot Y$

in X from b to a, then the directed edge in Y from a to c. Another comes from following the other directed edge in X from b to a, then the directed edge in Y from a to c. The third comes from following the directed edge in X from b to c, then the loop in Y at c. Note that we have effectively taken the dot product of the b-row of A_X with the c-column of A_Y—that is, $3 = 2 \cdot 1 + 0 \cdot 0 + 1 \cdot 1$. Figure 1.20 shows $X \cdot Y$. Note that to keep the picture simple, we wrote ßx3 over the directed edge from b to c in $X \cdot Y$. This is to indicate there are three directed edges from b to c. Observe that $X \cdot Y$ has adjacency matrix $A_{X \cdot Y} = \begin{pmatrix} 0 & 0 & 0 \\ 0 & 2 & 3 \\ 0 & 0 & 1 \end{pmatrix}$ with respect to the ordering (a, b, c) and that $A_{X \cdot Y} = A_X \cdot A_Y$.

Proposition 1.97
Let X_1 and X_2 be two finite graphs, each with the same vertex set V. Choose an ordering of V. Let A_1 (respectively, A_2) be the adjacency matrix of X_1 (respectively, X_2) with respect to this ordering. Then the adjacency matrix of $X_1 \cdot X_2$ with respect to this ordering is $A_1 \cdot A_2$.

The proof of Prop. 1.97 follows immediately from the definition; we leave it as an exercise for the reader. Use Example 1.96 as your guide.

Definition 1.98 Let X be a finite graph with vertex set V. Define X^n to be be the graph with vertex set V, where the multiplicity of the edge from v_1 to v_2 equals the number of walks of length n from v_1 to v_2. That is,

$$X^n = X \cdot X \cdot \cdots \cdot X,$$

where there are n copies of X in the product.

Proposition 1.99
Let X be a finite graph with vertex set V. Choose an ordering of V. Let A be the adjacency matrix of X with respect to this ordering. Then the adjacency matrix of X^n with respect to this ordering is A^n.

Proof
Apply Prop. 1.97.

Proposition 1.100
Let X be a finite graph with eigenvalues $\lambda_0, \ldots, \lambda_{n-1}$. Then the eigenvalues of X^j are $\lambda_0^j, \ldots, \lambda_{n-1}^j$.

Proof
Let A be an adjacency matrix of X. Let O such that OAO^{-1} is the diagonal matrix with diagonal entries $\lambda_0, \ldots, \lambda_{n-1}$, as in Corollary A.56. Then OA^jO^{-1} is the diagonal matrix with diagonal entries $\lambda_0^j, \ldots, \lambda_{n-1}^j$. The result follows from Lemma A.61 (2).

Definition 1.101 Let G be a group, and Γ be a multi-subset of G. The directed Cayley graph of G with respect to Γ, denoted by $\text{Cay}(G, \Gamma)$, is defined as follows. The vertices of $\text{Cay}(G, \Gamma)$ are the elements of G. For all $x, y \in G$, the multiplicity of the directed edge from x to y in $\text{Cay}(G, \Gamma)$ equals the number of elements γ in Γ, counted with multiplicity, such that $x = y\gamma$.

Proposition 1.102
Let G be a group. Let $\Gamma_1, \Gamma_2 \subseteq G$. Let $X_1 = \text{Cay}(G, \Gamma_1)$ and $X_2 = \text{Cay}(G, \Gamma_2)$. Let

$$\Gamma_1\Gamma_2 = \{\gamma_1\gamma_2 \mid (\gamma_1, \gamma_2) \in \Gamma_1 \times \Gamma_2\}.$$

Then $X_1 \cdot X_2 = \text{Cay}(G, \Gamma_1\Gamma_2)$.

Prop. 1.102 follows directly from the definitions, so we leave it as an exercise for the reader.

9. AN UPPER BOUND ON THE ISOPERIMETRIC CONSTANT

In this section we give a proof of the second part of Proposition 1.84. That is, we show that if X is a d-regular graph, then

$$h(X) \leq \sqrt{2d(d - \lambda_1(X))}.$$

Let A be the adjacency operator of X, and let Δ be the associated Laplacian. Let E be the edge multiset of X.

Let $g \in L_0^2(V, \mathbb{R})$ be a real-valued eigenfunction of the adjacency operator of X associated with the second largest eigenvalue $\lambda_1(X)$. Define $V^+ = \{x \in V \mid g(x) \geq 0\}$. Because $\sum_{x \in V} g(x) = 0$ we know that V^+ is not all of V. Since $-g$ is also an eigenvector associated with $\lambda_1(X)$, we may assume that $\left|V^+\right| \leq |V|/2$. Define the function $f \in L^2(V, \mathbb{R})$ as

$$f(x) = \begin{cases} g(x) & \text{if } x \in V^+. \\ 0 & \text{otherwise.} \end{cases}$$

The gist of the argument is as follows. Decompose the vertex set of X into "level sets" corresponding to the values taken on by the function f. The larger the isoperimetric constant $h(X)$ is, the more edges there must be that cross between level sets. The more such edges there are, the larger $\langle \Delta f, f \rangle$ must be, because $\langle \Delta f, f \rangle$ in some sense measures the total change in f across edges. The Rayleigh-Ritz theorem provides the final link to the spectral gap $d - \lambda_1(X)$. We now present the details of the proof.

We break the proof of Proposition 1.84 into two lemmas: Lemma 1.103 and Lemma 1.104. These two lemmas imply Proposition 1.84.

Lemma 1.103

$$\frac{\langle \Delta f, f \rangle_2}{\langle f, f \rangle_2} \leq d - \lambda_1(X).$$

Proof
If $x \in V^+$, then by Lemma 1.58 we have that

$$\begin{aligned} (\Delta f)(x) &= d \cdot f(x) - \sum_{y \in V^+} A_{x,y} f(y) = d \cdot g(x) - \sum_{y \in V^+} A_{x,y} g(y) \\ &\leq d \cdot g(x) - \sum_{y \in V} A_{x,y} g(y) = (\Delta g)(x). \end{aligned}$$

Thus

$$\begin{aligned} \langle \Delta f, f \rangle_2 &= \sum_{v \in V^+} (\Delta f)(v) f(v) \leq \sum_{v \in V^+} (\Delta g)(v) g(v) \\ &= \sum_{v \in V^+} (d - \lambda_1(X)) g(v)^2 = (d - \lambda_1(X)) \sum_{v \in V^+} f(v)^2 \\ &= (d - \lambda_1(X)) \langle f, f \rangle_2 . \end{aligned}$$

Lemma 1.104

$$\frac{h(X)^2}{2d} \leq \frac{\langle \Delta f, f \rangle_2}{\langle f, f \rangle_2}.$$

Proof
Orient the edges of X such that $f(e^+) \geq f(e^-)$ for all $e \in E$. Define

$$B_f = \sum_{e \in E} \left(f(e^+)^2 - f(e^-)^2 \right).$$

We prove this lemma in two steps. We first show that

$$B_f \leq \sqrt{2d \langle \Delta f, f \rangle_2 \langle f, f \rangle_2}.$$

We then show that

$$h(x) \langle f, f \rangle_2 \leq B_f.$$

This completes the proof of this lemma.

By the Cauchy-Schwarz inequality (Proposition A.20), Proposition 1.60, and and the fact that $(a+b)^2 \leq 2(a^2+b^2)$ for all $a, b \in \mathbb{R}$, we have that

$$\begin{aligned} B_f &= \sum_{e \in E} (f(e^+) + f(e^-))(f(e^+) - f(e^-)) \\ &\leq \sqrt{\sum_{e \in E} (f(e^+) + f(e^-))^2} \sqrt{\sum_{e \in E} (f(e^+) - f(e^-))^2} \\ &\leq \sqrt{2 \sum_{e \in E} (f(e^+)^2 + f(e^-)^2)} \sqrt{\langle \Delta f, f \rangle_2} \\ &= \sqrt{2d \sum_{v \in V} f(v)^2} \sqrt{\langle \Delta f, f \rangle_2} = \sqrt{2d \langle \Delta f, f \rangle_2 \langle f, f \rangle_2}. \end{aligned}$$

To prove the other inequality involving B_f, we break the vertices of X into level sets of f. Let $0 = \beta_0 < \beta_1 < \ldots < \beta_r$ be the values of f on the vertices V, and

$$L_i = \{x \in V \mid f(x) \geq \beta_i\}.$$

Note that

$$L_r \subset L_{r-1} \subset \cdots \subset L_1 \subset L_0 = V$$

and $L_i \subset V^+$ for $i \geq 0$.

Let us give an example to illustrate the decomposition of X. Consider the graph X in Figure 1.21. The spectrum of X is approximately

$$\text{Spec}(X) \approx \begin{pmatrix} -2.83424 & -2.38849 & -1.65662 & -1 & 0 & 0.812716 & 1.49086 & 2.57577 & 3 \\ 1 & 1 & 1 & 1 & 4 & 1 & 1 & 1 & 1 \end{pmatrix}.$$

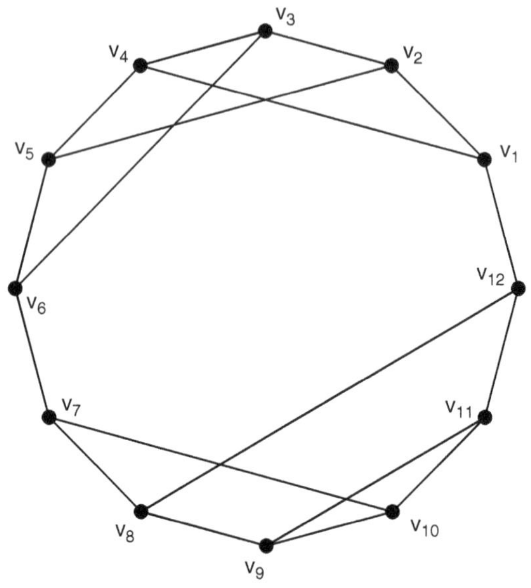

Figure 1.21 X

If we label the vertices with the ordering $v_1, \ldots, v_{12}$, as given in the figure for X, an eigenvector g corresponding to $\lambda_1(X) \approx 2.57577$ is approximately given by

$$g \approx (-1.11382, -1.93447, -1.93447, -1.93447, -1.93447, -1.11382,$$
$$1, 1.64131, 2.22763, 2.04828, 2.04828, 1)^t.$$

A corresponding function f with a level curves diagram for X is shown in Figure 1.22. The vertices are labeled with the approximate values of f.

Suppose that e is an edge such that $f(e^+) - f(e^-) \neq 0$. Then $f(e^-) = \beta_j$ and $f(e^+) = \beta_i$ with $j < i$, so

$$e \in \partial L_{j+1} \cap \partial L_{j+2} \cap \cdots \cap \partial L_i$$

and $e \notin \partial L_0, \partial L_1, \ldots, \partial L_j, \partial L_{i+1}, \ldots \partial L_r$. Furthermore,

$$\begin{aligned} f(e^+)^2 - f(e^-)^2 &= \beta_i^2 - \beta_j^2 \\ &= (\beta_i^2 - \beta_{i-1}^2) + (\beta_{i-1}^2 - \beta_{i-2}^2) + \cdots + (\beta_{j+1}^2 - \beta_j^2). \end{aligned}$$

Thus,

$$B_f = \sum_{\substack{e\in E \\ f(e^-)=\beta_j \\ f(e^+)=\beta_i \\ j<i}} \sum_{k=j+1}^{i} (\beta_k^2 - \beta_{k-1}^2) = \sum_{k=1}^{r} \left|\partial L_k\right| (\beta_k^2 - \beta_{k-1}^2).$$

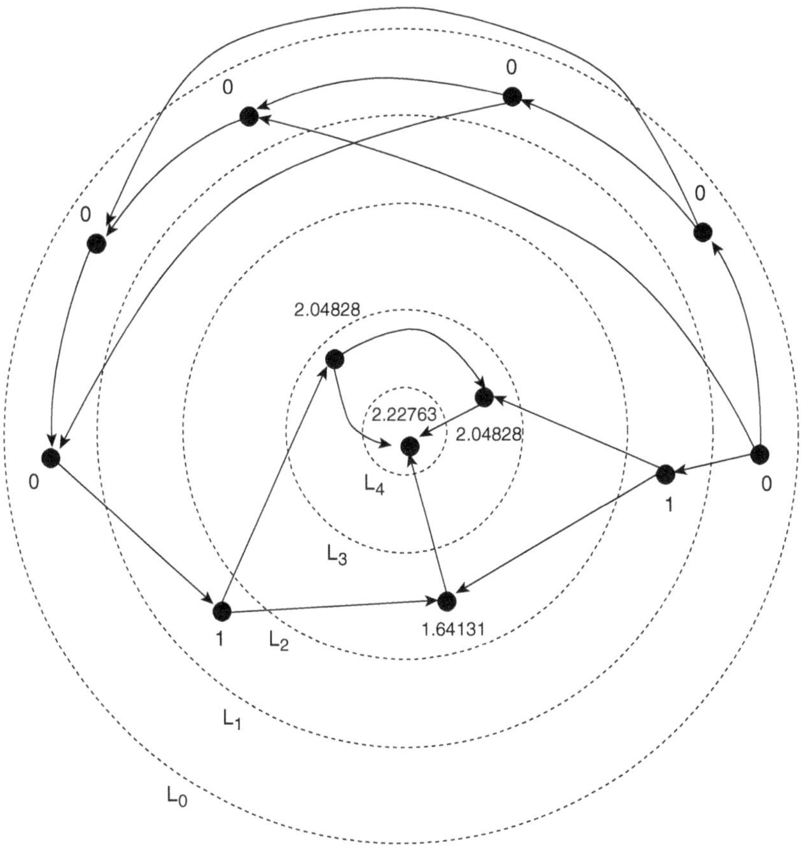

Figure 1.22 A level set decomposition

Because $L_i \subset V^+$ for $i > 0$ and $\left|V^+\right| \leq |V|/2$ we have that $\left|\partial L_i\right| \geq h(X)\left|L_i\right|$. Thus,

$$\begin{aligned} B_f &\geq h(X) \sum_{k=1}^{r} \left|L_k\right| (\beta_k^2 - \beta_{k-1}^2) \\ &= h(X) \left[\left|L_1\right| (\beta_1^2 - \beta_0^2) + \left|L_2\right| (\beta_2^2 - \beta_1^2) + \right. \\ &\quad \left. \cdots + \left|L_{r-1}\right| (\beta_{r-1}^2 - \beta_{r-2}^2) + \left|L_r\right| (\beta_r^2 - \beta_{r-1}^2) \right] \\ &= h(X) \left[\left|L_r\right| \beta_r^2 + \sum_{k=1}^{r-1} \beta_k^2 (\left|L_k\right| - \left|L_{k+1}\right|) \right]. \end{aligned}$$

Recall that $x \in L_k \setminus L_{k+1}$ if and only if $f(x) = \beta_k$. Therefore,

$$B_f \geq h(X) \sum_{i=1}^{r} \sum_{\substack{x \in V^+ \\ f(x) = \beta_i}} f(x)^2 = h(X) \left\langle f, f \right\rangle_2 .$$ ⍟

NOTES

1. Each of [40], [45], [87], and [129] cover different aspects of expander families. Sarnak [122] gives an excellent, concise, and readable two-page summary of the theory of expander families. Two textbooks on algebraic graph theory that do not touch on expanders are Biggs [23] and Godsil and Royle [66].
2. Prop. 1.84 was proved by Dodziuk [50], and independently by Alon and Milman [7], and Alon [4]. Several other sources discuss this proposition, its history, and its relationship to the geometry of manifolds: [87, pp. 46–48], [45, p. 13], and [70, p. 474].
3. A graph contains an Euler circuit iff every vertex has even degree. For this theorem to hold for graphs with loops, we must have each loop count 2 toward the degree.
4. In 1973, Pinsker [108] used a probabalistic argument to demonstrate that for any integer $d \geq 3$, there exists an d-regular expander family. See Lubotzky [87, pp. 5–6] for a proof. The first explicit construction of an expander family was given by Margulis [93]. However, Margulis did not provide any specific bound on the isoperimetric constants of his graphs. Later, Gabber and Galil [64] considered a similar family of graphs. Using classical analysis, they gave an explicit upper bound on the isoperimetric constants of their graphs. Jimbo and Maruoka [74] gave a different proof that used Fourier analysis on the group $\mathbb{Z}_m \times \mathbb{Z}_m$. Also see [92] and [94].

   As an example, construct a family (X_m) of 8-regular graphs as follows. The vertex set of X_m is given by $\mathbb{Z}_m \times \mathbb{Z}_m$. The neighbors of the vertex (x, y) are $(x, y + 2x)$, $(x + 2y, y)$, $(x, y + 2x + 1)$, $(x + 2y + 1, y)$, $(x, y - 2x)$, $(x - 2y, y)$, $(x, y - 2x - 1)$, and $(x - 2y - 1, y)$. Hoory, Linial, and Wigderson [70, pp. 503–8] and Xiao [133, pp. 24–29] follow the papers cited to show that $\lambda(X_m) \leq 5\sqrt{2} < 8$.
5. By the classification of finite simple groups, every nonabelian finite simple group is either an alternating group, a group of Lie type, or of one of finitely many sporadic groups. In 2006, Kassabov, Lubotzky, and Nikolov [77] proved the following result: "There exist $k \in \mathbb{N}$ and $0 < \epsilon \in \mathbb{R}$ such that every non-abelian finite simple group G, which is not a Suzuki group, has a set S of k generators for which $\text{Cay}(G, S)$ is an ϵ-expander." The definition of expander that they are using is given in the exercises (see Definition 1.107). They state that $k < 1000$ and $\epsilon > 10^{-10}$.
6. This note gives a construction of a 3-regular expander family (X_p) where p indexes over the odd primes. For details see Hoory, Linial, and Wigderson [70, pp. 453 and 529]. Let p be an odd prime and $\mathbb{Z}_p$ be the vertex set of X_p. Let $x \in \mathbb{Z}_p$. If $x \neq 0$, then put edges between x and the vertices $x + 1, x - 1$, and x^{-1} (the inverse of x under multiplication). If $x = 0$, then put edges between x and the vertices $x + 1$, $x - 1$, and 0. Note that X_p is nonbipartite since it has a loop at 0. Using the Selberg 3/16 theorem, one can show that (X_p) is an expander family.
7. Let S be a finite set of elements from $\text{SL}(2, \mathbb{Z})$ which is closed under taking inverses. Suppose further that S does not possess a solvable subgroup of finite index. For each prime p one may reduce S modulo p to obtain a finite set S_p of elements of $\text{SL}(2, \mathbb{F}_p)$. Bourgain and Gamburd [28] show that the sequence of

Cayley graphs $(\mathrm{Cay}(SL(2, \mathbb{F}_p), S_p))$ form an expander family. In [29], Bourgain and Gamburd generalize their results to $SL(2, \mathbb{F}_{p^n})$.

8. An article that discusses the isoperimetric constant of a graph is Mohar [99]. Mohar begins by discussing basic results about $h(X)$. For example, let X be a graph with n vertices. If X is k-edge-connected, then $h(X) \geq 2k/n$. (Suppose that X is connected. An edgecut set of X is a set of edges whose deletion makes X disconnected. X is k-edge-connected if the minimum number of edges in an edgecut set of X is k. There is a more general definition if X is a disconnected graph. See [66, p. 37].) Now suppose that X is a graph with n vertices and m edges. Mohar shows that

$$h(X) \leq \begin{cases} \frac{m}{n-1} & \text{if } n \text{ is even.} \\ \frac{m(n+1)}{n(n-1)} & \text{if } n \text{ is odd.} \end{cases}$$

If X is d-regular, then $m = dn/2$. Hence

$$h(X) \leq \begin{cases} \frac{d}{2}\left[\frac{n}{n-1}\right] & \text{if } n \text{ is even.} \\ \frac{d}{2}\left[\frac{n+1}{n-1}\right] & \text{if } n \text{ is odd.} \end{cases}$$

Note that $h(X) \approx d/2$.

Mohar gives the following upper bound for $h(X)$. Suppose that X is a d-regular graph with n vertices. Recall that $d - \lambda_1(X)$ is the second smallest eigenvalue of the Laplacian Δ for X (see Prop. 1.60). If X is not equal to K_1, K_2, or K_3, then $h(X) \leq \sqrt{(d - \lambda_1(X))(d + \lambda_1(X))}$. In Prop. 1.84 it was shown that $h(X) \leq \sqrt{2d(d - \lambda_1(X))}$. Since $h(X) \leq d$ (see Exercise 3), $\sqrt{(d - \lambda_1(X))(d + \lambda_1(X))} \leq \sqrt{2d(d - \lambda_1(X))}$. Thus Mohar's result is tighter than the result that is given in Prop. 1.84.

Mohar gives the values of $h(X)$ for various families of graphs and gives an algorithm for computing $h(X)$. He notes that for graphs with multiple edges, the computation of the isoperimetric number is NP-hard. Let X be a graph with n vertices and maximal vertex degree δ. He shows that

$$\mathrm{diam}(X) \leq \frac{\log(n/2)}{\log((\delta + h(X))/(\delta - h(X)))}.$$

See Note 6 of Chapter 3 for information on the growth rate of $h(X)$.

9. Let

$$D_n = \langle r, s | r^n = s^n = 1, sr = r^{-1}s \rangle$$

be the dihedral group of order $2n$. In Rosenhouse [119], it is shown that $h(\mathrm{Cay}(D_n, \{r, r^{-1}, s\}) = 4/n$. Hence, they do not yield an expander family. See Exercise 14 of this chapter. Rosenhouse ends with conjectures about Cayley graphs on generalized dihedral groups.

10. Let G be the group $\mathbb{Z}_n \times \mathbb{Z}_n \setminus \{(0,0)\}$. The nth Platonic graph, π_n, has vertex set $G/\{\pm 1\}$. Two vertices (a,b) and (c,d) are adjacent if and only if $\det\begin{pmatrix} a & b \\ c & d \end{pmatrix} = \pm 1 (\text{mod } n)$. The graph π_n is n-regular. Lanphier and Rosenhouse [83] show that when n is a prime, π_n can be viewed as a complete multigraph where each vertex is itself a wheel on $n+1$ vertices. They give a similar theorem when n is a power of a prime. If p is a prime, they prove that

$$h(\pi_{p^r}) \leq \begin{cases} \frac{p^r(p-1)}{2(p+1)} & \text{if } p \not\equiv 3(\text{mod } 4), \\ \frac{p^{2r}-2p^{2r-1}+5p^{2r-2}-4p^{r-1}+4}{2(p^r-2p^{r-1}-3p^{r-2}+4p^{-1})} & \text{if } p \equiv 3(\text{mod } 4). \end{cases}$$

This extends a result of Brooks, Perry, and Petersen [31], which states that if p is a prime and $p \equiv 1(\text{mod } 4)$ then

$$\frac{p^2-2p+5}{4(p-1)} \leq h(\pi_p) \leq \frac{(p-1)p}{2(p+1)}.$$

11. Using several models of random regular graphs (including the one described in the next paragraph), Broder and Shamir [30] showed that the second-largest eigenvalue of a random $2d$-regular graph is concentrated in an interval of width $O(\sqrt{d})$ around its mean, and its mean is $O(d^{3/4})$. This result holds for three different models of random regular graphs.

    We now describe one of the models that Broder and Shamir [30] use. One can randomly construct an undirected, $2d$-regular graph G_{2d} with n vertices as follows. Suppose the vertex set of G_{2d} is $1, 2, \ldots, n$. First, independently choose d permutations $\pi_1, \ldots, \pi_d$ from S_n uniformly at random (that is, each of the d permutations is chosen independently of one another, and when choosing a permutation from S_n you make each permutation equally likely to be chosen, so the probability of choosing any given permutation is $1/n!$). For each permutation π that is chosen, and each vertex v of the graph, add the edge $\{v, \pi(v)\}$ to G_{2d}. Hence, the graph G_{2d} has nd edges. Note that loops and multiple edges are allowed. Also, note that there may be several ways to get the same graph. The probability space of this random construction is the space $S_n^d = S_n \times \cdots \times S_n$. There are $(n!)^d$ elements in this space, and each element is assigned equal probability.

12. By applying the ideas from Broder and Shamir [30] to Cayley graphs, Alon and Roichman [8] showed the following: "For every $1 > \epsilon > 0$ there exists $c(\epsilon) > 0$ such that the following holds. Let G be a group of order n, and let S be a random set of $c(\epsilon)\log_2(n)$ elements of G, then the Cayley graph $X(G,S)$ is an ϵ-expander almost surely. That is, the probability it is an expander tends to 1 as $n \to \infty$." Here they use the notion of expander given in the exercises (see Definition 1.106). Loh and Schulman [86] and Landau and Russell [82] use representation theory to give proofs of similar results. The main idea behind the proofs of all of these results lies in constructing a random walk in a randomly chosen graph and bounding the number of closed walks of a given length in the graph at a given vertex. See also [57].

13. Let $r \geq 3$ and $n > r$. Let $G(n, r - \text{reg})$ denote the set of all r-regular graphs with vertex set $V = \{1, 2, \ldots, n\}$. It is assumed that rn is even. The set $G(n, r - \text{reg})$ is turned into a probability space by assigning equal probability to each graph in the set. A random element of $G(n, r - \text{reg})$ is denoted by $G_{r-\text{reg}}$. We say that almost every $G_{r-\text{reg}}$ has property Q if the probability that $G_{r-\text{reg}}$ has property Q tends to 1 as $n \to \infty$. For more on this probability space of graphs see Bollobás [26].

    Bollobás [25] proves the following result about the isoperimetric constant of a random regular graph using the probabilistic model $G(n, r - \text{reg})$. Let $r \geq 3$ and $0 < \eta < 1$ be such that $2^{4/r} < (1 - \eta)^{(1-\eta)}(1 + \eta)^{(1+\eta)}$. Then almost every r-regular graph has isoperimetric constant at least $(1 - \eta)r/2$. For example, set $\eta = 0.5$. Then $(1 - 0.5)^{(1-0.5)}(1 + 0.5)^{(1+0.5)} \approx 1.299038$. For $r \geq 11$, we have that $2^{4/r} < 1.299038$. Hence, for $r \geq 11$, almost every r-regular graph has isoperimetric constant at least $0.5r/2 = r/4$.
14. Morally speaking, the results summarized in Note 13 and Note 9 of Chapter 3 imply that for $d \geq 3$, if you choose a sequence of d-regular graphs "at random," then it is almost certainly an expander family.

EXERCISES

1. Show that if A_1 and A_2 are two adjacency matrices of a graph X using different orderings of the vertices, then A_1 and A_2 have the same eigenvalues. (Hint: Use Prop. A.44 and Lemma A.61.)

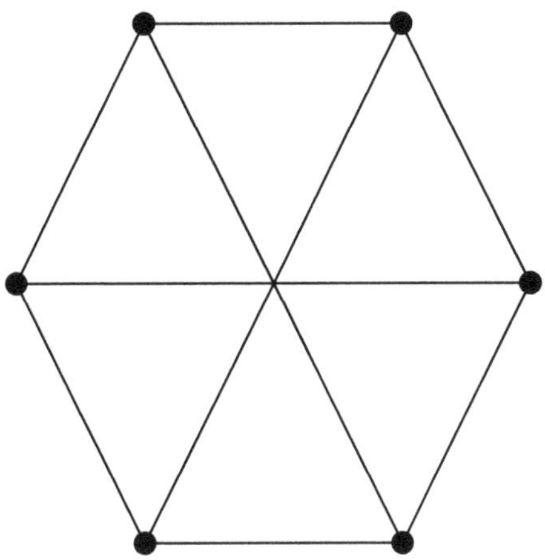

Figure 1.23

2. Let X be the graph shown in Figure 1.23.
    (a) Compute the isoperimetric constant $h(X)$.
    (b) Calculate the eigenvalues of the graph. (Hint: The matrix is circulant.) Use your answer to compute the spectral gap of the graph. Plug your results into Proposition 1.84 and see how good the approximation is.

3. Suppose that X is a d-regular graph. Prove that $0 \leq h(X) \leq d$.
4. Prove that $h(X) = 0$ if and only if X is a disconnected graph.
5. Prove that the sum of the eigenvalues of the graph is the number of loops in the graph counted with multiplicity. (Hint: Use Lemma A.60.)
6. Prove that a graph is bipartite if and only if it has no closed walks of odd length. (A walk is closed if it begins and ends at the same vertex.)

We use the following definition for Exercises 7 and 8.

Definition 1.105 Let X be a graph with vertex set V. Define an equivalence relation on V where $v \sim w$ iff there is a path from v to w. Let $V_1, \ldots, V_m$ be the equivalence classes of $\sim$. Define the graphs $X_1, \ldots, X_m$ as follows. The vertex set of X_i is V_i. There is an edge incident to the vertices a and b of V_i iff there is an edge incident to a and b in V. The graphs $X_1, \ldots, X_m$ are called the *connected components* of X.

7. Let X, V, and $V_1, \ldots, V_m$ be as in Def. 1.105. Prove the following:
   (a) $V = V_1 \cup V_2 \cup \cdots \cup V_m$.
   (b) $V_i \cap V_j = \emptyset$ if $i \neq j$.
   (c) If $x, y \in V_i$ for some i, then there is a path in the graph connecting x to y.
   (d) If $x \in V_i$ and $y \in V_j$ for some $i \neq j$, then there is no path in the graph connecting x to y.
8. Let X, V, and $V_1, \ldots, V_m$ be as in Def. 1.105. Prove that the adjacency matrix of X is the direct sum of the adjacency matrices of the X_i's. (You will need to choose the ordering on the vertices correctly.) Prove that this implies that the spectrum of X is the union of the spectra of its connected components. Show that this implies that the multiplicity of d as an eigenvalue of A is equal to m.
9. Let G be a finite group, and let $\Gamma \Subset G$ such that $\mathrm{Cay}(G, \Gamma)$ is a cycle graph. Prove that G is either a cyclic group or a dihedral group.
10. Let X be a graph. Show that an eigenvalue of X is rational if and only if it is an integer. (Hint: Think about the characteristic polynomial of an adjacency matrix of X, and use the rational roots theorem.)
11. Let X be a d-regular graph with adjacency matrix A.
    (a) Show that if $|\mu| > d$, then $\mu I + A$ is an invertible matrix.
    (b) Show that if X is nonbipartite, then $dI + A$ is an invertible matrix.
12. Prove Propositions 1.97 and 1.102.
13. Let X be a finite graph. Let X' be the graph obtained by adding a single loop to each vertex of X. Prove that the eigenvalues of X' are precisely $1 + \lambda_0(X), \ldots, 1 + \lambda_{n-1}(X)$.
14. Let

$$D_n = \langle r, s \,|\, r^n = s^n = 1, sr = r^{-1}s \rangle$$

be the dihedral group of order $2n$. Show that

$$h(\mathrm{Cay}(D_n, \{r, r^{-1}, s\}) \leq 4/n.$$

Hence, $h(\mathrm{Cay}(D_n, \{r, r^{-1}, s\})) \to 0$ as $n \to \infty$. (Hint: Draw a few of the graphs. They look like ladders. See Note 9.)

15. Show that $X_n = \mathrm{Cay}(\mathbb{Z}_{2n}, \{1, n, 2n-1\})$ does not yield an expander family. (Hint: The adjacency matrices are circulant.)
16. Let $n \geq 3$. Define the graph $K_{n,n}$ as follows. The vertices of $K_{n,n}$ are given by $V \cup W$ where $|V| = |W| = n$. Every $v \in V$ is adjacent to every $w \in W$ via an edge of multiplicity 1. There are no other edges. Prove that the spectrum of $K_{n,n}$ is given by

$$\mathrm{Spec}(K_{n,n}) = \begin{pmatrix} -n & 0 & n \\ 1 & 2n-2 & 1 \end{pmatrix}.$$

17. This exercise is a bit like graph eigenvalue "Sudoku." Let X be a 4-regular graph with 10 vertices and no loops. In each case, *partial* information is given about the spectrum of X. Your task is to use the given information, along with Prop. 1.48, to deduce additional information about the rest of the spectrum of X and then use that information to answer the following questions: Is X connected? Is X bipartite? (Hint: Use Exercise 5.)

    (a) -4 is an eigenvalue of X with multiplicity 2.
    (b) -3 is an eigenvalue of X with multiplicity 3.
    -2 is an eigenvalue of X with multiplicity 2.
    0 is an eigenvalue of X with multiplicity 1.
    3 is an eigenvalue of X with multiplicity 3.
    (c) -2 is an eigenvalue of X with multiplicity 4.
    2 is an eigenvalue of X with multiplicity 4.
    (d) -3 is an eigenvalue of X with multiplicity 2.
    -2 is an eigenvalue of X with multiplicity 1.
    -1 is an eigenvalue of X with multiplicity 4.

18. The following exercise is taken from [66]. Let X be a finite graph with vertex set V. Prove that there exists $S \subset V$ such that $\frac{|\partial S|}{|S|} = h(X)$ and the subgraphs induced by S and $V \setminus S$ are both connected. (Hint: First show that if a, b, c, d are positive real numbers, then $\frac{a+b}{c+d}$ lies between $\frac{a}{c}$ and $\frac{b}{d}$. To do so, consider the slopes of the vectors (c, a) and (d, b), and draw the diagram that corresponds to vector addition.)

We use the following definition for Exercises 19–21.

Definition 1.106 Let X be a d-regular graph with vertex set V and $|V| = n$. Then X is called an (n, d, c)-expander if for every subset A of V,

$$\left|\partial' A\right| \geq c\left(1 - \frac{|A|}{n}\right)|A|,$$

where $\partial' A = \{y \in V \mid \mathrm{dist}(y, A) = 1\}$ is the *vertex boundary* of A and $\mathrm{dist}(y, A) = \min_{x \in A} \mathrm{dist}(y, x)$. We sometimes simply write that X is a *c-expander.*

Let (X_n) be a sequence of d-regular graphs with $|X_n| \to \infty$ as $n \to \infty$. We say that (X_n) is a *family of vertex expanders* if there exists a $c > 0$ such that each X_n is a c-expander.

19. Prove that K_n is a $(n, n-1, 1)$-expander.
20. Prove that (C_n) is not a vertex expander family.
21. Let X be a d-regular graph with n vertices. Prove the following:
    (a) If X is an (n, d, c)-expander, then $h(X) \geq c/2$.
    (b) X is an $(n, d, h(X)/d)$-expander.

We use the following definition for Exercise 22.

Definition 1.107 Let X be a d-regular graph with vertex set V and $|V| = n$. Then X is called an ϵ-expander if for every subset A of V with $|A| \leq |X|/2$ we have

$$\left|\partial' A\right| \geq \epsilon \, |A| \, .$$

Here $\partial' A$ is as in Def. 1.106.

22. Suppose that X is a d-regular graph. Prove the following:
    (a) If X is an ϵ-expander, then $h(X) \geq \epsilon$.
    (b) X is an $(h(X)/d)$-expander.
    (c) If (X_n) is a sequence of d-regular graphs with $|X_n| \to \infty$, then (X_n) is an expander family iff there exists ϵ such that each X_n is an ϵ-expander.

2

Subgroups and Quotients

One question that recurs throughout this book is: Which sequences of finite groups yield expander families? Many positive answers to this question are known—symmetric groups work, for example—but alas, the techniques involved are too advanced for this simple text. We content ourselves with negative results, that is, theorems that show that certain sequences of finite groups never yield expander families.

In this chapter, we lay the groundwork for extending such negative results. In particular, we prove two statements we've dubbed the Quotients Nonexpansion Principle and the Subgroups Nonexpansion Principle. Together, these state that if a sequence (G_n) of finite groups admits a nonexpanding sequence of quotients or bounded-index subgroups, then (G_n) does not yield an expander family.

1. COVERINGS AND QUOTIENTS

In this section, we study how the isoperimetric constant and spectral gap of a Cayley graph on a group G compare to those of a corresponding Cayley graph on a quotient of G. We state the results in the more general setting of graph coverings. Ultimately, we obtain the Quotients Nonexpansion Principle (Prop. 2.20), which states that (G_n) cannot yield an expander family if any of its quotients do not. Of course, it's equivalent to say that if (G_n) yields an expander family, then so does any sequence of quotients (Q_n). However, our main use of Prop. 2.20 will be to produce negative results—see, for example, Theorem 4.47, Example 4.27, and Exercise 4.9. We have therefore chosen to accentuate the negative.

Definition 2.1 Let X and X' be graphs with vertex sets V and V' and edge multisets E and E', respectively. A graph *homomorphism* between X and X' is a pair of maps $\phi_V : V \to V'$ and $\phi_E : E \to E'$ where whenever $e \in E$ has endpoints $a, b \in V$ and $\phi_E(e)$ has endpoints $a', b' \in V'$, then $\phi_V(\{a, b\}) = \{a', b'\}$. To simplify matters, we usually write ϕ for both ϕ_E and ϕ_V. If both ϕ_V and ϕ_E are bijective (one-to-one and onto), then we say that ϕ is an *isomorphism*.

Let v be a vertex of a graph X. Let E_v be the set of edges incident to v. (Regard E_v as a set, not a multiset, by regarding multiple edges as distinct elements of E_v.) Note that if ϕ is a homomorphism from X to Y, then ϕ maps E_v to $E_{\phi(v)}$.

Definition 2.2 Let X, Y be graphs. Let ϕ be a homomorphism from X to Y. Let v be a vertex of X. We say that ϕ is *bijective at* v if the map from E_v to $E_{\phi(v)}$ induced by ϕ is bijective. We say that ϕ is *locally bijective* if ϕ is bijective at v for all vertices v of X. We say that ϕ is a *covering from* X *to* Y if ϕ is locally bijective and ϕ maps the vertex set of X surjectively onto the vertex set of Y. If there is a covering from X to Y, then we say that X *covers* Y.

Remark 2.3
In the parlance of topology, we would say that ϕ is a *covering map* and X is a *covering space*.

Example 2.4
Let X, Y be the graphs depicted in Figure 2.1. There is a covering ϕ from X to Y such that $\phi(v_1) = \phi(v_4) = w_1$, $\phi(v_2) = \phi(v_3) = w_2$, and $\phi(v_5) = \phi(v_6) = w_3$. Because X has no multiple edges, what ϕ does to edges is determined by what it does to vertices, so we have completely specified the map ϕ. The edges labeled a, b, c, for example, map to the edges labeled a', b', c', respectively. Note that $E_{v_1} = \{a, b, c\}$, and $E_{w_1} = \{a', b', c'\}$, so ϕ is bijective at v_1. The reader should verify that ϕ is bijective at all other vertices of X and therefore is indeed a covering. We see this example again in Example 2.23.

Example 2.5
Let X, Y be the graphs depicted in Figure 2.2. Let ϕ_1 be the graph homomorphism from X to Y that sends a to a' and b to b'. Let ϕ_2 be the graph homomorphism from X to Y that sends both a and b to a'. Then ϕ_1 is a covering, but ϕ_2 is not.

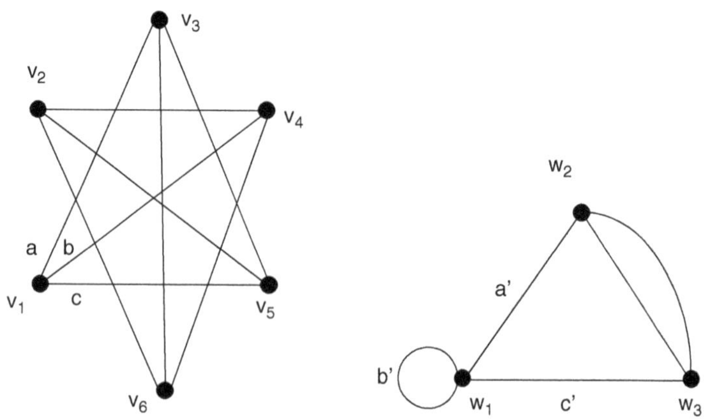

Figure 2.1 X (the left graph) covers Y (the right graph)

Figure 2.2 One homomorphism from X (the left graph) to Y (the right graph) is a covering, one is not

Remark 2.6
If X covers Y, then X is d-regular iff Y is d-regular.

Lemma 2.7
If X covers Y and X is connected, then Y is connected.

Proof
Let w_1, w_2 be two vertices of Y. Let ϕ be a covering from X to Y. Because ϕ surjects onto the vertices of Y, we know that there exist vertices v_1, v_2 of X such that $\phi(v_1) = w_1$ and $\phi(v_2) = w_2$. Since X is connected, we know there exists a walk from v_1 to v_2 using edges $e_1, e_2, \ldots, e_n$. Since ϕ is a homomorphism, it follows that the edges $\phi(e_1), \ldots, \phi(e_n)$ give us a walk in Y from w_1 to w_2. ⊛

Remark 2.8
Lemma 2.7 admits a quick and easy topological proof: the forward image of a connected space under a continuous map is connected.

Definition 2.9 Let ϕ be a covering from X to Y. Let w be a vertex of Y. Then the *fibre of ϕ at w*, denoted $\phi^{-1}(w)$, is the set of all vertices v of X such that $\phi(v) = w$.

Our next lemma states that if Y is connected, then all fibers of a covering of Y have the same size.

Lemma 2.10
Suppose that X, Y are finite graphs and ϕ is a covering from X to Y. If Y is connected, then $|\phi^{-1}(w_1)| = |\phi^{-1}(w_2)|$ for all vertices w_1, w_2 of Y.

Proof
First observe that by Def. 2.2, we have that if $v \in \phi^{-1}(w_1)$, then ϕ induces a bijection from between the set of edges in E_v incident to a vertex in $\phi^{-1}(w_2)$, and the set of edges between w_1 and w_2.

We prove the desired result by induction on $\text{dist}(w_1, w_2)$. We know this distance is finite, because Y is connected. For the base case, assume w_1 and w_2 are distinct and adjacent. Let m be the number of edges between w_1 and w_2. Since w_1 and w_2 are adjacent, we know that $m > 0$. The number of edges between a vertex in $\phi^{-1}(w_1)$ and a vertex in $\phi^{-1}(w_2)$ is $m \cdot |\phi^{-1}(w_1)|$. Reversing the roles of w_1 and w_2, we see that this number also equals $m \cdot |\phi^{-1}(w_2)|$. Because $m > 0$, we therefore have $|\phi^{-1}(w_1)| = |\phi^{-1}(w_2)|$.

This same line of reasoning similarly shows that the fiber at a vertex whose distance from w_1 is j has the same size as the fiber at a vertex whose distance from w_1 is $j + 1$. The inductive step follows. ⌂

Next we show that the isoperimetric constant of a covering is never better than that of any graph it covers.

Lemma 2.11
If X and Y are finite graphs such that X covers Y, then $h(X) \leq h(Y)$.

Proof
If Y is not connected, then by Lemma 2.7 we know that X is not connected, so $h(X) = 0 = h(Y)$.

Now suppose that Y is connected. Let S be a set of vertices of Y such that $h(Y) = \frac{|\partial S|}{|S|}$ and $|S| \leq \frac{1}{2}|Y|$. Let

$$\phi^{-1}(S) = \{v \in V_X \mid \phi(v) \in S\}.$$

Let w be any vertex of Y, and let $a = |\phi^{-1}(w)|$. By Lemma 2.10, we have that $|\phi^{-1}(S)| = a|S|$ and $|X| = a|Y|$, so $|\phi^{-1}(S)| \leq \frac{1}{2}|X|$. By Def. 2.2, we see that every edge in Y between a vertex in S and a vertex in $V_Y \setminus S$ has exactly a preimages in X. So $|\partial[\phi^{-1}(S)]| = a|\partial S|$. So

$$h(X) \leq \frac{|\partial[\phi^{-1}(S)]|}{|\phi^{-1}(S)|} = \frac{a|\partial S|}{a|S|} = h(Y).$$ ⌂

Given a symmetric subset Γ of a group G and a subgroup H of G, we construct a "coset graph" $\mathrm{Cos}(H\backslash G, \Gamma)$ which is covered by the Cayley graph $\mathrm{Cay}(G, \Gamma)$.

Definition 2.12 Let G be a group. Let $\Gamma \Subset G$. Let H be a subgroup of G. Define the *coset graph* $\mathrm{Cos}(H\backslash G, \Gamma)$ to be the graph whose vertex set is the set $H\backslash G$ of right cosets of H in G, so that the multiplicity between two vertices Hx and Hy equals the number of γ in Γ, counted with multiplicity, such that $Hx = Hy\gamma$.

Remark 2.13
Note that $\mathrm{Cos}(H\backslash G, \Gamma)$ is regular with degree $|\Gamma|$.

Example 2.14
See the Notations and Conventions section for our conventions for dihedral groups. Let $G = D_6$, let $\Gamma = \{s, s, r, r^2\}$, and let $H = \{1, s\}$. Then $H\backslash D_6 = \{H, Hr, Hr^2\}$. The graphs $\mathrm{Cay}(D_6, \Gamma)$ and $\mathrm{Cos}(H\backslash D_6, \Gamma)$ are shown in Figures 2.3 and 2.4. The three edges between Hr and Hr^2, for example, come from the elements s, s, and r of Γ.

Let G/H, respectively $H\backslash G$, denote the set of left, respectively right, cosets of H in G. Recall that if $H \lhd G$, then $Ha = aH$ for all $a \in G$. So $H\backslash G = G/H$ and G/H

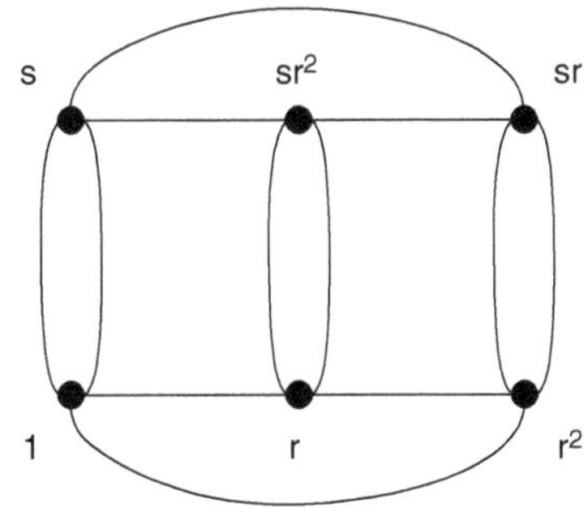

Figure 2.3 Cay(D_6, Γ)

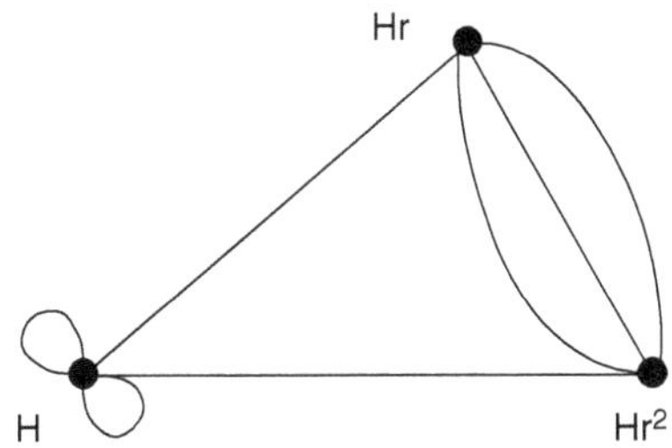

Figure 2.4 Cos$(H\backslash D_6, \Gamma)$

forms a group with $(aH) \cdot (bH) = (ab)H$ as the group operation. The map $a \mapsto aH$ from G to G/H is called the *canonical homomorphism*. As the name suggests, it is indeed a homomorphism.

Remark 2.15
Note that if $H \lhd G$, then $\text{Cos}(H\backslash G, \Gamma) = \text{Cay}(G/H, \overline{\Gamma})$, where $\overline{\Gamma} = \{\gamma H \mid \gamma \in \Gamma\}$ denotes the image of Γ under the canonical homomorphism.

Lemma 2.16
Let G be a finite group, let $H < G$, and let $\Gamma \Subset G$. Then $\text{Cay}(G, \Gamma)$ covers $\text{Cos}(H\backslash G, \Gamma)$.

Proof
For $g \in G$ and $\gamma \in \Gamma$, denote by $e(g, \gamma)$ the edge in $\text{Cay}(G, \Gamma)$ between g and $g\gamma$ induced by γ. Similarly, denote by $e(Hg, \gamma)$ the edge in $\text{Cos}(H\backslash G, \Gamma)$ between Hg and $Hg\gamma$ induced by γ. Let ϕ be the map from $\text{Cay}(G, \Gamma)$ to $\text{Cos}(H\backslash G, \Gamma)$ that takes g to Hg and $e(g, \gamma)$ to $e(Hg, \gamma)$. It is then straightforward to verify that ϕ is a covering. (See Exercise 1.) ◬

Corollary 2.17
Let G, H, Γ be as in Lemma 2.16. Then $h(\text{Cay}(G, \Gamma)) \leq h(\text{Cos}(H\backslash G, \Gamma))$.

Proof

This follows from Lemmas 2.11 and 2.16. ◈

Definition 2.18 Let (G_n) and (Q_n) be sequences of finite groups. We say that (G_n) *admits* (Q_n) *as a sequence of quotients* if for each n there exists $H_n \lhd G_n$ such that $G_n/H_n \cong Q_n$.

Definition 2.19 Let (G_n) be a sequence of finite groups. We say that (G_n) yields an expander family if for some positive integer d there exists a sequence (Γ_n), where for each n we have that $\Gamma_n \subseteq_s G_n$ with $|\Gamma_n| = d$, so that the sequence of Cayley graphs $(\mathrm{Cay}(G_n, \Gamma_n))$ is an expander family.

Proposition 2.20 (Quotients Nonexpansion Principle)
Let (G_n) be a sequence of finite groups. Suppose that (G_n) admits (Q_n) as a sequence of quotients. If (Q_n) does not yield an expander family, then (G_n) does not yield an expander family.

Proof

Assume temporarily that there exists a positive integer d and for each n a symmetric subset $\Gamma_n \subseteq_s G_n$ such that $(\mathrm{Cay}(G_n, \Gamma_n))$ is an expander family. Let $\overline{\Gamma_n}$ be the image of Γ_n under the canonical homomorphism. Then by Remark 2.15 and Corollary 2.17, we have that $(\mathrm{Cay}(Q_k, \overline{\Gamma_k}))$ is an expander family, which is a contradiction. ◈

Example 2.21
Let p_n be the nth prime number. (So $p_1 = 2, p_2 = 3, p_3 = 5$, and so on.) Let

$$G_n = \left\{ \begin{pmatrix} 1 & a & b \\ 0 & 1 & c \\ 0 & 0 & 1 \end{pmatrix} : a, b, c \in \mathbb{Z}_{p_n} \right\}.$$

Note that G_n is a group under matrix multiplication. (G_n is called the *group of* 3×3 *unipotent matrices over* $\mathbb{Z}_{p_n}$.)

Define a map $\phi : G_n \to \mathbb{Z}_{p_n} \times \mathbb{Z}_{p_n}$ by

$$\phi \left(\begin{pmatrix} 1 & a & b \\ 0 & 1 & c \\ 0 & 0 & 1 \end{pmatrix} \right) = (a, c).$$

The reader can verify that ϕ is a surjective homomorphism with kernel

$$K_n = \left\{ \begin{pmatrix} 1 & 0 & b \\ 0 & 1 & 0 \\ 0 & 0 & 1 \end{pmatrix} : b \in \mathbb{Z}_{p_n} \right\}.$$

So $G_n/K_n \cong \mathbb{Z}_{p_n} \times \mathbb{Z}_{p_n}$, by the first isomorphism theorem. Hence, (G_n) yields $(\mathbb{Z}_{p_n} \times \mathbb{Z}_{p_n})$ as a sequence of quotients. Later, in Example 4.27, we use

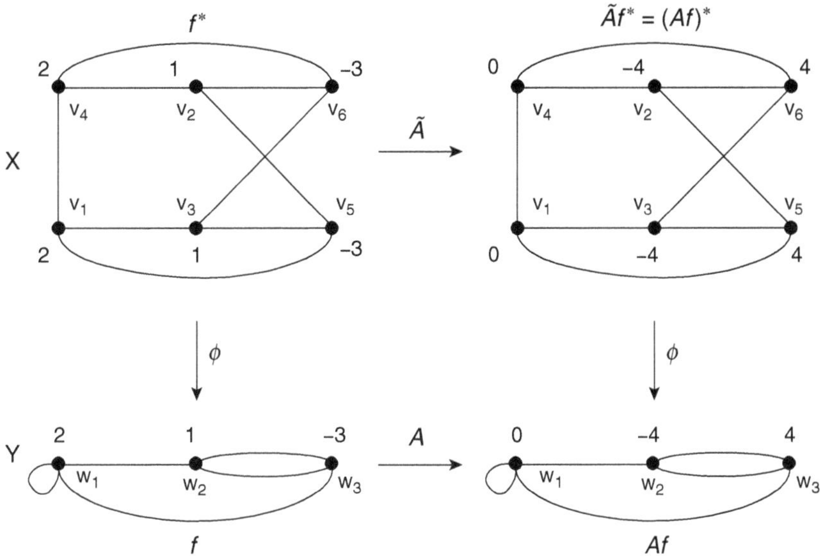

Figure 2.5 Pullbacks commute with adjacency operators

this fact, in conjunction with Prop. 2.20, to show that (G_n) does not yield an expander family.

From Corollary 1.87, we know that for a sequence of d-regular graphs, $h \to 0$ iff $\lambda_1 \to d$. Hence, we should be able to recover Prop. 2.20 by studying eigenvalues of coverings. Indeed, we can do just that, as we now demonstrate.

Definition 2.22 Let ϕ be a homomorphism from X to Y. Let $f \in L^2(Y)$. Define $f^* \in L^2(X)$ by $f^* = f \circ \phi$. We say that f^* is the *pullback of f via ϕ*.

Example 2.23
Take X, Y, ϕ as in Example 2.4.

Define $f \in L^2(Y)$ by $f(w_1) = 2, f(w_2) = 1, f(w_3) = -3$, as in Figure 2.5 (lower left). Then $f^*(v_1) = f(\phi(v_1)) = f(w_1) = 2$. The other values of f^* are shown in Figure 2.5 (upper left).

Let $A, \tilde{A}$ be the adjacency operators of Y, X, respectively. Observe, as shown on the right-hand side of Figure 2.5, that it makes no difference whether we first pull back via ϕ and then apply the adjacency operator, or if we first apply the adjacency operator and then pull back via ϕ.

Example 2.23 illustrates the following lemma.

Lemma 2.24
Let X, Y be finite graphs such that X covers Y. Let $A, \tilde{A}$ be the adjacency operators of Y, X, respectively. Let $f \in L^2(Y)$. Then

$$(Af)^* = \tilde{A}f^*.$$

Proof
Let ϕ be a covering from X to Y. Let v be a vertex of X. Recall that E_v is the set of all edges incident to v. If e is an edge incident to v, then let $e(v)$ denote the other vertex incident to e. Then

$$(Af)^*(v) = (Af)(\phi(v)) = \sum_{e \in E_{\phi(v)}} f(e(\phi(v))), \text{ and}$$

$$(\tilde{A}f^*)(v) = \sum_{\tilde{e} \in E_v} f^*(\tilde{e}(v)) = \sum_{\tilde{e} \in E_v} f(\phi(\tilde{e}(v))).$$

Equality follows because ϕ is a covering: it defines a bijection between E_v and $E_{\phi(v)}$, and it preserves incidence, that is, $e(\phi(v)) = \phi(\tilde{e}(v))$ where $e = \phi(\tilde{e})$. ⊛

Lemma 2.25
Let X and Y be finite graphs such that X covers Y. Then every eigenvalue of Y is an eigenvalue of X.

Proof
Let $A, \tilde{A}$ be the adjacency operators of Y, X, respectively. Let μ be an eigenvalue of A with corresponding eigenfunction f. Then

$$\tilde{A}f^* = (Af)^* = (\mu f)^* = \mu f^*,$$

by Lemma 2.24. So μ is an eigenvalue of $\tilde{A}$. ⊛

Proposition 2.26
Suppose X and Y are finite d-regular graphs such that X covers Y. Then $\lambda_1(X) \geq \lambda_1(Y)$.

Proof
This follows from Lemma 2.25 and Remark 2.6. ⊛

Remark 2.27
Recall from Remark 1.85 that roughly speaking the smaller λ_1 is, the better the graph is as a communications network. So Prop. 2.26, like Lemma 2.11, tells us that the quality of a graph is no better than that of any graph it covers. Maintaining speed and reliability, after all, becomes more difficult as the graphs become larger.

Remark 2.28
We could have proved Prop. 2.20 (the Quotients Nonexpansion Principle) in an alternate way, using Prop. 2.26, Lemma 2.16, and Corollary 1.87. See Exercise 2.

2. SUBGROUPS AND SCHREIER GENERATORS

In this section, we define a standard way in which a pair (G, Γ) with $\Gamma \Subset G$ and a subgroup H of G induces a pair $(H, \hat{\Gamma})$ with $\hat{\Gamma} \Subset G$. The set $\hat{\Gamma}$ is called the set of *Schreier generators*. We then show that the isoperimetric constant of $\text{Cay}(G, \Gamma)$ cannot be better than that of $\text{Cay}(H, \hat{\Gamma})$, divided by the index of H in G. As a consequence, we prove the Subgroups Nonexpansion Principle: If a sequence (G_n) of finite groups admits a nonexpanding bounded-index sequence of subgroups, then (G_n) does not yield an expander family. As with quotients, we show that working with eigenvalues gives us the same sorts of results as working with isoperimetric constants.

Definition 2.29 Let G be a finite group, and let $H < G$. Let $T \subset G$ such that T contains exactly one element from each right coset of H in G. Then T is a *set of transversals for H in G*.

With G, H, and T as in Def. 2.29, for any $x \in G$, we denote by $\overline{x}$ the unique element of T such that $Hx = H\overline{x}$. If $t, \gamma \in G$, we introduce the notation

$$\widehat{(t, \gamma)} = t\gamma(\overline{t\gamma})^{-1}.$$

Let $t, \gamma \in G$. Then there exists a unique $h \in H$ so that $t\gamma = h\overline{t\gamma}$ (see Lemma 2.34). Because $\widehat{(t, \gamma)} = t\gamma(\overline{t\gamma})^{-1} = h$, we see that $\widehat{(t, \gamma)}$ is an element of H. Hence $t\gamma = \widehat{(t, \gamma)}\overline{t\gamma}$ is the unique expression of $t\gamma$ in the form $h\overline{t\gamma}$.

Definition 2.30 Let G, H, and T be as in Def. 2.29. Let $\Gamma \subset G$. Define

$$\hat{\Gamma} = \{\widehat{(t, \gamma)} \mid (t, \gamma) \in T \times \Gamma\}.$$

We call $\hat{\Gamma}$ the set of Schreier generators for H in G with respect to Γ.

Remark 2.31
The meticulous reader will observe that we have not defined the Cartesian product $T \times \Gamma$, because Γ is not a set but a *multiset*. In cases like this where we have failed to extend set-theoretic definitions to multisets, simply apply the usual definition but be careful to count with multiplicity. So $T \times \Gamma$ is defined to be the multiset whose elements are all ordered pairs (t, γ) such that $t \in T, \gamma \in \Gamma$, where the multiplicity of the element (t, γ) in $T \times \Gamma$ equals the multiplicity of $\gamma \in \Gamma$.

More generally, for multisets A, B, the multiplicity of (a, b) in $A \times B$ equals the multiplicity of a in A times the multiplicity of b in B.

Remark 2.32
Note that $\left|\hat{\Gamma}\right| = [G : H] \cdot |\Gamma|$.

Example 2.33
Let G be the dihedral group $D_3 = \{1, r, r^2, s, sr, sr^2\}$. (See Notations and conventions for our notations for the dihedral group.) Let $H = \{1, r, r^2\}$, and

let $\Gamma = \{s, sr\}$. The right cosets of H in G are H and $Hs = \{s, sr, sr^2\}$. Let $T = \{1, s\}$ and note that T is a set of transversals for H in G. For example, we have $\overline{sr} = s$. Taking $t = 1$ and $\gamma = sr$, we have

$$\widehat{(1, sr)} = sr \cdot (\overline{sr})^{-1} = sr \cdot s^{-1} = srs = r^2 \in \hat{\Gamma}.$$

The complete multiset of Schreier generators is

$$\hat{\Gamma} = \{\widehat{(1, s)}, \widehat{(s, s)}, \widehat{(1, sr)}, \widehat{(s, sr)}\} = \{1, 1, r, r^2\}.$$

We return to this example later in Example 2.38.

The following lemma lists a handful of computational tricks that we employ frequently.

Lemma 2.34
Let G, H, and T be as in Def. 2.29. Then:

1. For all $x \in G$, there exists a unique $h \in H$ such that $x = h\overline{x}$.
2. For all $h \in H$, $a \in G$, we have $\overline{ha} = \overline{a}$.
3. For all $a, b \in G$, we have $\overline{\overline{a}b} = \overline{ab}$.
4. For all $t \in T$, we have $\overline{t} = t$.

Proof
The proofs are left as an exercise.

Lemma 2.35
Let G, H, and T be as in Def. 2.29. If $\Gamma \subset G$, then $\hat{\Gamma} \subset H$. Moreover, if $\Gamma \circledS G$, then $\hat{\Gamma} \circledS H$.

Proof
First we show that if $\Gamma \subset G$, then $\hat{\Gamma} \subset H$. Let $t \in T$, $\gamma \in \Gamma$. Let $x = t\gamma$. By Lemma 2.34, there exists a unique $h \in H$ such that $x = h\overline{x}$. So $\widehat{(t, \gamma)} = x\overline{x}^{-1} = h$.

Now we show that if Γ is symmetric, then $\hat{\Gamma}$ is symmetric. To do so, we define a map $\phi : T \times \Gamma \to T \times \Gamma$ by $\phi(t, \gamma) = (\overline{t\gamma}, \gamma^{-1})$. Observe that by Lemma 2.34, we have

$$(\phi \circ \phi)(t, \gamma) = \phi\left(\overline{t\gamma}, \gamma^{-1}\right) = \left(\overline{\overline{t\gamma}\gamma^{-1}}, \gamma\right) = (t, \gamma).$$

So ϕ equals its own inverse map. In particular, ϕ is bijective.

Now notice that again by Lemma 2.34, we have

$$\widehat{(t, \gamma)}^{-1} = \overline{t\gamma}\gamma^{-1}t^{-1} = \overline{t\gamma}\gamma^{-1}\left(\overline{\overline{t\gamma}\gamma^{-1}}\right)^{-1} = \widehat{\phi(t, \gamma)}.$$

Because ϕ is bijective, it follows that $\widehat{(t, \gamma)}^{-1}$ appears in $\hat{\Gamma}$ with the same multiplicity as $\widehat{(t, \gamma)}$.

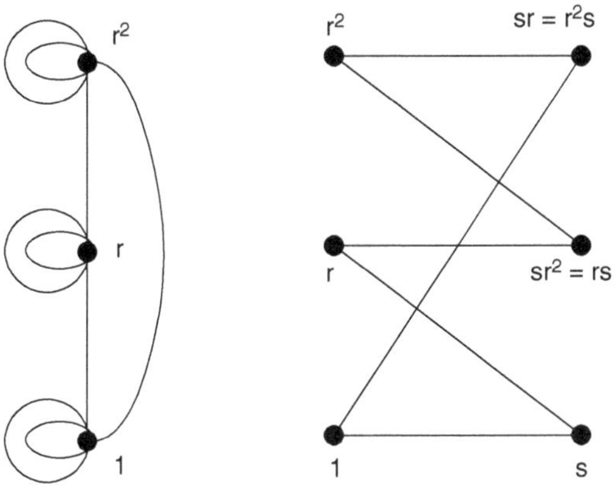

Figure 2.6 On the left, Cay$(H, \hat{\Gamma})$; on the right, Cay(G, Γ)

Remark 2.36
As in Remark 2.31, we have that $T \times \Gamma$ is a multiset, not a set. To be completely rigorous when we define ϕ in the proof of Lemma 2.35, we should extend the definition of a *function* to include those whose domains and codomains are multisets. We should then extend the definitions of *injective, surjective,* and *bijective.* We leave it to the reader to attend to these details. The guiding principle is to pretend that the multiset is a set but to remember to count with multiplicity.

Remark 2.37
Lemma 2.35 tells us that if Cay(G, Γ) is an undirected graph, then Cay$(H, \hat{\Gamma})$ is also an undirected graph.

Example 2.38
We continue Example 2.33. Because every element in G can be expressed uniquely in the form ht with $h \in H$ and $t \in T$, we can arrange the vertices of Cay(G, Γ) in a grid, with H as the vertical axis and T as the horizontal axis, as in Figure 2.6.

Observe in Figure 2.6 that every edge of Cay$(H, \hat{\Gamma})$ seems to come from an edge of Cay(G, Γ), by collapsing horizontally. Note that the edge in Cay(G, Γ) between 1 and s becomes *two* loops at 1 in Cay$(H, \hat{\Gamma})$. The right way to think about this is to regard the single edge as two directed edges, one from 1 to s, and one from s to 1. Then regard the two loops as the two corresponding directed edges from $1 \in H$ to itself. We return to this example later in Example 2.42.

Remark 2.39
Let G, H, and T be as in Definition 2.29. Every element of G can be expressed uniquely in the form ht with $h \in H$ and $t \in T$. This gives us a surjective map from G onto H defined by $ht \mapsto h$. This map "does" the horizontal collapsing that is discussed in Example 2.38.

To make more precise the observation in Example 2.38, if $g \in G$ and $\gamma \in \Gamma$, we denote by $e(g, \gamma)$ the directed edge in $\mathrm{Cay}(G, \Gamma)$ from g to $g\gamma$ induced by γ.

Lemma 2.40
There is a one-to-one correspondence between the set of directed edges in $\mathrm{Cay}(G, \Gamma)$ and the set of directed edges in $\mathrm{Cay}(H, \hat{\Gamma})$. The correspondence is given by the map $e(ht, \gamma) \mapsto e(h, \widehat{(t, \gamma)})$.

Proof
The proof is a matter of unwinding definitions; we leave it to the reader as an exercise. ⓐ

The next lemma establishes an upper bound on the isoperimetric constant of a Cayley graph on a group in terms of the isoperimetric constant of a Cayley graph on a subgroup.

Lemma 2.41
Let G, H, T be as in Def. 2.29. Let $\Gamma \subseteq G$. Then

$$h(\mathrm{Cay}(G, \Gamma)) \leq \frac{h(\mathrm{Cay}(H, \hat{\Gamma}))}{[G : H]}.$$

Proof
Let $S \subset H$ such that $|S| \leq \frac{1}{2}|H|$ and $\frac{|\partial S|}{|S|} = h(\mathrm{Cay}(H, \hat{\Gamma}))$. Let $\tilde{S} = \{ht \mid h \in S, t \in T\}$. Then

$$|\tilde{S}| = |S| \cdot |T| \leq \frac{1}{2}|H| \cdot |T| = \frac{1}{2}|G|.$$

Note that the map $ht \mapsto h$ is the same as the map $g \mapsto g(\overline{g})^{-1}$. So if $h \in H$, $t \in T$, $\gamma \in \Gamma$, then $ht\gamma \in \tilde{S}$ iff $ht\gamma(\overline{ht\gamma})^{-1} \in S$ iff $h\widehat{(t, \gamma)} \in S$. By Lemma 2.40 we have

$$\begin{aligned}
|\partial\tilde{S}| &= |\{e(g, \gamma) \ : \ g \in \tilde{S}, \gamma \in \Gamma, g\gamma \notin \tilde{S}\}| \\
&= |\{e(ht, \gamma) \ : \ h \in S, t \in T, \gamma \in \Gamma, ht\gamma \notin \tilde{S}\}| \\
&= |\{e(h, \widehat{(t, \gamma)}) \ : \ h \in S, t \in T, \gamma \in \Gamma, ht\gamma \notin \tilde{S}\}| \\
&= |\{e(h, \widehat{(t, \gamma)}) \ : \ h \in S, t \in T, \gamma \in \Gamma, h\widehat{(t, \gamma)} \notin S\}| \\
&= |\partial S|.
\end{aligned}$$

Therefore $h(\mathrm{Cay}(G, \Gamma)) \leq \dfrac{|\partial\tilde{S}|}{|\tilde{S}|} = \dfrac{|\partial S|}{|S| \cdot [G : H]} = \dfrac{h(\mathrm{Cay}(H, \hat{\Gamma}))}{[G : H]}$. ⓐ

The following example illustrates the proof of Lemma 2.41.

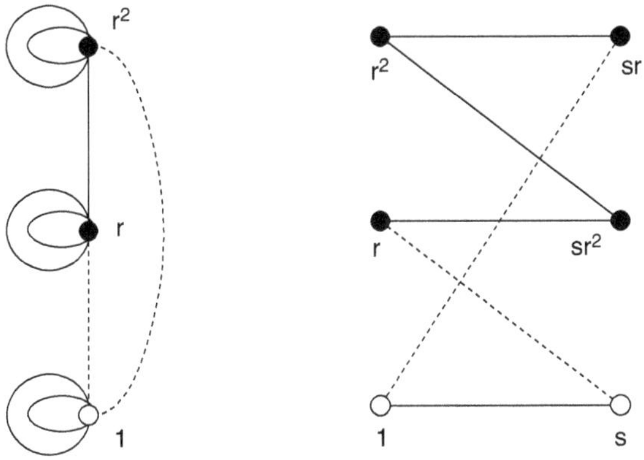

Figure 2.7 On the left, ∂S. On the right, $\partial\tilde{S}$. Observe that the figure on the left results from collapsing the figure on the right horizontally

Example 2.42
We continue Example 2.38. Suppose $S = \{1\} \subset H$. Then $\frac{|\partial S|}{|S|} = 2 = h(\mathrm{Cay}(H, \hat{\Gamma}))$. Also, $\tilde{S} = \{1, s\}$. Now, ∂S consists of the two dotted edges shown in the left-hand side of Figure 2.7. $S, \tilde{S}$ consists of the white vertices on the left and on the right sides, respectively, of Figure 2.7.

These two edges then “become” the two dotted edges shown in the right-hand side of Figure 2.7, which are precisely the edges in $\partial\tilde{S}$.

Lemma 2.41 tells us that $h(\mathrm{Cay}(G, \Gamma)) \leq \frac{h(\mathrm{Cay}(H,\hat{\Gamma}))}{[G:H]} = 1$. In fact, $\mathrm{Cay}(G, \Gamma)$ is a 6-cycle graph, so $h(\mathrm{Cay}(G, \Gamma))$ is precisely $\frac{2}{3}$, which is reassuring.

We return yet again to these graphs in Example 2.51.

Remark 2.43
As a corollary of Lemma 2.41, we obtain a group-theoretic fact known as the *Schreier subgroup lemma,* which states that if Γ generates G, then $\hat{\Gamma}$ generates H. For

if Γ generates G,
then $\mathrm{Cay}(G, \Gamma)$ is connected,
which implies that $h(\mathrm{Cay}(G, \Gamma)) > 0$,
which by Lemma 2.41 implies that $h(\mathrm{Cay}(H, \hat{\Gamma})) > 0$,
which implies that $\mathrm{Cay}(H, \hat{\Gamma})$ is connected,
which implies that $\hat{\Gamma}$ generates H.

The converse of this statement is false, however (see Exercise 4).

Definition 2.44 Let (G_n) and (H_n) be sequences of finite groups. We say that (G_n) *admits* (H_n) *as a bounded-index sequence of subgroups* if $H_n < G_n$ for all n and the sequence $([G_n : H_n])$ is bounded.

Example 2.45
Let G_n be the dihedral group D_n. (See Notations and conventions for our notations for the dihedral group.) For each n, let $H_n = \langle r \rangle < G_n$. Then (G_n) admits (H_n) as a bounded-index sequence of subgroups, because $[G_n : H_n] = 2$ for all n.

Proposition 2.46 (Subgroups Nonexpansion Principle)
Let (G_n) be a sequence of finite groups. Suppose that (G_n) admits (H_n) as a bounded-index sequence of subgroups. If (H_n) does not yield an expander family, then (G_n) does not yield an expander family.

Proof
Assume temporarily that there is a positive integer d and symmetric subsets $\Gamma_n \subseteq_S G_n$ with $|\Gamma_n| = d$ for all n such that $(\mathrm{Cay}(G_n, \Gamma_n))$ is an expander family. Let $\epsilon > 0$ such that $h((\mathrm{Cay}(G_n, \Gamma_n))) \geq \epsilon$ for all n.

Let M be a positive integer such that $[G_n : H_n] \leq M$ for all n. For each n, let T_n be a set of transversals for H_n in G_n. Let

$$\Lambda_n = \hat{\Gamma}_n \cup \{(M - [G_n : H_n])d \cdot e_n\}.$$

(So Λ_n is essentially the set of Schreier generators, but with enough copies of the identity thrown in so that $|\Lambda_n| = M \cdot d$ for all n.) Then by Lemma 2.41, we have that

$$h(\mathrm{Cay}(H_n, \Lambda_n)) = h(\mathrm{Cay}(H_n, \hat{\Gamma}_n)) \geq h((\mathrm{Cay}(G_n, \Gamma_n))) \geq \epsilon$$

for all n. Thus $(\mathrm{Cay}(H_n, \Lambda_n))$ is an expander family, which is a contradiction. ⊛

We now turn our attention to eigenvalues, whereupon we obtain a result similar to Lemma 2.41. The basic idea is to start with an eigenfunction on $\mathrm{Cay}(H, \hat{\Gamma})$ and spread it out horizontally to $\mathrm{Cay}(G, \Gamma)$, arranging the vertices as in Figure 2.7. Although we do not necessarily obtain an eigenfunction of $\mathrm{Cay}(G, \Gamma)$ in this manner (unlike coverings, where the pullback of an eigenfunction is an eigenfunction), we can however apply the Rayleigh-Ritz theorem to get a bound on the second-largest eigenvalue.

Recall the definition of λ_1 given in Def. 1.38.

Lemma 2.47
Let G, H, T as in Def. 2.29. Let $\Gamma \subseteq_S G$. Then

$$\lambda_1(\mathrm{Cay}(G, \Gamma)) \geq \frac{\lambda_1(\mathrm{Cay}(H, \hat{\Gamma}))}{[G : H]}.$$

Proof
Let A_G, A_H be the adjacency operators of $\mathrm{Cay}(G, \Gamma), \mathrm{Cay}(H, \hat{\Gamma})$, respectively. Let $\lambda_H = \lambda_1(\mathrm{Cay}(H, \hat{\Gamma}))$. Let $f \in L^2_0(H, \mathbb{R})$ such that $A_H f = \lambda_H f$.

Define $\tilde{f} \in L^2(G)$ by $\tilde{f}(ht) = f(h)$ for all $h \in H, t \in T$. Note that $\tilde{f} \in L^2_0(G, \mathbb{R})$, because

$$\sum_{g \in G} \tilde{f}(g) = \sum_{t \in T} \sum_{h \in H} \tilde{f}(ht) = \sum_{t \in T} \sum_{h \in H} f(h) = 0.$$

So by the Rayleigh-Ritz theorem (Prop. 1.82), we have that

$$\lambda_1(\mathrm{Cay}(G, \Gamma)) \geq \frac{\langle A_G \tilde{f}, \tilde{f} \rangle}{\langle \tilde{f}, \tilde{f} \rangle}.$$

We compute that

$$\begin{aligned}
\langle \tilde{f}, \tilde{f} \rangle &= \sum_{g \in G} \tilde{f}(g)^2 \\
&= \sum_{t \in T} \sum_{h \in H} \tilde{f}(ht)^2 \\
&= \sum_{t \in T} \sum_{h \in H} f(h)^2 \\
&= \sum_{t \in T} \langle f, f \rangle \\
&= [G : H] \langle f, f \rangle.
\end{aligned}$$

Recall that the map $ht \mapsto h$ is the same as the map $g \mapsto g(\overline{g})^{-1}$. So, by Lemma 2.34, for all $h \in H, t \in T, \gamma \in \Gamma$,

$$\tilde{f}(ht\gamma) = f(ht\gamma(\overline{ht\gamma})^{-1}) = f(h\widehat{(t, \gamma)}).$$

So

$$\begin{aligned}
\langle A_G \tilde{f}, \tilde{f} \rangle &= \sum_{g \in G} A_G \tilde{f}(g) \cdot \tilde{f}(g) \\
&= \sum_{g \in G} \sum_{\gamma \in \Gamma} \tilde{f}(g\gamma) \cdot \tilde{f}(g) \\
&= \sum_{h \in H} \sum_{t \in T} \sum_{\gamma \in \Gamma} \tilde{f}(ht\gamma) \cdot \tilde{f}(ht) \\
&= \sum_{h \in H} \sum_{t \in T} \sum_{\gamma \in \Gamma} f(h\widehat{(t, \gamma)}) \cdot f(h) \\
&= \sum_{h \in H} (A_H f)(h) \cdot \overline{f(h)} \\
&= \langle A_H f, f \rangle \\
&= \lambda_H \langle f, f \rangle.
\end{aligned}$$

So

$$\lambda_1(\mathrm{Cay}(G,\Gamma)) \geq \frac{\lambda_H\langle f,f\rangle}{[G:H]\langle f,f\rangle} = \frac{\lambda_1(\mathrm{Cay}(H,\hat{\Gamma}))}{[G:H]}.$$

Remark 2.48
Lemma 2.47 and Corollary 1.87 together provide an alternate route to proving the Subgroups Nonexpansion Principle. See Exercise 3.

Remark 2.49
Lemma 2.47 can be used in conjunction with Prop. 1.48 to give yet another proof of the Schreier subgroup lemma (see Remark 2.43).

Remark 2.50
Lemma 2.47 can be stated a bit more cleanly in terms of *normalized* adjacency operators. For a d-regular graph X with adjacency operator A, define $\mu_1(X) = \frac{\lambda_1(X)}{d}$. Then Lemma 2.47 is equivalent to $\mu_1(\mathrm{Cay}(G,\Gamma)) \geq \mu_1(\mathrm{Cay}(H,\hat{\Gamma}))$.

Note that $\mu_1(X)$ is the second-largest eigenvalue of $\frac{A}{d}$, which is the transition matrix for a random walk on X. In other words, suppose that for each vertex v, the probability that our random walker is at v is $f(v)$. Then the walker chooses at random, that is, uniformly, an edge incident to v and traipses along that edge to its other endpoint. Then the new probability vector is precisely $\frac{1}{d}Af$.

Example 2.51
We continue Example 2.42. From our previous computations, we can show that $\mathrm{Cay}(H,\hat{\Gamma})$ consists of a 3-cycle graph with two loops added at each vertex. It is then straightforward to compute that $\lambda_1(\mathrm{Cay}(H,\hat{\Gamma})) = 1$. By Lemma 2.47, then, we have that $\lambda_1(\mathrm{Cay}(G,\Gamma)) \geq \frac{1}{2}$. In fact, from Example 1.53 we know that $\lambda_1(\mathrm{Cay}(G,\Gamma)) = \cos(\frac{2\pi}{6}) = \frac{1}{2}$, so it turns out that we have equality in this case (see Exercise 5).

NOTES

1. Exercise 11 comes from [24], in which Bilu and Linial use this result to construct expander families by starting with a base graph and recursively taking 2-lifts. More precisely, for any $d \geq 3$, they explicitly construct a family of d-regular graphs with $\lambda_1 \leq O(\sqrt{d(\log d)^3})$. They conjecture that every d-regular Ramanujan graph admits a Ramanujan 2-lift.
2. Eigenvalues and related invariants of random coverings of graphs (also called random lifts or k-lifts) are studied in many publications, including [10], [11], [12], [58], [85], and [91].
3. Many articles discuss zeta functions of graph coverings; these include [55], [71], [98], [123], [130], [131], and [132]. See Note 17 in Chapter 3 for more about zeta functions of graphs.
4. See, for example, [125] for a discussion of Schreier generators.

EXERCISES

1. In the proof of Lemma 2.16, prove that ϕ is a covering.
2. Prove Prop. 2.20 by using Prop. 2.26 in lieu of Lemma 2.11.
3. Prove Prop. 2.46 by using Lemma 2.47 in lieu of Lemma 2.41.
4. Find G, H, Γ, T such that $\text{Cay}(G, \Gamma)$ is disconnected but both $\text{Cay}(H, \hat{\Gamma})$ and $\text{Cos}(H\backslash G, \Gamma)$ are connected.
5. Give an example where equality does not hold in Lemma 2.47.
6. Prove Lemma 2.34.
7. Give an example to show that Lemma 2.10 can fail if Y is not connected.
8. Let X be a graph. An *automorphism* of X is a bijective homomorphism of X. Let $\text{Aut}(X)$ be the set of all automorphisms of X. Prove that $\text{Aut}(X)$ is a group under composition. The set $\text{Aut}(X)$ is called the *automorphism group* of X.
9. Let $X = \text{Cay}(G, \Gamma)$. Prove that X is *vertex transitive*. That is, given two vertices v and w of X, prove that there is an automorphism of the graph ϕ such that $\phi(v) = \phi(w)$. (See Exercise 8 for a definition of automorphism.) (Hint: First show that if $g \in G$, then the left multiplication map given by $\phi_g : G \to G$, where $\phi_g(h) = gh$, is an automorphism of the graph X.)
10. Prove that the graph Y in Figure I.1 is not a Cayley graph. (Hint: Use Exercise 9. Observe that deleting any edge incident to the center vertex leaves behind a disconnected graph.)
11. Suppose that X, Y are finite connected graphs, and that ϕ is a covering from X to Y such that every fiber of ϕ contains exactly two vertices. We say that X is a *2-lift* of Y—see Note 1. Let A be the adjacency operator of Y. For every vertex w of Y, let ψ_w be a bijective map from $\phi^{-1}(w)$ to $\{-1, 1\}$.

    (a) Let e be an edge of Y. Show that there are exactly two edges $\tilde{e}$ in X such that $\phi(\tilde{e}) = e$.

    (b) Let $\tilde{e}$ be an edge of X such that $\phi(\tilde{e}) = e$. Let v_0, v_1 be the endpoints of $\tilde{e}$. Let $w_0 = \phi(v_0)$, $w_1 = \phi(v_1)$. Define $\sigma(e)$ to be 1 if $\psi_{w_0}(v_0) = \psi_{w_1}(v_1)$ and -1 otherwise. Show that σ is well defined.

    (c) Define a linear operator $A_s : L^2(Y) \to L^2(Y)$ by

    $$(A_s f)(w) = \sum_{e \in E_w} \sigma(e) f(e(w)),$$

    where E_w is the set of edges incident to w, and $e(w)$ is w if e is a loop, and $e(w)$ is the endpoint of e other than w, if e is not a loop. If f is an eigenvector of A_s with eigenvalue λ, then define a vector f_s in $L^2(X)$ by $f_s(v) = \psi_{\phi(v)}(v) \cdot f(\phi(v))$. Let $\tilde{A}$ be the adjacency operator of X. Show that f_s is an eigenvector of $\tilde{A}$ with eigenvalue λ.

    (d) Show that A_s is symmetric.

    (e) Show that if f, f' are orthogonal, then f_s, f_s' are orthogonal.

    (f) Show that if $f, g \in L^2(Y)$, then f_s is orthogonal to g^*.

    (g) Show that the spectrum of X equals the union of the spectrum of Y and the spectrum of A_s.

STUDENT RESEARCH PROJECT IDEAS

1. Investigate the following partial converse of the Subgroups Nonexpansion Principle union the Quotients Nonexpansion Principle. Suppose (G_k) is a sequence of finite groups. For each k, suppose $N_k \lhd G_k$ and $G_k/N_k \cong Q_k$. Moreover, suppose that (N_k) and (Q_k) each yield a family of expanders. Does it follow that (G_k) yields a family of expanders?
2. In a related vein, suppose that (Y_n) and (F_n) are two expander families (or at least, two sequences with a uniform lower bound on the isoperimetric constant such that the number of vertices of at least one of them goes to infinity). For each n, suppose that X_n covers Y_n and that the number of vertices in each fiber is $|F_n|$. Form a new graph X'_n by "adding in" a copy of F_n to each fiber of X_n. (We would very much like to call X'_n a "cloud covering" of Y_n where the copies of F_n are the "clouds.") Is (X'_n) an expander family?

   The strategy might be something like the following. Let S be a subset of the vertex set of X'_n containing no more than half the vertices. If S contains an equal number of vertices in each fiber, then $h(F_n)$ provides a lower bound on $|\partial S|$. On the other hand, if S is a union of fibers, then $h(Y_n)$ provides a lower bound on $|\partial S|$. Then interpolate between these two extremes. The paper [99], in which isoperimetric constants of graph products are studied, may contain some ideas relevant to this question.

3

The Alon-Boppana Theorem

1. STATEMENT AND CONSEQUENCES

Let X be a d-regular graph. Recall from Proposition 1.84 that

$$\frac{d - \lambda_1(X)}{2} \leq h(X) \leq \sqrt{2d(d - \lambda_1(X))}.$$

Hence, if the spectral gap $d - \lambda_1(X)$ is large, the isoperimetric constant $h(X)$ is large. If we want graphs with large isoperimetric constant (e.g., to construct fast, reliable communications networks), then our goal is to find graphs with small λ_1. The main result of this chapter is Prop. 3.1, which places a constraint on how small λ_1 can be.

Proposition 3.1
If (X_n) is a sequence of connected d-regular graphs with $|X_n| \to \infty$ as $n \to \infty$, then

$$\liminf_{n\to\infty} \lambda_1(X_n) \geq 2\sqrt{d-1}.$$

That is, for every $\epsilon > 0$, there exists an $N > 0$ such that $\lambda_1(X_n) > 2\sqrt{d-1} - \epsilon$ for all $n > N$.

Figure 3.1 illustrates this proposition. For each graph X_n, plot a point with n on the horizontal axis and $\lambda_1(X_n)$ on the vertical axis. For any $\epsilon > 0$, draw a horizontal line at $2\sqrt{d-1} - \epsilon$. As you move from left to right, after you cross some threshold value (N), all the plotted points must lie above that line.

Prop. 3.1 tells us that if we fix d and pick a large d-regular graph X, then $\lambda_1(X)$ is at best a little bit smaller than $2\sqrt{d-1}$. Thus, asymptotically the best spectral gap for a large d-regular graph is $d - 2\sqrt{d-1}$.

We prove Prop. 3.1 in Section 2 of this chapter. In fact, for a d-regular graph X one can give a lower bound for $\lambda_1(X)$ in terms of d and the diameter of X. We present this proposition and its proof, both due to Nilli [103], in Section 2. One then derives Proposition 3.1 as a consequence.

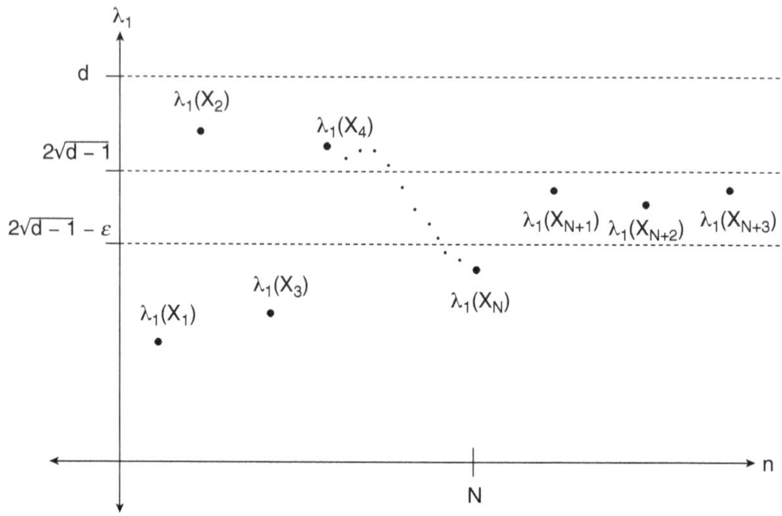

Figure 3.1

Definition 3.2 Suppose that X is a d-regular graph with n vertices. If X is not bipartite, then $\lambda_0(X) = d$ is called the *trivial eigenvalue* of X. If X is bipartite, then $\lambda_0(X) = d$ and $\lambda_{n-1}(X) = -d$ are called the trivial eigenvalues of X.

Definition 3.3 Let X be a d-regular graph with n vertices. If X is nonbipartite, let $\lambda(X) = \max\{|\lambda_1(X)|, |\lambda_{n-1}(X)|\}$. If X is bipartite, let $\lambda(X) = \max\{|\lambda_1(X)|, |\lambda_{n-2}(X)|\}$.

Remark 3.4
Suppose X is d-regular graph. If X is disconnected, then $\lambda(X) = d$. So $\lambda(X)$ "detects" disconnected graphs, just like $\lambda_1(X)$. If X is connected, then

$$\lambda(X) = \max\ \{|\lambda| \ : \ \lambda \text{ is a nontrivial eigenvalue of } X\}.$$

Remark 3.5
Some authors define $\lambda(X) = \max\{|\lambda_1(X)|, |\lambda_{n-1}(X)|\}$. For nonbipartite graphs this gives the same value as our definition, but for bipartite graphs this gives $\lambda(X) = d$.

Let X be a d-regular graph. Note that $\lambda(X) \geq \lambda_1(X)$. Hence, bounding $\lambda_1(X)$ from below bounds $\lambda(X)$ from below. Therefore, Proposition 3.1 gives us the following theorem, due to Alon and Boppana. This is not the only way to prove the Alon-Boppana theorem. We give another proof of this theorem using a completely different method in Section 3.

Proposition 3.6 (Alon-Boppana)
If (X_n) is a sequence of connected d-regular graphs with $|X_n| \to \infty$ as $n \to \infty$, then

$$\liminf_{n\to\infty} \lambda(X_n) \geq 2\sqrt{d-1}.$$

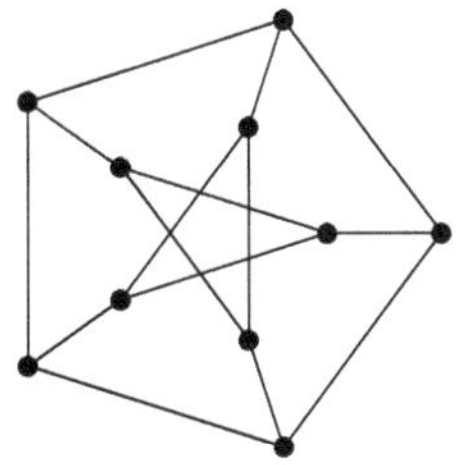

Figure 3.2 The Petersen graph

Suppose that (X_n) is a sequence of d-regular graphs with $|X_n| \to \infty$. Note that $d - \lambda(X_n) \leq d - \lambda_1(X_n)$ for all n. Hence, (X_n) is an expander family if $d - \lambda(X_n)$ is bounded below by some fixed positive constant. Thus, information about $\lambda(X)$ yields information about the isoperimetric constant of X. The Alon-Boppana theorem tells us that for a large, d-regular graph X the strongest upper bound for $\lambda(X)$ is $2\sqrt{d-1}$. This leads us to the following definition.

Definition 3.7 We say that a d-regular graph X is *Ramanujan* if $\lambda(X) \leq 2\sqrt{d-1}$.

Remark 3.8
Note that a d-regular Ramanujan graph must be connected, because otherwise $\lambda(X) = d$.

Example 3.9
The 3-regular graph X in Figure 3.2 is called the Petersen graph. The spectrum of X is $\begin{pmatrix} -2 & 1 & 3 \\ 4 & 5 & 1 \end{pmatrix}$. Therefore, the Petersen graph is Ramanujan because $\lambda(X) = 2 < 2.828 \approx 2\sqrt{3-1}$.

Example 3.10
Note that the cycle graph C_n and the complete graph K_n are Ramanujan graphs for $n \geq 3$. See Examples 1.52 and 1.53.

Example 3.11
The 3-regular graphs shown in Figure 3.3 are not Ramanujan. Recall that $2\sqrt{3-1} \approx 2.828$. Using software we computed the spectra of X_1 and X_2. All values are approximate. The spectrum of X_1 is

$$\{-2.83424,\ -2.38849,\ -1.65662,\ -1,\ 0,\ 0,\ 0,\ 0,\ 0.812716,\ 1.49086,\ 2.57577,\ 3\}.$$

So $\lambda(X_1) \approx 2.83424 > 2\sqrt{3-1}$. The spectrum of X_2 is

$$\{-2.3574,\ -1.88145,\ -1.61803,\ -1,\ -1,\ -1,\ 0,\ 0.501401,\ 0.618034,\ 1.90519,\ 2.83226,\ 3\}.$$

So $\lambda(X_2) \approx 2.83226 > 2\sqrt{3-1}$.

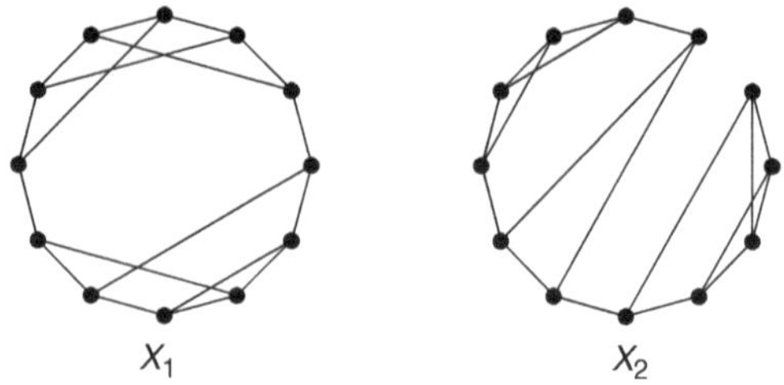

Figure 3.3

Remark 3.12

Suppose (X_n) is a sequence of d-regular Ramanujan graphs. Then $\lambda_1(X_n) \leq \lambda(X) \leq 2\sqrt{d-1}$. If $d \geq 3$, then

$$h(X_n) \geq \frac{d - \lambda_1(X_n)}{2} \geq \frac{d - 2\sqrt{d-1}}{2} > 0.$$

If $d \geq 3$, then any sequence of d-regular Ramanujan graphs is an expander family. As d increases, the lower bound $(d - 2\sqrt{d-1})/2$ on $h(X_n)$ becomes larger. This makes sense, because we are allowing more edges at each vertex of X_n. Note also that we needed $d \geq 3$ to get an expander family. For example, (C_n) is a family of 2-regular Ramanujan graphs, but it is not an expander family. (See Example 1.77 or Example 1.88.)

The Rayleigh-Ritz theorem (Prop. 1.82) told us that we can find $\lambda_1(X)$ by maximizing $\langle Af, f \rangle$ over the unit sphere in $L^2_0(X, \mathbb{R})$. The following proposition gives a similar method for computing $\lambda(X)$.

Proposition 3.13

Let X be a nonbipartite, d-regular graph with vertex set V. Let A be the adjacency operator for X. Then

$$\lambda(X) = \max_{\substack{f \in L^2_0(V) \\ \|f\|_2 = 1}} \left| \langle Af, f \rangle_2 \right| = \max_{f \in L^2_0(V)} \frac{\left| \langle Af, f \rangle_2 \right|}{\langle f, f \rangle_2}.$$

Proof

Let $n = |V|$. By Theorem A.53, there exists an orthonormal basis $\{f_0, f_1, f_2, \ldots, f_{n-1}\}$ for $L^2(V)$, such that f_i is an eigenfunction of A associated with the eigenvalue $\lambda_i = \lambda_i(X)$. Recall that f_0 is constant on V.

Let $f \in L^2_0(V)$ with $\|f\|_2 = 1$. As in Proposition 1.82, $f = c_1 f_1 + \cdots + c_{n-1} f_{n-1}$ for some $c_i \in \mathbb{C}$. Hence,

$$\left| \langle Af, f \rangle_2 \right| = \left| \left\langle \sum_{i=1}^{n-1} c_i \lambda_i f_i, \sum_{j=1}^{n-1} c_j f_j \right\rangle_2 \right|$$

$$\leq \sum_{i=1}^{n-1}\sum_{j=1}^{n-1} c_i\overline{c_j}\left|\lambda_i\left\langle f_i, f_j\right\rangle_2\right| = \sum_{i=1}^{n-1} c_i\overline{c_i}\left|\lambda_i\right|$$

$$\leq \lambda(X)\sum_{i=1}^{n-1} c_i\overline{c_i} = \lambda(X)\left\|f\right\|_2^2 = \lambda(X).$$

So

$$\lambda(X) \geq \max_{\substack{f\in L_0^2(V)\\ \|f\|_2=1}} \left|\langle Af, f\rangle_2\right|.$$

Now we prove the reverse inequality. If $\lambda(X) = \lambda_1(X)$, let $f = f_1$. Otherwise, let $f = f_{n-1}$. Then $f \in L_0^2(V)$ and $\|f\|_2 = 1$, and $\left|\langle Af, f\rangle_2\right| = \lambda(X)$. ⊛

Remark 3.14
One can show that

$$\lambda(X) = \max_{f\in L_0^2(V,\mathbb{R})} \frac{\left|\langle Af, f\rangle_2\right|}{\langle f, f\rangle_2}.$$

We invite the reader to fill in the details.

Remark 3.15
Let X be a bipartite d-regular graph with vertex set V. We leave it to the reader to show that

$$\lambda(X) \leq \max_{\substack{f\in L_0^2(V)\\ \|f\|_2=1}} \left|\langle Af, f\rangle_2\right| = d,$$

with equality if and only if X is disconnected.

2. FIRST PROOF: THE RAYLEIGH-RITZ METHOD

In this section, we give a proof of Proposition 3.1. We begin by providing a lower bound for $\lambda_1(X)$ for any d-regular graph X. This result is due to Nilli [103]. We follow the presentation given in Murty [101].

Proposition 3.16
Recall the notation for the floor function $\lfloor\cdot\rfloor$ given in Notations and conventions. Let X be a connected d-regular graph. If the $\operatorname{diam}(X) \geq 4$, then

$$\lambda_1(X) > 2\sqrt{d-1} - \frac{2\sqrt{d-1}-1}{\left\lfloor \frac{1}{2}\operatorname{diam}(X) - 1\right\rfloor}. \tag{5}$$

Before proving Proposition 3.16, we first show that it implies Proposition 3.1.

Consider a connected d-regular graph X. Fix a vertex v of X. The number of walks of length 1 starting at v is d. The number of walks of length 2 starting at v is d^2.

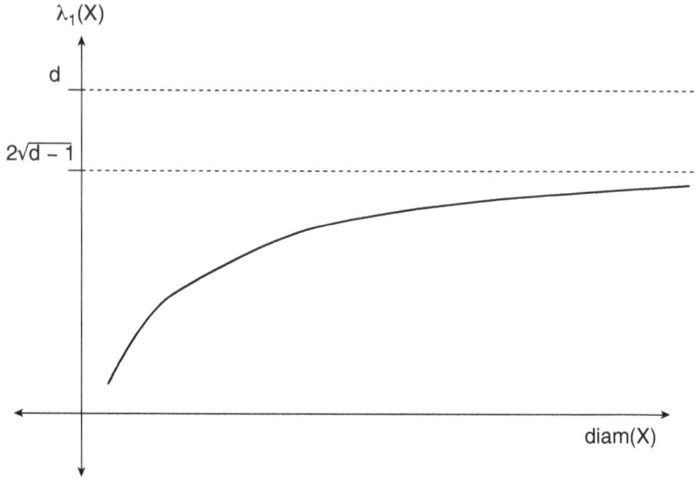

Figure 3.4

In general, the number of walks of length a starting at v is d^a. Note that a walk of length a contains at most $a + 1$ vertices.

We can cover the entire graph (possibly several times over) by taking all walks of length $\mathrm{diam}(X)$ from the fixed vertex v. There are $d^{\mathrm{diam}(X)}$ such walks, each containing at most $\mathrm{diam}(X) + 1$ distinct vertices. Hence,

$$|X| \leq (\mathrm{diam}(X) + 1)d^{\mathrm{diam}(X)}. \tag{6}$$

Let (X_n) be a sequence of connected, d-regular graphs such that $|X_n| \to \infty$ as $n \to \infty$. By Equation 6, $\mathrm{diam}(X_n) \to \infty$ as $n \to \infty$. Hence the term

$$\frac{2\sqrt{d-1} - 1}{\left\lfloor \frac{1}{2}\mathrm{diam}(X) - 1 \right\rfloor}$$

in Equation 5 approaches 0 as $n \to \infty$. Prop. 3.1 follows. See Figure 3.4—the curve shows the right-hand side of Equation 5. For any graph X, Prop. 3.16 tells us that $\lambda_1(X)$ must lie above this curve.

We now give the proof of Proposition 3.16.

Proof of Proposition 3.16

Let $b = \left\lfloor \frac{1}{2}\mathrm{diam}(X) - 1 \right\rfloor$. Throughout this proof, let $q = d - 1$; V be the vertex set of X; and E be the edge multiset of X.

The idea of the proof is as follows. We pick two vertices v_1 and v_2 that are as far apart as possible in the graph. Then, we carefully construct a function $f \in L_0^2(V, \mathbb{R})$ that has local maxima at v_1 and v_2 and decays rapidly as one moves away from v_1 and v_2. By Proposition 1.82, $\lambda_1(X) \geq d - \langle \Delta f, f \rangle_2 / \langle f, f \rangle_2$. We then calculate $\langle f, f \rangle_2$ and give an upper bound on $\langle \Delta f, f \rangle_2$. This will provide a lower bound for $\lambda_1(X)$. We break the proof into four steps.

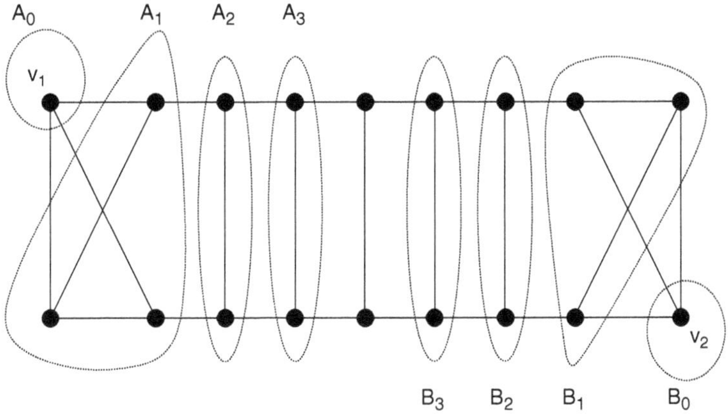

Figure 3.5 A 3-regular graph X

Step 1: In this step we construct an f in $L_0^2(V, \mathbb{R})$ as described. Let v_1 and v_2 be two vertices of X such that $\operatorname{dist}(v_1, v_2) \geq 2b + 2$. Define the sets

$$A_i = \{v \in V \mid \operatorname{dist}(v, v_1) = i\}$$

and

$$B_i = \{v \in V \mid \operatorname{dist}(v, v_2) = i\}$$

for $i = 0, 1, \ldots, b$.

As an example, consider the graph X in Figure 3.5. The diameter of X is 8. So $b = \lfloor 4 - 1 \rfloor = 3$. The sets A_0, A_1, A_2, A_3, and B_0, B_1, B_2, B_3 are shown in Figure 3.5.

If $x \in A_i \cap B_j$ for some $0 \leq i, j \leq b$, then $\operatorname{dist}(v_1, v_2) \leq \operatorname{dist}(v_1, x) + \operatorname{dist}(x, v_2) \leq 2b < 2b + 2$. This can't happen. Thus, $A_0, A_1, \ldots, A_b, B_0, B_1, \ldots, B_b$ are disjoint sets.

Let

$$A = \bigcup_{i=0}^{b} A_i \quad \text{and} \quad B = \bigcup_{i=0}^{b} B_i.$$

Suppose that $x \in A$ and $y \in B$ are adjacent. Then, $\operatorname{dist}(v_1, v_2) \leq \operatorname{dist}(v_1, x) + \operatorname{dist}(x, y) + \operatorname{dist}(y, v_2) \leq 2b + 1 < 2b + 2$. This can't happen. So, no vertex of A is adjacent to a vertex of B.

Define the function $f \in L_0^2(V, \mathbb{R})$ as

$$f(x) = \begin{cases} \alpha & \text{if } x \in A_0 \\ \alpha\, q^{-(i-1)/2} & \text{if } x \in A_i, \text{ for } i \geq 1 \\ 1 & \text{if } x \in B_0 \\ q^{-(i-1)/2} & \text{if } x \in B_i, \text{ for } i \geq 1 \\ 0 & \text{otherwise,} \end{cases}$$

where $\alpha \in \mathbb{R}$ will be chosen next.

If f_0 is the function that is equal to 1 on all vertices of V, then

$$\begin{aligned}\langle f, f_0\rangle_2 &= \alpha\left(|A_0| + \sum_{i=1}^{b} q^{-(i-1)/2}|A_i|\right) + \left(|B_0| + \sum_{i=1}^{b} q^{-(i-1)/2}|B_i|\right)\\ &= \alpha c_0 + c_1.\end{aligned}$$

for some real numbers $c_0, c_1 > 0$. Let $\alpha = -c_1/c_0$ for the remainder of this proof. Then $\langle f, f_0\rangle_2 = 0$.

Step 2: We now compute $\langle f, f\rangle_2$. Note that

$$\begin{aligned}\langle f, f\rangle_2 &= \sum_{x\in V} f(x)\overline{f(x)} = \sum_{i=0}^{b}\sum_{x\in A_i}|f(x)|^2 + \sum_{i=0}^{b}\sum_{x\in B_i}|f(x)|^2\\ &= S_A + S_B,\end{aligned}$$

where

$$S_A = \alpha^2 + \sum_{i=1}^{b}|A_i|\,\alpha^2 q^{-(i-1)} \quad\text{and}\quad S_B = 1 + \sum_{i=1}^{b}|B_i|\,q^{-(i-1)}.$$

Step 3: In this step we find an upper bound for $\langle \Delta f, f\rangle_2$. This step takes some work. Orient the edges of the graph X. That is, for each edge $e \in E$, label one endpoint e^+ and the other e^-. Recall that the Laplacian of X is independent of this labeling. From Proposition 1.60, we have that

$$\langle \Delta f, f\rangle_2 = C_A + C_B,$$

where

$$C_A = \sum_{\substack{e\in E\\ e^+\text{ or }e^-\in A}} \left(f(e^+) - f(e^-)\right)^2 \quad\text{and}\quad C_B = \sum_{\substack{e\in E\\ e^+\text{ or }e^-\in B}} \left(f(e^+) - f(e^-)\right)^2.$$

We see that

$$C_A = \sum_{i=0}^{b-1}\sum_{x\in A_i}\sum_{y\in A_{i+1}} A_{x,y}\left(f(x) - f(y)\right)^2 + \sum_{x\in A_b}\sum_{y\notin A} A_{x,y}\left(f(x) - 0\right)^2.$$

(Note that for $i = 0, f(x) - f(y) = 0$ in the sum.) For each $x \in A_i$, there are at most q elements y in A_{i+1} that are adjacent to x. Thus,

$$C_A \le \sum_{i=1}^{b-1} q\,|A_i|\left(q^{-(i-1)/2} - q^{-i/2}\right)^2\alpha^2 + q\,|A_b|\,q^{-(b-1)}\alpha^2.$$

Note that $\left(q^{-(i-1)/2} - q^{-i/2}\right)^2 = (q^{1/2} - 1)^2 q^{-i}$ and $q = (q^{1/2} - 1)^2 + 2\,q^{1/2} - 1$. Thus,

$$C_A \le \alpha^2 \sum_{i=1}^{b-1} q\left|A_i\right|(q^{1/2}-1)^2 q^{-i} + \alpha^2\left((q^{1/2}-1)^2 + 2q^{1/2} - 1\right)\left|A_b\right| q^{-(b-1)}$$

$$= \alpha^2 (q^{1/2}-1)^2 \left(\sum_{i=1}^{b} \left|A_i\right| q^{-(i-1)}\right) + \alpha^2\left(2q^{1/2}-1\right)\left|A_b\right| q^{-(b-1)}.$$

Note that $S_A - \alpha^2 = \alpha^2 \sum_{i=1}^{b} \left|A_i\right| q^{-(i-1)}$. Hence,

$$C_A \le (q^{1/2}-1)^2\left(S_A - \alpha^2\right) + \alpha^2\left(\frac{2\sqrt{q}-1}{b}\right) b\left|A_b\right| q^{-(b-1)}.$$

If $x \in A_i$ where $1 \le i \le b-1$, then there is at least one vertex from A_{i-1} that is adjacent to x, and at most q vertices from A_{i+1} that are adjacent to x. Hence, $\left|A_{i+1}\right| \le q\left|A_i\right|$ for $1 \le i \le b-1$. Similarly, $\left|B_{i+1}\right| \le q\left|B_i\right|$ for $1 \le i \le b-1$. So

$$\left|A_1\right| \ge q^{-1}\left|A_2\right| \ge q^{-2}\left|A_3\right| \ge \cdots \ge q^{-(b-2)}\left|A_{b-1}\right| \ge q^{-(b-1)}\left|A_b\right|.$$

In particular,

$$\alpha^2 b\left|A_b\right| q^{-(b-1)} = \alpha^2 \sum_{i=1}^{b} \left|A_b\right| q^{-(b-1)} \le \alpha^2 \sum_{i=1}^{b} \left|A_i\right| q^{-(i-1)} = S_A - \alpha^2. \tag{7}$$

Because X is connected and $\operatorname{diam}(X) \ge 4$, we have that $d \ge 2$ and $(2\sqrt{q} - 1)/b > 0$. Also $0 < (q^{1/2}-1)^2 = q + 1 - 2\sqrt{q}$. Thus

$$C_A \le (q^{1/2}-1)^2\left(S_A - \alpha^2\right) + \left(\frac{2\sqrt{q}-1}{b}\right)\left(S_A - \alpha^2\right) \qquad \text{(by (7))}$$

$$= \left(q + 1 - 2\sqrt{q} + \frac{2\sqrt{q}-1}{b}\right)\left(S_A - \alpha^2\right)$$

$$< \left(q + 1 - 2\sqrt{q} + \frac{2\sqrt{q}-1}{b}\right) S_A.$$

Similarly,

$$C_B < \left(q + 1 - 2\sqrt{q} + \frac{2\sqrt{q}-1}{b}\right) S_B.$$

Putting the pieces together, we see that

$$\langle \Delta f, f \rangle_2 = C_A + C_B < \left(q + 1 - 2\sqrt{q} + \frac{2\sqrt{q} - 1}{b} \right) (S_A + S_B). \quad (8)$$

Step 4: By the Rayleigh-Ritz theorem (Prop. 1.82)

$$\begin{aligned} d - \lambda_1(X) &= \min_{\substack{g \in L_0^2(V) \\ \|g\|_2 = 1}} \langle \Delta g, g \rangle_2 \\ &\leq \frac{\langle \Delta f, f \rangle_2}{\langle f, f \rangle_2} \\ &= \frac{C_A + C_B}{S_A + S_B} \\ &< q + 1 - 2\sqrt{q} + \frac{2\sqrt{q} - 1}{b} \qquad \text{(by Equation 8)} \\ &= d - 2\sqrt{d-1} + \frac{2\sqrt{d-1} - 1}{b}. \end{aligned}$$

Solving for $\lambda_1(X)$ gives the desired result.

3. SECOND PROOF: THE TRACE METHOD

In this section, we present the combinatorial proof of Proposition 3.6 that is given by Lubotzky, Phillips, and Sarnak [89]. Recall the statement of the proposition: if (X_n) is a sequence of connected d-regular graphs with $|X_n| \to \infty$ as $n \to \infty$, then $\liminf_{n\to\infty} \lambda(X_n) \geq 2\sqrt{d-1}$. That is, for any $\epsilon > 0$, there exists an $N > 0$ such that $\lambda(X_n) \geq 2\sqrt{d-1} - \epsilon$ for all $n > N$.

Before embarking on the proof, we give a short outline. Let X be a d-regular graph with vertices $v_1, v_2, \ldots, v_n$. Let A be the adjacency matrix of X with respect to this ordering on the vertices. It can be shown that

$$\sum_{i=0}^{n-1} \lambda_i(X)^k = \operatorname{tr}(A^k) = (A^k)_{1,1} + \ldots (A^k)_{n,n}, \quad (9)$$

where k is any positive integer. Furthermore, it can be shown that $(A^k)_{i,i}$ equals the number of walks of length k from vertex v_i to vertex v_i. One can use Equation 9 to bound $\lambda(X)$ in terms of the number of closed walks (see Definition 3.28) of a given length in X. For example, if X is connected and nonbipartite, and k is a positive integer, then

$$\lambda(X) \geq \left(\frac{(\#\text{ closed walks of length } 2k) - d^{2k}}{n-1} \right)^{1/2k}.$$

This turns the problem of finding a lower bound on $\lambda(X)$ into a problem of counting the number of closed walks in X of length $2k$. (The reason for changing k into $2k$ will become apparent later.)

This is exactly the tactic we use in this section; however, there is one slight wrinkle. To get a lower bound on the number of closed walks of length $2k$ in the graph X, we count the number of closed walks that start and end at a fixed base point (which never return to that base point in the middle of the walk) in the universal covering graph of X.

When counting the number of such walks, combinatorial quantities called Catalan numbers arise. Hence, we begin by discussing the basic properties of Catalan numbers.

3.1 Catalan Numbers

In this subsection, we introduce the Catalan numbers. These numbers occur in various counting problems. We use them in the proof of the Alon-Boppana theorem.

Definition 3.17 Let $a = (a_1, a_2, \ldots, a_{2k})$ be a sequence where $a_i = \pm 1$ for $i = 1, \ldots, 2k$. The *value* of a is $a_1 + \cdots + a_{2k}$. The *length* of a is $2k$. We say that a is *balanced* if a has value 0 and $a_1 + \cdots + a_i \geq 0$ for $i = 1, \ldots, 2k$. We say that a is *unbalanced* if a is not balanced.

Throughout this subsection, we only consider sequences that consist of the numbers 1 and -1.

Example 3.18
The sequence $(1, 1, 1, -1)$ has value $1 + 1 + 1 - 1 = 2$.

The sequence $a_1 = (1, -1, 1, 1, -1, -1)$ has value 0. The sequence a_1 is balanced because the values $1, 1-1, 1-1+1, 1-1+1+1, 1-1+1+1-1$, and $1-1+1+1-1-1$ are all greater than or equal to 0.

The sequence $a_2 = (1, -1, -1, 1)$ has value 0. The sequence a_2 is unbalanced because $1 - 1 - 1 = -1 < 0$.

Definition 3.19 Let n be a positive integer. The nth Catalan number is the number of balanced sequences of length $2n$ consisting of n positive ones and n negative ones. By convention, $C_0 = 1$.

Example 3.20
By Table 3.1, $C_1 = 1$, $C_2 = 2$, and $C_3 = 5$.

Table 3.1 The First few Catalan Numbers

n	**balanced sequences of length $2n$**	C_n
1	$(1, -1)$	1
2	$(1, 1, -1, -1), (1, -1, 1, -1)$	2
3	$(1, 1, 1, -1, -1, -1), (1, 1, -1, 1, -1, -1), (1, -1, 1, -1, 1, -1),$ $(1, -1, 1, 1, -1, -1), (1, 1, -1, -1, 1, -1)$	5

Remark 3.21
A string consisting of n open parentheses (and n closed parentheses) is called balanced if while reading the string from left to right there is never a point where the number of previously read open parentheses is stricly less than the number of previously read closed parentheses. For example, the string ()(()) is a balanced set of parentheses. The string ())(is not balanced. One can define the nth Catalan number as the number of balanced strings consisting of n open parentheses and n closed parentheses. Again, by convention, $C_0 = 1$.

Remark 3.22
One can show that Def. 3.19 is equivalent to the following recurrence relations:

$$C_{n+1} = \sum_{i=0}^{n} C_i C_{n-i} \text{ and } C_0 = 1.$$

Lemma 3.23
The nth Catalan number is $C_n = \dfrac{1}{n+1}\dbinom{2n}{n}$.

Proof
The number of sequences of length $2n$ composed of n positive ones and n negative ones is $\dbinom{2n}{n}$, because we are choosing locations for the n positive ones out of $2n$ possible locations. To count the number of balanced sequences of length $2n$, we subtract the number of unbalanced sequences of length $2n$ from $\dbinom{2n}{n}$.

Given an unbalanced sequence, $s = (a_1, a_2, \ldots, a_{2n})$, let k_0 be the smallest integer such that $a_1 + \cdots + a_{k_0} < 0$ and

$$\hat{s} = (a_1, a_2, \cdots, a_{k_0}, -a_{k_0+1}, -a_{k_0+2}, \cdots, -a_{2n}).$$

For example, if $s = (1, -1, -1, 1, 1, -1, -1, 1)$, then $\hat{s} = (1, -1, -1, -1, -1, 1, 1, -1)$.

If $s = (a_1, \ldots, a_{2n})$ is an unbalanced sequence with value 0 and k_0 is the smallest integer such that $a_1 + \cdots + a_{k_0} < 0$, then $\hat{s}$ has value

$$\sum_{i=1}^{k_0} a_i - \left(\sum_{i=k_0+1}^{2n} a_i \right) = -1 - 1 = -2.$$

Conversely, given a sequence $\hat{s} = (a_1, \ldots, a_{2n})$ of value -2, let k_0 be the smallest integer such that $a_1 + \cdots + a_{k_0} = -1$. Then, we may reverse the process and get an unbalanced sequence

$$s = (a_1, a_2, \ldots, a_{k_0}, -a_{k_0+1}, -a_{k_0+2}, \ldots, -a_{2n})$$

that has value 0. Thus, the number of unbalanced sequences of value 0 corresponds to the number of sequences with value -2. A sequence of value -2

consists of $n-1$ positive ones and $n+1$ negative ones. Therefore, there are $\binom{2n}{n+1}$ unbalanced sequences with value 0.

Hence,

$$\begin{aligned} C_n &= \binom{2n}{n} - \binom{2n}{n+1} \\ &= \binom{2n}{n} - \frac{(2n)!}{(n+1)!(n-1)!} \\ &= \binom{2n}{n} - \frac{n}{n+1}\binom{2n}{n} \\ &= \frac{1}{n+1}\binom{2n}{n}. \end{aligned}$$

3.2 The Universal Covering Graph

In this subsection, we define the universal covering graph T of a regular graph X. We give a count for the number of closed walks that start and end at a base point v in T, but do not hit v at any point in the middle of the walk. This count will be used in the proof of the Alon-Boppana theorem to give a lower bound for $\lambda_1(X)$.

Definition 3.24 Let X be a graph. Let

$$w = (v_0, e_0, v_1, e_1, \ldots, v_{n-1}, e_{n-1}, v_n)$$

be a walk in X. We say that w is *nonbacktracking* if $e_i \neq e_{i+1}$ for $i = 0, 1, \ldots, n-2$.

Definition 3.25 Let X be a connected d-regular graph with vertex set V. Let $v_0 \in V$ be a fixed vertex of X. The *universal covering graph* T_{v_0} of X using v_0 as a base point is constructed as follows. Each vertex of T_{v_0} is a nonbacktracking walk of X that begins at v_0. Two vertices are adjacent, via an edge of multiplicity 1, if one is $(v_0, e_0, v_1, \ldots, e_{n-2}, v_{n-1})$ and the other is $(v_0, e_0, v_1, \ldots, e_{n-2}, v_{n-1}, e_{n-1}, v_n)$. (That is, if one walk "extends" the other by a single step.) A pair of vertices not of this form are not adjacent.

Remark 3.26

The words *backtrackless* or *irreducible* are sometimes used instead of *nonbacktracking*. A nonbacktracking walk is sometimes called a *trek*.

Example 3.27

Consider the graph X in Figure 3.6 with the given labeling. The universal covering graph T_a of X using a as a base point is given in Figure 3.7. Note that T_a is an infinite graph.

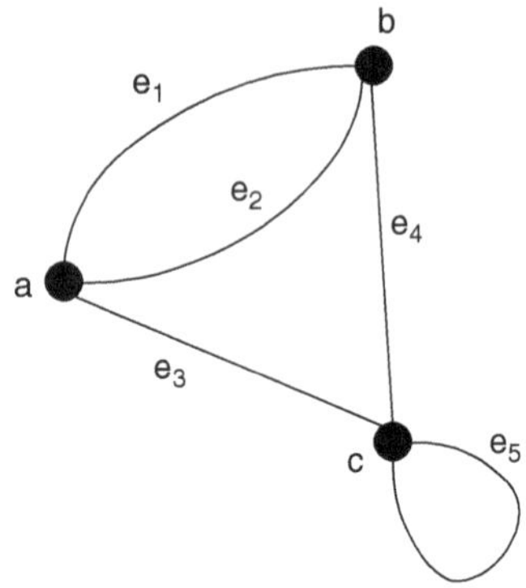

Figure 3.6 A 3-regular graph X

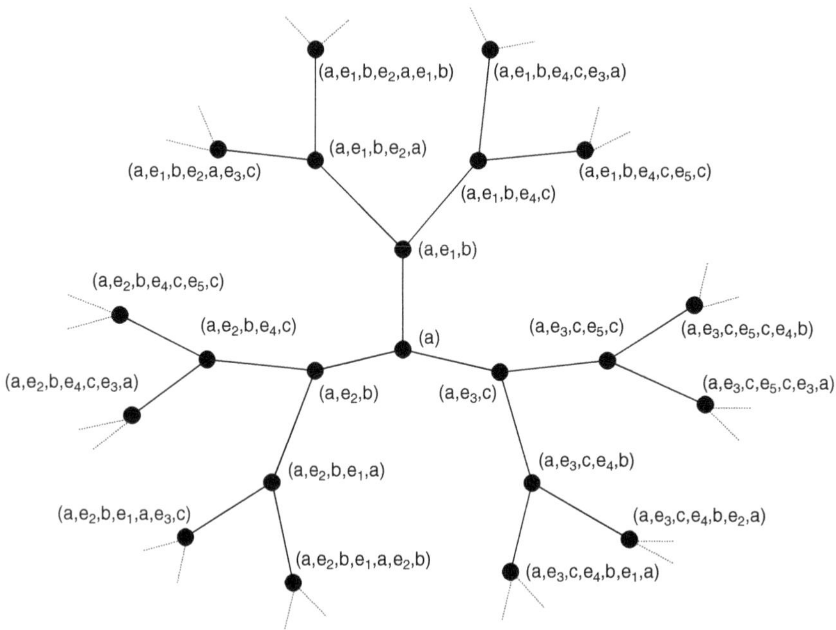

Figure 3.7 The universal covering graph of X

Definition 3.28 Let X be a graph. A walk $(v_0, e_0, v_1, \ldots, e_{n-1}, v_n)$ in X is said to be *closed* if $v_0 = v_n$. A closed walk is said to be a *circuit* if it has no repeated edges, that is, if $e_i \neq e_j$ if $i \neq j$.

We say that X is a *tree* if X is connected and has no circuits.

Example 3.29
The graph T in Figure 3.8 is a tree. The graph X in Figure 3.6 is not a tree since, for example, the closed walk $(a, e_1, b, e_4, c, e_3, a)$ is a circuit. For that matter, X has a loop, and no graph with a loop can be a tree, because a loop is a circuit of length 1.

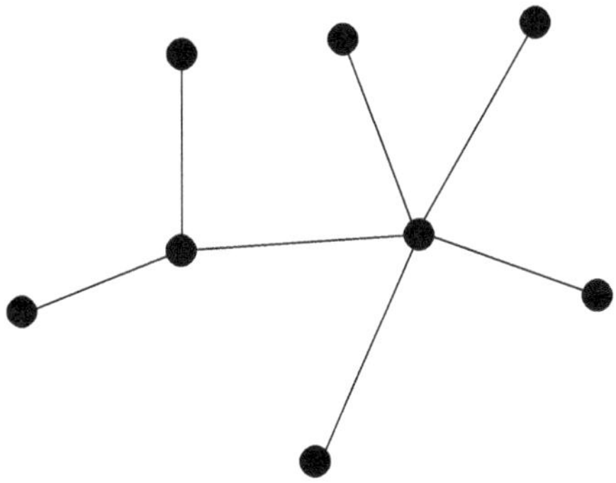

Figure 3.8 A tree

Let e denote the edge that lies between the vertices (a) and (a, e_1, b) in the graph T_a in Figure 3.7. Consider the following closed walk

$$((a), e, (a, e_1, b), e, (a)).$$

Note that this walk is closed but is not a circuit because it has a repeated edge. In fact, T_a has many closed walks; we prove later that none of them are circuits—that is, T_a is a tree.

Definition 3.30 Let X be a graph, and let v be a vertex of X. Let $C = (v, e_0, v_1, \ldots, v_{n-1}, e_{n-1}, v)$ be a closed walk starting and ending at v. We say that C is *unfactorable* if $v_i \neq v$ for $i = 1, \ldots, n-1$. Otherwise, we say that C is *factorable.*

That is, C is unfactorable if the walk C encounters v at the beginning and the end of the walk, but never in the middle of the walk.

Example 3.31
Consider the graph X in Figure 3.6. The walks (a, e_1, b, e_2, a) and $(a, e_1, b, e_4, c, e_5, c, e_4, b, e_2, a)$ are unfactorable. The walk $(a, e_1, b, e_2, a, e_1, b, e_2, a)$ is factorable.

Remark 3.32
Suppose that X is a graph and that v is some fixed vertex of X. Let C and D be closed walks that start and end at v. One can define the "product" of these walks, denoted by CD, as the walk that first goes along C and then goes along D. A closed walk E that starts and ends at v is factorable iff it can be written $E = CD$ for some closed walks C and D, each of length ≥ 1, that start and end at v. In other words, E is factorable if it can be nontrivially "factored." So unfactorable walks are sort of like prime numbers.

You should be aware that although we are using these terms for now, "factorable" and "unfactorable" are not standard terms in graph theory.

Lemma 3.33

Let X be a connected d-regular graph with vertex set V and edge multiset E. Let $T = T_{v_0}$ be the universal cover of X constructed using some fixed vertex $v_0 \in V$ as a base point. Then:

1. T is d-regular.
2. T is connected.
3. T is a tree.
4. The distance in T between $(v_0, e_0, \ldots, e_{n-1}, v_n)$ and (v_0) is n.
5. Let k be a fixed positive integer. The number of unfactorable walks of length $2k$ in T that begin and end at (v_0) is equal to

$$\frac{1}{k}\binom{2k-2}{k-1}d(d-1)^{k-1}.$$

Proof

1. The vertex (v_0) of T has the d neighbors $(v_0, e_1, v_1), \ldots, (v_0, e_d, v_d)$, where $e_1, \ldots, e_d$ are the d edges incident to v_0 in X. Because T is by definition a simple graph (i.e., T has no multiple edges and no loops), the degree of (v_0) is d. Let $t = (v_0, e_0, v_1, \ldots, v_{n-1}, e_{n-1}, v_n)$ be a vertex of T where $n \geq 1$. The vertex v_n of X has d edges incident to it in X. One of these edges is e_{n-1}. Call the remaining edges $e'_1, e'_2, \ldots, e'_{d-1}$. Let w_i be the other vertex of X (perhaps equal to v_n if e'_i is a loop) that is incident to e'_i for $i = 1, \ldots, d-1$. Then, t has d neighbors, namely,

$$(v_0, e_0, v_1, \ldots, v_{n-1})$$
$$\text{and} \quad (v_0, e_0, v_1, \ldots, v_{n-1}, e_{n-1}, v_n, e'_i, w_i) \text{ for } i = 1, \ldots d-1.$$

   So $\deg(t) = d$.
2. Let $(v_0, e_1, v_1, \ldots, v_{n-2}, e_{n-1}, v_n)$ be a vertex of T. Then the sequence of vertices

$$(v_0), (v_0, e_1, v_1), \ldots, (v_0, e_1, v_1, \ldots, v_{n-2}), (v_0, e_1, v_1, \ldots, v_{n-2}, e_{n-1}, v_n)$$

   gives a walk from (v_0) to $(v_0, e_1, v_1, \ldots, v_{n-2}, e_{n-1}, v_n)$.
3. Suppose that T has a circuit of the form

$$(w_0), (w_0, e_0, w_1), \ldots, (w_0, e_0, w_1, \ldots, e_{n-1}, w_n), (w_0).$$

   But then, (w_0) would be adjacent to $(w_0, e_0, w_1, \ldots, e_{n-1}, w_n)$. This can't happen unless $n = 1$, in which case, we don't have a circuit.
4. Consider a vertex $t = (v_0, e_0, v_1, \ldots, v_{n-1}, e_{n-1}, v_n)$ of X. First we show that $\text{dist}(t, (v_0)) \geq n$. Let $(t_k, t_{k-1}, \ldots, t_1, t_0)$ be a walk in T_{v_0} from $t = t_k$ to $(v_0) = t_0$. Let $l(t_i)$ denote the length of t_i as a walk in X. By definition of adjacency in T_{v_0}, for each i we have $l(t_i) = l(t_{i-1}) \pm 1$. Note that (v_0) is a walk of length 0 in X, so $l(t_0) = 0$. By induction, it follows that $l(t_i) \geq i$ for all i. But $l(t_k) = l(t) = n$, so $n \leq k$. That is,

the length of any walk in T_{v0} from t to (v_0) is at least n. Therefore, $\text{dist}(t, (v_0)) \geq n$. To see that $\text{dist}(t, (v_0)) \leq n$, observe that

$$(v_0, e_0, v_1, \ldots, v_{n-1}, e_{n-1}, v_n), (v_0, e_0, v_1, \ldots, v_{n-1}), \ldots, (v_0, e_0, v_1), (v_0)$$

is a walk of length n from t to (v_0).

5. Consider an unfactorable walk of length $2k$ that starts and ends at (v_0). By part (4) of this lemma, at each step of the walk our distance from (v_0) will either increase by 1 or decrease by 1. Hence, we may associate such a walk with a sequence $(a_1, a_2, \ldots, a_{2k})$ of 1s and -1s such that $a_1 = 1, a_{2k} = -1$, and $\sum_{i=2}^{m} a_i \geq 0$ for $m = 2, \ldots, 2k-1$. That is, the first step of the walk takes us away from (v_0), the last step of the walk takes us to closer to (v_0), and at each step in between we are at least distance 1 from (v_0). By Lemma 3.23, there are $C_{k-1} = \frac{1}{k}\binom{2k-2}{k-1}$ such sequences.

   How many walks correspond to such a sequence? Each such walk takes k steps away from (v_0) and k steps toward (v_0). Because T_{v0} is a d-regular tree, from a fixed vertex of T_{v0} there are exactly $d-1$ edges that move us farther away from (v_0) and only one edge that moves us closer to (v_0). In the first step of the walk, there are d choices for which edge to go along. Afterward, each step toward (v_0) is uniquely determined, and each step away from (v_0) can be chosen in $d-1$ ways.

   Thus, the number of walks of length $2k$ that start at (v_0) and end at (v_0) for the first time is $\frac{1}{k}\binom{2k-2}{k-1}d(d-1)^{k-1}$. ⓐ

Let X be a d-regular graph, v_0 a vertex of X, and $T = T_{v0}$ the universal cover of X using v_0 as a base point. In the proof of the Alon-Boppana theorem, in the next subsection, we make use of the fact that one can project closed walks in T that start and end at (v_0) to closed walks in X that start and end at v_0. To do so, we need to describe the "covering map" associated with T.

Definition 3.34 Let X be a d-regular graph, v_0 a vertex of X, and $T = T_{v0}$ the universal cover of X using v_0 as a base point. Define the *covering map* $\phi_{v0} : T \to X$ of T as follows. For a vertex $(v_0, e_0, v_1, e_1, \ldots, e_{n-1}, v_n)$ of T, define $\phi_{v0}(v_0, e_0, \ldots, e_{n-1}, v_n) = v_n$. Let e be the edge of T that is incident to $(v_0, e_0, v_1, e_1, \ldots, e_{n-1}, v_n)$ and $(v_0, e_0, v_1, e_1, \ldots, e_{n-1}, v_n, e_n, v_{n+1})$. Define $\phi_{v0}(e) = e_n$.

Remark 3.35
The space T_{v0} is the "universal covering space" of X, and ϕ_{v0} is a covering map in the sense of algebraic topology, provided we topologize graphs appropriately.

Example 3.36
Consider the graph X in Figure 3.6 with covering graph T_a in Figure 3.7. Let $\phi_a : T_a \to X$ be the covering map. Then $\phi_a((a)) = a$ and $\phi_a((a, e_3, c, e_4, b)) = b$.

Also, ϕ_a maps the edge in T_a that is incident to (a, e_3, c) and (a, e_3, c, e_5, c) to the loop e_5 in X. Consider the walk

$$w = ((a), (a, e_1, b), (a, e_1, b, e_4, c), (a, e_1, b, e_4, c, e_3, a))$$

in T_a. (We do not include the edges of the walk because T_a has no multiple edges or loops.) Applying ϕ_a to each vertex and edge in w, we get the walk

$$(a, e_1, b, e_4, c, e_3, a)$$

in X.

Lemma 3.37
Let X, v_0, T, and ϕ_{v_0} be as in Definition 3.34. The number of closed walks of length $2k$ in X beginning and ending at v_0 is greater than or equal to the number of closed walks of length $2k$ in T_{v_0} beginning and ending at (v_0).

Proof
We leave the proof to the reader. See Exercise 1. ⊗

3.3 A Combinatorial Proof of the Alon-Boppana Theorem

In this subsection, we complete our second proof of the Alon-Boppana theorem. First, we establish a couple of technical lemmas.

Lemma 3.38

$$\lim_{k\to\infty} \binom{2k-2}{k-1}^{1/2k} = 2.$$

Proof
Let n be a positive integer. Because $\ln(x)$ is an increasing function, left-handed Riemann sums underestimate integrals of the logarithm and right-handed Riemann sums overestimate integrals of the logarithm. Hence,

$$\ln(n!) = \ln(2) + \cdots + \ln(n) \geq \int_1^n \ln(x)\, dx = n\ln(n) - n + 1,$$

and similarly

$$\ln(n!) \leq \int_2^{n+1} \ln(x)\, dx = (n+1)\ln(n+1) - (n+1) - 2\ln(2) + 2.$$

Thus,

$$\begin{aligned}
\lim_{k\to\infty}\frac{1}{2k}\ln\binom{2k-2}{k-1} &= \lim_{k\to\infty}\frac{1}{2k}\ln\frac{(2k-2)!}{[(k-1)!]^2}\\
&= \lim_{k\to\infty}\frac{\ln[(2k-2)!]-2\ln[(k-1)!]}{2k}\\
&\le \lim_{k\to\infty}\frac{(2k-1)\ln(2k-1)-(2k-1)-2\ln(2)+2-2[(k-1)\ln(k-1)-(k-1)+1]}{2k}\\
&= \lim_{k\to\infty}\ln(2k-1)-\ln(k-1)\\
&=\ln(2).
\end{aligned}$$

Similarly,

$$\begin{aligned}
\lim_{k\to\infty}\frac{1}{2k}\ln\binom{2k-2}{k-1} &= \lim_{k\to\infty}\frac{\ln[(2k-2)!]-2\ln[(k-1)!]}{2k}\\
&\ge \lim_{k\to\infty}\frac{(2k-2)\ln(2k-2)-(2k-2)+1-2[k\ln(k)-k-2\ln(2)+2]}{2k}\\
&= \ln(2).
\end{aligned}$$

Hence,

$$\binom{2k-2}{k-1}^{1/2k} = \exp\left(\ln\binom{2k-2}{k-1}^{1/2k}\right)\to 2$$

as $k\to\infty$. ◈

Lemma 3.39
Let A and B be be real numbers with $A\ge B\ge 0$. If k is a positive integer, then $(A-B)^{1/2k}\ge A^{1/2k}-B^{1/2k}$.

Proof
Define the function $f(x)=(x-B)^{1/2k}+B^{1/2k}-x^{1/2k}$ where $x\ge B$. The derivative of f is given by

$$f'(x)=\frac{1}{2k}(x-B)^{1/2k-1}-\frac{1}{2k}x^{1/2k-1}=\frac{1}{2k}\left(\frac{x^{1-1/2k}-(x-B)^{1-1/2k}}{(x-B)^{1-1/2k}x^{1-1/2k}}\right).$$

Note that $f'(x)>0$ for all $x>B$. Since $f(B)=0$, this implies that $f(x)\ge 0$ for all $x\ge B$. The result follows by plugging in A for x. ◈

We now give a combinatorial proof of the Alon-Boppana theorem. We follow the exposition given by Lubotzky, Phillips, and Sarnak [89].

Theorem 3.40 (Alon-Boppana)
Let $d \geq 2$ be a fixed integer and $(X_n)_{n=1}^{\infty}$ a family of connected d-regular graphs with $|X_n| \to \infty$ as $n \to \infty$. Then

$$\liminf_{n\to\infty} \lambda(X_n) \geq 2\sqrt{d-1}.$$

That is, given an $\epsilon > 0$, there exists an $N \geq 1$ such that $\lambda(X_n) \geq 2\sqrt{d-1} - \epsilon$ for all $n \geq N$.

Proof
Consider a d-regular graph X with vertex set V ordered as $v_1, v_2, \ldots, v_n$. Assume that $n \geq 3$. Let A be the adjacency matrix of X with this ordering of the vertices. It follows from Prop. 1.99 that $(A^{2k})_{i,j}$ equals the number of walks of length $2k$ from vertex v_i to vertex v_j. Taking the trace of A^{2k} gives

$$\sum_{i=0}^{n-1} \lambda_i(X)^{2k} = \operatorname{tr}(A^{2k}) = \sum_{i=1}^{n} (A^{2k})_{i,i} = w(2k),$$

where $w(2k)$ is the number of closed walks of length $2k$ in X. Here we use Prop. 1.100 and Lemma A.60. Given a vertex $v \in V$, let $\rho_v(2k)$ be the number of walks of length $2k$ beginning and ending at (v) in the covering graph T_v. By Lemma 3.37, we have

$$\sum_{i=0}^{n-1} \lambda_i(X)^{2k} = w(2k) \geq \sum_{i=1}^{n} \rho_{v_i}(2k). \tag{10}$$

Let $\rho'_v(2k)$ denote the number of unfactorable closed walks of length $2k$ in the covering graph T_v beginning and ending at (v). Since $\rho_v(2k)$ counts *all* closed walks of length $2k$ based at v but $\rho'_v(2k)$ counts only the unfactorable ones, $\rho_v(2k) \geq \rho'_v(2k)$. By Lemma 3.33(5),

$$\rho'_v(2k) = \frac{1}{k}\binom{2k-2}{k-1} d(d-1)^{k-1}. \tag{11}$$

Hence, $\rho'_v(2k)$ is independent of the choice of v. Henceforth, we denote $\rho'_v(2k)$ by $\rho'(2k)$. So, from Equation 10, we have

$$\sum_{i=0}^{n-1} \lambda_i(X)^{2k} \geq \sum_{i=1}^{n} \rho'(2k) = n\rho'(2k).$$

If X is bipartite, then $\lambda_0(X) = d$ and $\lambda_{n-1}(X) = -d$, so

$$(n-2)\lambda(X)^{2k} \geq \sum_{i=1}^{n-2} \lambda_i(X)^{2k} \geq n\rho'(2k) - 2d^{2k},$$

and

$$\lambda(X)^{2k} \geq \frac{n}{n-2}\rho'(2k) - \frac{2d^{2k}}{n-2} \geq \rho'(2k) - \frac{2d^{2k}}{n-2}.$$

(Recall that $n \geq 3$, so dividing by $n-2$ is okay.) If X is not bipartite, then $\lambda_0(X) = d$, so

$$(n-1)\lambda(X)^{2k} \geq \sum_{i=1}^{n-1} \lambda_i(X)^{2k} \geq n\rho'(2k) - d^{2k},$$

and

$$\lambda(X)^{2k} \geq \frac{n}{n-1}\rho'(2k) - \frac{d^{2k}}{n-1} \geq \rho'(2k) - \frac{2d^{2k}}{n-2}.$$

In either case, we have

$$\lambda(X)^{2k} \geq \rho'(2k) - \frac{2d^{2k}}{n-2}.$$

From Equation 11, we get

$$\begin{aligned}\lambda(X)^{2k} &\geq \frac{1}{k}\binom{2k-2}{k-1}d(d-1)^{k-1} - \frac{2d^{2k}}{n-2}\\ &\geq \frac{1}{k}\binom{2k-2}{k-1}(d-1)^{k} - \frac{2d^{2k}}{n-2}\\ &= \frac{1}{k}\binom{2k-2}{k-1}(\sqrt{d-1})^{2k} - \frac{2d^{2k}}{n-2}.\end{aligned}$$

By Lemma 3.39,

$$\lambda(X) \geq \frac{1}{k^{1/2k}}\binom{2k-2}{k-1}^{1/2k}\sqrt{d-1} - \frac{2^{1/2k}d}{(n-2)^{1/2k}}.$$

Now let (X_n) be a sequence of connected d-regular graphs. From the foregoing arguments, we have

$$\liminf_{n\to\infty} \lambda(X_n) \geq \frac{1}{k^{1/2k}}\binom{2k-2}{k-1}^{1/2k}\sqrt{d-1}$$

for all $k \geq 1$. Recall that $k^{1/2k} \to 1$ as $k \to \infty$ by taking logarithms and applying L'Hôpital's rule. Now let $k \to \infty$ and use Lemma 3.38 to get the desired result. ◬

NOTES

1. The Catalan numbers are named after Belgian mathematician Eugène Charles Catalan (1814–1894).
2. Several proofs of the Alon-Boppana theorem have appeared in the literature. Our first proof of the theorem follows the exposition in Murty [101] based on a proof by Nilli [103]. Our second proof follows the one given by Lubotzky, Phillips, and Sarnak [89]. Davidoff, Sarnak, and Valette [45] present a proof based on Serre's theorem (see Note 14). Hoory, Linial, and Wigderson [70] present several proofs of the theorem.
3. Each of [45], [87], and [129] cover different aspects of Ramanujan graphs. [101] is a survey paper on Ramanujan graphs.
4. The first construction of Ramanujan graphs was given in 1988 by Lubotzky, Phillips, and Sarnak [89]. This article is indisputably one of the most important and seminal in the field. Their construction is as follows: let p and q be odd primes with $p, q \equiv 1 (\bmod 4)$, and i be an integer such that $i^2 \equiv -1 (\bmod q)$. There are $p + 1$ integral solutions to the equation $p = a^2 + b^2 + c^2 + d^2$ where $a > 0$ is odd and b, c, d are even. For each solution construct the matrix $\begin{pmatrix} a+ib & c+id \\ -c+id & a-ib \end{pmatrix}$ in $PGL(2, \mathbb{Z}_q)$. Let $\left(\frac{p}{q}\right)$ denote the Legendre symbol. If $\left(\frac{p}{q}\right) = -1$, then $X^{p,q}$ is the Cayley graph $\text{Cay}(PGL(2, \mathbb{Z}_2), S)$ where S is the set of $p + 1$ matrices constructed as before. If $\left(\frac{p}{q}\right) = 1$, then the matrices all lie in the index two subgroup $PSL(2, \mathbb{Z}_q)$. In this case, construct the graph $X^{p,q}$ using $PSL(2, \mathbb{Z}_q)$ as the vertex set. It is shown in [89] that $X^{p,q}$ is a $(p + 1)$-regular Ramanujan graph. The book by Davidoff, Sarnak, and Valette [45] gives a simplifed proof that the $X^{p,q}$ graphs yield expander families.
5. The Ramanujan graphs constructed by Lubotzky, Phillips, and Sarnak (see Note 4) are $(p + 1)$-regular where p is an odd prime. In 1992, Chiu [37] constructed a family of 3-regular Ramanujan graphs. In 1994, Morgenstern [100] constructed families of $(p^e + 1)$-regular Ramanujan graphs for any prime p. Note that the problem of constructing families of d-regular Ramanujan graphs is open for d that is not of the above form.
6. In [33], Buser shows that for any $n > 0$ there exists a cubic graph X (i.e., a 3-regular graph) with $|X| \geq n$ and $h(X) \geq 1/128$. Mohar [99] defines $F(n, k) = \max\{h(X) \mid X \text{ is } k - \text{regular with } n \text{ vertices}\}$ and

$$f(k) = \limsup_{n \to \infty} F(n, k).$$

Note that Buser's result shows that $f(3) \geq 1/128$. Mohar uses the Ramanujan graphs constructed by Lubotzky, Phillips, Sarnak graphs (see Note 4) to show that if p is a prime with $p \equiv 1 (\bmod 4)$ then $f(p + 1) \geq \frac{1}{2}(p + 1) - \sqrt{p}$. Using this result he shows that $f(k) \geq \frac{k}{2} + O(k^{1-\epsilon})$ for some $\epsilon > 0$. This is asymptotically the best possible result because $f(k) \leq k/2$. He conjectures that there are constants l and u such that for each $k \geq 3$, $f(k) = k/2 - c_k(k - 1)^{1/2}$ where $1/2 < l < c_k < u < 1$. See Note 8 of Chapter 1 for more results from [99].

7. The *chromatic number* of a graph X, denoted by $\chi(X)$, is the minimal number of colors necessary to color the vertices of X in such a way that no two adjacent vertices have the same color. For example, a bipartite graph has chromatic number equal to 2. (A bipartition $V_1 \cup V_2$ of the graph, let the vertices in V_1 be colored red, and the vertices in V_2 be colored blue.) Let X be a connected nonbipartite d-regular graph on n vertices, without loops. Then

$$\chi(X) \geq \frac{d}{\max\{|\lambda_1|, |\lambda_{n-1}|\}} = \frac{d}{\lambda(X)}.$$

See [45, p. 31] for a proof.

8. The explicit construction of Ramanujan graphs by Lubotzky, Phillips, and Sarnak [89] gives a solution to a famous extremal problem in graph theory: the explicit construction of graphs with arbitrarily large girth and chromatic number. The *girth* of a connected graph X, denoted by $g(X)$ is the length of the shortest cycle in X. If there is no shortest cycle (that is, the graph is a tree), then we say that $g(X) = \infty$. The chromatic number of a graph X, denoted by $\chi(X)$, is defined in Note 7. Given two large constants a and b, is it possible to construct a graph X with $\chi(X) \geq a$ and $g(X) \geq b$? Using probabilistic methods, Erdös [53] proved that such graphs exist but did not show how to explicitly construct them. One can show that if X is a connected, d-regular, nonbipartite, Ramanujan graph without loops, then

$$\chi(X) \geq \frac{d}{2\sqrt{d-1}} \sim \frac{\sqrt{d}}{2}. \tag{12}$$

(See Note 7.) Using Equation 12 and approximations on the girth of their Ramanujan graphs, Lubotzky, Phillips, and Sarnak produced explicit graphs with arbitrarily large girth and chromatic number.

9. Friedman [59] proved that for any $\epsilon > 0$ and d, the second-largest eigenvalue of "most" random d-regular graphs is at most $2\sqrt{d-1} + \epsilon$. More explicitly, Friedman shows the following for even $d \geq 4$. Consider a random d-regular graph model formed by $d/2$ uniform, independent permutations on $\{1, \ldots, n\}$. He shows that for any $\epsilon > 0$, all the eigenvalues besides $\lambda_0 = d$ are bounded above by $2\sqrt{d-1} + \epsilon$ with probability $1 - O(n^{-\tau})$, where $\tau = \lceil(\sqrt{d-1} + 1)/2\rceil - 1 > 0$. He proves related theorems for other models of random graphs, including some results with d odd.

10. Cioabă and Murty [44] consider infinite families of k-regular graphs where $k-1$ is not a prime power. By perturbing known Ramanujan graph families and using results about gaps between consecutive primes, they are able to construct infinite families of "almost" Ramanujan graphs for almost every value of k. That is, for every $\epsilon > 0$ and for almost every value of k (what they mean by "almost every" is described in their paper) they show that there exist infinitely many k-regular graphs such that all the nontrivial eigenvalues of these graphs have absolute value less that $(2+\epsilon)\sqrt{d-1}$.

11. Experimental evidence gathered by Miller and Novikoff [97] seems to indicate that Ramanujan graphs exist in great abundance. Their computer-generated data set suggests that about 52 percent of all regular bipartite graphs and about 27 percent of regular nonbipartite graphs are Ramanujan (that is, in the limit as the number of vertices goes to infinity). Moreover, they conjecture that the distribution of $\lambda(X)$ converges to a Tracy-Widom distribution.
12. Jakobson, Miller, Rivin, and Rudnick [73] carried out a numerical study on the fluctuations in the spectrum of a regular graph. Their experiments indicate that the level spacing distribution of a generic k-regular graph approaches the Gaussian orthogonal ensemble of random matrix theory as they increase the number of vertices in the graphs. The paper also gives a brief survey of quantum chaos for graph theorists.
13. There are several papers that generalize the Alon-Boppana theorem. In [61], Friedman and Tillich show that the Alon-Boppana bound can be generalized to finite quotients of a large class of graphs G. That article goes on to discuss applications to error-correcting codes. Ceccherini-Silberstein, Scarabotti, and Tolli [35] generalize the Alon-Boppana theorem to edge-weighted graphs. Hoory [69] uses a generalization of the Alon-Boppana bound to prove that the spectral radius of the universal cover of a finite connected graph G with average degree $d \geq 2$ is greater than or equal to $2\sqrt{d-1}$.
14. The following result is known as Serre's theorem. For every $\epsilon > 0$ there exists a constant $c = c(\epsilon, d) > 0$ such that for any finite, connected d-regular graph X, the number of eigenvalues λ such that $\lambda > (2-\epsilon)\sqrt{k-1}$ is at least $c\,|X|$. There are several proofs of Serre's theorem. See [45] for a proof involving Chebyshev polynomials. Cioabă presents an elementary proof of Serre's theorem in [43]. In his Ph.D. dissertation, Greenberg [67] proves a version of Serre's theorem for arbitrary (not necessarily regular) graphs. Cioabă uses a path-counting argument in [42] to give a simpler proof of Greenberg's result.
15. Cioabă [41] shows that a Cayley graph on an abelian group contains many closed walks of even length. He uses this result to give the following analogue of Serre's theorem for such graphs. Let $d \geq 3$ and $\epsilon > 0$. There exists a constant $C = C(\epsilon, d) > 0$ such that if $X = \text{Cay}(G, \Gamma)$, where G is a finite abelian group and Γ is a symmetric subset of G of size d that does not contain the identity of G, then the number of eigenvalues λ_i of X that satisfy the equation $\lambda_i \geq d - \epsilon$ is at least $C \cdot |G|$.
16. In the expository article [52], Dsouza and Krebs discuss the implications of the path-counting method we employed in our combinatorial proof of the Alon-Boppana theorem; Exercise 4 comes from that paper. Additional details can be found in [51]. An alternate graph-theoretic proof of the result in Exercise 4 can be found in [21].
17. Throughout this note let X be a fixed graph. Let $C = (v_0, e_0, v_1, \ldots, e_{n-1}, v_n)$ be a closed walk in X. That is, $v_0 = v_n$. We say that C has a *tail* if $e_0 = e_{n-1}$. Otherwise, we say that C is *tailless*. The *equivalence class* of C is given by

$$
\begin{aligned}
[C] = \{&(v_0, e_0, v_1, \ldots, e_{n-1}, v_n), \\
&(v_1, e_1, \ldots, e_{n-1}, v_n, e_0, v_1), \\
&\ldots, (v_{n-1}, e_{n-1}, v_n, e_0, v_1, \ldots, e_{n-2}, v_{n-1})\}.
\end{aligned}
$$

Thus, two closed walks in X are equivalent if they are the same up to the starting point. For any positive integer n, define the closed walk C^n to be the walk that repeats C a total of n times.

We say that a closed walk P in X is *prime* if it is nonbacktracking, tailless, and not of the form C^n, where C is some other closed walk in X.

The *Ihara zeta function* of X is given by

$$\zeta_X(u) = \prod_{[P]} \left(1 - u^{\mu(P)}\right)^{-1}$$

where the product is over all equivalence classes of prime walks in X, u is a complex variable, and $\mu(P)$ is the length of P.

Suppose that X is a $(q+1)$-regular graph. One can show that the radius of convergence of ζ_X is $1/q$. We say that Z_X satisfies the Riemann hypothesis if $0 < Re(s) < 1$ and $Z_X(q^{-s}) = 0$ implies that $Re(s) = 1/2$. One can show that Z_X satisfies the Riemann hypothesis if and only if X is a Ramanujan graph.

For more details on the Ihara zeta function of a graph, we refer the reader to the articles by Terras and Stark [130], [131], [132], and Kotani and Sunada [80].

In [115], Reeds defines the Kronecker product of finite graphs and explores the following question. Given a pair of graphs with equal zeta functions, if we take the Kronecker products of the two graphs with a third graph, is the equality of zeta functions preserved? This work was done as an Research Experience for Undergraduates (REUs) project.

EXERCISES

1. Prove Lemma 3.37. Break your proof into two steps:

   (a) ϕ_{v_0} maps walks in T down to walks in X and closed walks based at (v_0) to closed walks based at v_0.
   (b) If w_1 and w_2 are distinct walks in T, then ϕ_{v_0} maps w_1 and w_2 to distinct walks in X. (Hint: Look at the first place where the distinct walks in the covering graph differ.)

2. Let (G_n) be a sequence of finite groups with $|G_n| \to \infty$. Let d be a positive integer. For each n, let $\Gamma_n \subseteq_s G_n$ such that $|\Gamma_n| = d$ and Γ_n contains the identity element of G_n. Let $X_n = \text{Cay}(G_n, \Gamma_n)$. Prove that at most finitely many of the graphs X_n are Ramanujan. (Hint: Use Exercise 13 of Chapter 1.)
3. Let $X = \text{Cay}(G, \Gamma)$ be a Cayley graph where $n = |G|$. Let A be the adjacency operator for X and 1 the identity element of G. Assume that X is connected. For each $g \in G$ and positive integer k, let $N_g(k)$ denote the number of closed walks of length k that start and end at the vertex g.

   (a) Let $g_1, g_2 \in G$. Prove that $N_{g_1}(k) = N_{g_2}(k)$. (Hint: Use Exercise 9 of Chapter 2.)
   (b) Prove that $\text{tr}(A^k) = nN_1(k)$.
   (c) Prove that

$$\lambda(X)^{2k} \geq \frac{n}{n-1} N_1(2k) - \frac{2d^{2k}}{n-2}.$$

(d) Prove that

$$\lambda(X)^{2k} \le nN_1(2k) - d^{2k}.$$

(e) Prove that the covering map in Def. 3.34 is a covering. (See Def. 2.2.)

4. Let $C_n = \mathrm{Cay}(\mathbb{Z}_n, \{1, -1\})$ be the n-cycle graph (see Example 1.53). Let $N_0(k)$ be the number of closed walks of length k starting and ending at 0.

(a) Prove that a closed walk of length k in C_n is composed of x 1s and $k - x$ -1s such that n divides $2x - k$. Hence,

$$N_0(k) = \sum_{\substack{0 \le x \le k \\ n \mid 2x-k}} \binom{k}{x}.$$

(b) Use part 4a and Exercise 3 to show that the number of closed walks of length k in C_n is equal to

$$nN_0(k) = n \sum_{\substack{0 \le x \le k \\ n \mid 2x-k}} \binom{k}{x}.$$

(c) Use part 4b, Example 1.53, and Proposition 1.100 to show that

$$\sum_{\substack{0 \le x \le k \\ n \mid 2x-k}} \binom{k}{x} = \frac{1}{n} \sum_{j=0}^{n-1} \left(2\cos\left(\frac{2\pi j}{n}\right)\right)^k. \tag{13}$$

(It's interesting to note that Equation 13 gives us a formula for certain sums of "evenly spaced" entries in a row of Pascal's triangle.)

STUDENT RESEARCH PROJECT IDEAS

1. We can generalize Equation 9 to an arbitrary (not necessarily regular) finite graph X as follows:

($\star$) The sum of the kth powers of the eigenvalues of X equals the number of closed walks of length k in X.

Obtain a copy of [114]. Select a few small graphs from that book, and work out each side of the equation ($\star$) separately, along the lines of Exercise 4. Try to generalize your results to an infinite family of graphs.

2. [115] is a paper about zeta functions of graphs, based on work done by an undergraduate as part of the Research Experience for Undergraduates (REUs) program. (See Note 17 in this chapter.) Read this paper, and attempt to prove or disprove one of the conjectures in it.

PART TWO

Combinatorial Techniques

4

Diameters of Cayley Graphs and Expander Families

When we think of expander families as good communications networks, we expect messages in them to spread quickly. In other words, we expect them to have small diameters. In Section 1, we make this notion precise by showing that diameter growth in expander families is logarithmic, which is optimal. In Section 2, we discuss the diameter of a Cayley graph in terms of the underlying group structure. In Section 3, we show that a sequence of Cayley graphs on abelian groups cannot have logarithmic diameter and hence cannot be an expander family. In Section 4, we establish some results about the diameter of a Cayley graph vis à vis subgroups and quotients. In Section 5, we iteratively apply the Subgroups and Quotients Nonexpansion Principles to the base case of abelian groups to conclude that a sequence of solvable groups with bounded derived length cannot yield an expander family. The results of Sections 3, 4, and 5, then, provide some necessary conditions that a sequence of groups must satisfy if it is to yield an expander family. It is natural to attempt to refine such conditions until they are both necessary and sufficient. No such "if and only if" theorem currently exists. (In the Notes we speculatively offer some conjectures as to what such a theorem might look like.) In Section 7, we construct a single example (the sequence of cube-connected cycle graphs CCC_n) that demonstrates the falsity of many of the direct converses: the sequence (CCC_n) has logarithmic diameter, yet these graphs are Cayley graphs on solvable groups with derived length 2. The construction of CCC_n makes use of the wreath product, a special case of the semidirect product, both of which we define in Section 6.

1. EXPANDER FAMILIES HAVE LOGARITHMIC DIAMETER

In this section, we show that for a sequence (X_n) of d-regular finite graphs, the best possible diameter growth rate is $O(\log |X_n|)$, and expander families achieve this bound. (See Appendix B for basic facts about "big oh" notation.)

Definition 4.1 Let X be a graph. Let v be a vertex of X, and let r be a non-negative integer. Define $B_r[v] = \{w \in V_X \mid \text{dist}(v, w) \leq r\}$. That is, $B_r[v]$ is the set of all vertices of X whose distance from v is less than or equal to r. We call $B_r[v]$ the *closed ball of radius r centered at v*. Define $S_r[v] = \{w \in V_X \mid \text{dist}(v, w) = r\}$. That is, $S_r[v]$ is the set of all vertices of X whose distance from v equals r. We call $S_r[v]$ the *sphere of radius r centered at v*.

Remark 4.2
This terminology should sound familiar from the language of metric spaces—recall from Remark 1.20 that dist is a metric.

Example 4.3
The set of white vertices in Figure I.3 from the introduction is $B_4[1]$.

In the proofs that follow, we repeatedly use the following logic. If $|B_r[v]|$ grows quickly as a function of r, then we can get to many vertices in just a few steps—that is, the diameter is small. Likewise, if $|B_r[v]|$ grows slowly, the diameter is large. The next example illustrates a typical way we use this sort of argument.

Example 4.4
Let X be a finite graph, and let v be a vertex of X. Suppose that $|B_r[v]| \leq r^2$ for all $r \geq 1$. We now show that $\text{diam}(X) \geq |X|^{1/2}$. Let $k = \text{diam}(X)$. Then for any vertex w of X, we have $\text{dist}(v, w) \leq k$. In other words, $X = B_k[v]$. So $|X| = |B_k[v]| \leq k^2$. Hence $|X|^{1/2} \leq k = \text{diam}(X)$.

Let X be a finite d-regular graph. Assume that $d \geq 3$ and that $\text{diam}(X) \geq 3$. Let v be a vertex of X. Note that $|S_0[v]| = 1$ and $|S_1[v]| \leq d$. Note that if $j \geq 2$, then for any vertex w in $S_j[v]$, at least one edge incident to w is also incident to a vertex in $S_{j-1}[v]$. (Consider a path of length j from v to w.) Therefore, of the d edges incident to w, no more than $d - 1$ of them are also incident to vertices in $S_{j+1}[v]$. It follows that $|S_{j+1}[v]| \leq (d-1)|S_j[v]|$. By induction, then, we have that $|S_j[v]| \leq d(d-1)^{j-1}$. For for any r, we have that $B_r[v]$ is the union of the sets $S_0[v], S_1[v], \ldots, S_r[v]$. So $|B_r[v]| \leq 1 + d\left(\sum_{j=0}^{r-1}(d-1)^j\right)$ vertices. The right-hand side is a polynomial in d of degree r, so we expect it to be controlled by d^r. More precisely, we claim that $|B_r[v]| \leq d^r$ if $r \geq 3$. To prove this, first observe by elementary algebra that $0 \leq d^2 - 3d + 1$, because $d \geq 3$. Therefore $(d-1)^3 \leq d^2(d-2)$. It follows that $(d-1)^r = (d-1)^{r-3}(d-1)^3 \leq d^{r-3}d^2(d-2) = d^{r-1}(d-2)$. Hence, $d(d-1)^r - 2 \leq d(d-1)^r \leq d^r(d-2)$. Therefore

$$1 + d\left(\sum_{j=0}^{r-1}(d-1)^j\right) = 1 + d\left[\frac{(d-1)^r - 1}{d-2}\right] \leq d^r.$$

Let $k = \text{diam}(X)$. Then $|X| = |B_k(v)| \leq d^k$. So $\text{diam}(X) \geq \log_d |X|$.

The moral of the story is that if $d \geq 3$ is fixed and (X_n) is a sequence of d-regular graphs with $|X_n| \to \infty$, then $\text{diam}(X_n)$ grows at least logarithmically. In other words, logarithmic diameter growth is the best possible.

Lemma 4.5
Let X be a connected finite graph. Let $a > 1$. Suppose that for any vertex v of X, we have that $\left|B_r[v]\right| \geq a^r$ whenever $|B_{r-1}[v]| \leq \frac{1}{2}|X|$. Then

$$\operatorname{diam}(X) \leq \left(\frac{2}{\log a}\right) \log |X|.$$

Proof
Let w_1, w_2 be two vertices of X. Let r_1 be the smallest non-negative integer such that $|B_{r_1}[w_1]| > \frac{1}{2}|X|$. (We know that r_1 exists because X is connected, hence $B_k[w_1] = X$, where $k = \operatorname{diam}(X) < \infty$.) Then $|B_{r_1}[w_1]| \geq a^{r_1}$, since $r_1 - 1 < r_1$, and therefore $|B_{r_1-1}[v]| \leq \frac{1}{2}|X|$. Similarly, letting r_2 be the smallest non-negative integer such that $|B_{r_2}[w_2]| > \frac{1}{2}|X|$, we have $|B_{r_2}[w_2]| \geq a^{r_2}$. Now, $|B_{r_1}[w_1]| + |B_{r_2}[w_2]| > |X|$, so we must have that $B_{r_1}[w_1] \cap B_{r_2}[w_2] \neq \emptyset$. Let $w_3 \in B_{r_1}[w_1] \cap B_{r_2}[w_2]$. Then $\operatorname{dist}(w_1, w_3) \leq r_1$ and $\operatorname{dist}(w_2, w_3) \leq r_2$, so

$$\begin{aligned}\operatorname{dist}(w_1, w_2) &\leq r_1 + r_2 \\ &\leq \frac{\log |B_{r_1}[w_1]|}{\log a} + \frac{\log |B_{r_2}[w_2]|}{\log a} \\ &\leq \left(\frac{2}{\log a}\right) \log |X|.\end{aligned}$$

Because this inequality holds for any two vertices w_1 and w_2, we conclude that $\operatorname{diam}(X) \leq \left(\frac{2}{\log a}\right) \log |X|$.

The following proposition gives an upper bound on the diameter of a graph in terms of the number of vertices and the isoperimetric constant.

Proposition 4.6
Let X be a connected d-regular graph. Let $C = 1 + \frac{h(X)}{d}$. Then

$$\operatorname{diam}(X) \leq \left(\frac{2}{\log C}\right) \log |X|.$$

Proof
Let v be any vertex of X. Suppose $|B_{r-1}[v]| \leq \frac{1}{2}|X|$. By the definition of $h(X)$ (Def. 1.63), it follows that

$$|\partial B_{r-1}[v]| \geq h(X)|B_{r-1}[v]|.$$

Any edge in $\partial B_{r-1}[v]$ must be incident to a vertex in $S_r[v]$. Because X is d-regular, it follows that

$$|S_r[v]| \geq \frac{|\partial B_{r-1}[v]|}{d} \geq \frac{h(X)}{d}|B_{r-1}[v]|.$$

Note that $B_r[v]$ is the disjoint union of $B_{r-1}[v]$ and $S_r[v]$. So

$$|B_r[v]| = |B_{r-1}[v]| + |S_r[v]| \geq |B_{r-1}[v]| + \frac{h(X)}{d}|B_{r-1}[v]| = C|B_{r-1}[v]|.$$

By induction, therefore, we have that $|B_r[v]| \geq C^r$ whenever $|B_{r-1}[v]| \leq \frac{1}{2}|X|$. The result now follows from Lemma 4.5. ⍟

By Proposition 4.6, we see that if the isoperimetric constant of a sequence (X_n) of d-regular graphs is bounded away from zero, then the diameters of the graphs grow at most logarithmically as a function of $|X_n|$. From the discussion preceding Lemma 4.5, we know that this growth rate is the slowest possible.

Definition 4.7 Let (X_n) be a sequence of graphs. We say that (X_n) *has logarithmic diameter* if $\operatorname{diam}(X_n) = O(\log |X_n|)$. (See Appendix B for a refresher on "big oh" notation.)

Corollary 4.8
Let d be a non-negative integer. If (X_n) is a family of d-regular expanders, then (X_n) has logarithmic diameter.

Proof
Because (X_n) is a family of expanders, for some $\epsilon > 0$ we have that $h(X_n) \geq \epsilon$ for all n. Let $C_n = 1 + h(X_n)/d$, and let $C = 1 + \epsilon/d$. Since $\epsilon \leq h(X_n)$, we see that $2/\log C_n \leq 2/\log C$. Thus, by Proposition 4.6,

$$\operatorname{diam}(X_n) \leq \left(\frac{2}{\log C_n}\right) \log |X_n| \leq \left(\frac{2}{\log C}\right) \log |X_n|.$$

Hence, $\operatorname{diam}(X_n) = O(\log |X_n|)$. ⍟

Our main use of Corollary 4.8 will be to show that certain sequences of graphs are not expander families.

Example 4.9
Let $X_n = \operatorname{Cay}(\mathbb{Z}_{4n}, \{1, -1, 2n\})$. It is not too hard to show that $\operatorname{diam}(X_n) = n$. (See Exercise 1.) So $\operatorname{diam}(X_n) = \frac{1}{4}|X_n|$ is linear as a function of $|X_n|$. Hence $\operatorname{diam}(X_n) \neq O(\log |X_n|)$, by Lemma B.2. Therefore, by Corollary 4.8, we have that (X_n) is not an expander family.

This example is a special case of the results of Section 3, in which we see that abelian groups *never* yield expander families.

The following example shows that the converse of Corollary 4.8 is not true.

Example 4.10
We now construct a sequence $(X_n)_{n=3}^{\infty}$ of 3-regular graphs. We show that this sequence has logarithmic diameter but is not an expander family.

The graph X_n can roughly be described as follows. X_n has a vertex located at the "top" of the graph. From this vertex we have three subgraphs that are almost

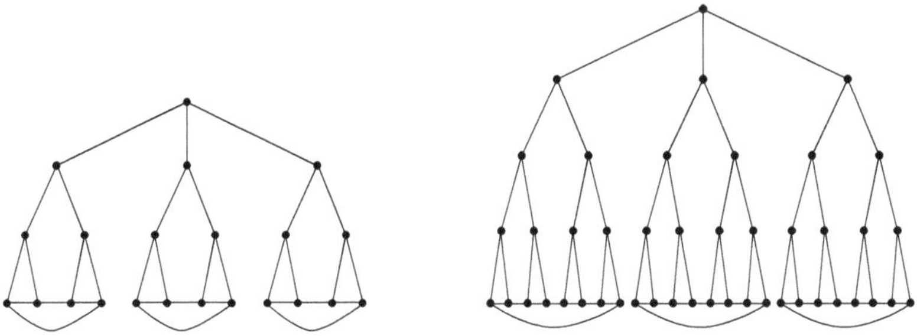

Figure 4.1 X_3 and X_4

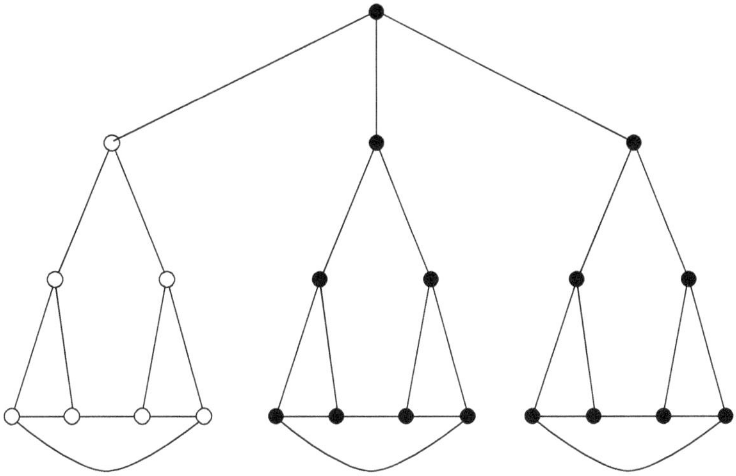

Figure 4.2 F_3 consists of the white vertices in X_3

binary trees, except for the fact that their "bottom" vertices are connected by a cycle. The graphs X_3 and X_4 are shown in Figure 4.1. (See Exercise 2 for a precise definition of a family of graphs that is similar to this one.)

Note that $|X_n| = 1 + 3(2^n - 1) = 3 \cdot 2^n - 2 \geq 2^n$. Hence, we have that $n \leq \log_2 |X_n|$. It's straightforward to see that $\mathrm{diam}(X_n) = 2n$; the "worst-case scenario" is traveling from a vertex at the bottom to another vertex that is also at the bottom but in a different cycle. So $\mathrm{diam}(X_n) = O(\log |X_n|)$.

Let us show that this is not an expander family. Let F_n consist of all the vertices in the left subgraph of X_n. For example, in Figure 4.2, the set F_3 consists of all the white vertices in the graph X_3.

Then, $h(X_n) \leq |\partial F_n| / |F_n| = 1/(2^n - 1) \to 0$ as $n \to \infty$. Therefore, (X_n) is not an expander family.

2. DIAMETERS OF CAYLEY GRAPHS

Definition 4.11 Let (G_n) be a sequence of finite groups. We say that (G_n) *has logarithmic diameter* if for some positive integer d there exists a sequence (Γ_n), where for each n we have that $\Gamma_n \Subset G_n$ with $|\Gamma_n| = d$, so that the sequence of Cayley graphs $(\text{Cay}(G_n, \Gamma_n))$ has logarithmic diameter.

Definition 4.12 Let Γ be a set, and let n be a positive integer. Then a *word of length n in Γ* is an element of the Cartesian product $\Gamma \times \cdots \times \Gamma = \Gamma^n$. If $\Gamma \subseteq G$ for some group G and $w = (w_1, \ldots, w_n)$ is a word in Γ, then *w evaluates to g* (or *g can be expressed as w*) if g equals the product $w_1 \cdot \cdots \cdot w_n$.

Example 4.13
Let $G = \mathbb{Z}$, and let $\Gamma = \{-1, 1, 2, 5\}$. Then $w_1 = (1, 2, 1)$ is a word of length 3 in Γ, and $w_2 = (2, 2)$ is a word of length 2 in Γ. Notice that both w_1 and w_2 evaluate to 4. So 4 can be expressed in many different ways as a word in Γ. Moreover, note that 2 is the minimal length of any word in Γ that evaluates to 4.

Example 4.14
Let $G = \mathbb{Z}_{10}$, and let $\Gamma = \{0, 2, 4, 6, 8\}$. Then 7 cannot be expressed as a word in Γ.

Definition 4.15 Let G be a group, let $\Gamma \subseteq G$, and let $g \in G$ such that g can be expressed as a word in Γ. We say that the *word norm of g in Γ* is the minimal length of any word in Γ which evaluates to g.

Remark 4.16
The standard convention is to say that the word of length 0 evaluates to the identity element. So the identity element has word norm 0.

Proposition 4.17
Let G be a finite group. Let $\Gamma \Subset G$. Let $X = \text{Cay}(G, \Gamma)$. Then:

1. X is connected iff every element of G can be expressed as a word in Γ.
2. If $a, b \in G$ and there is a walk in X from a to b, then the distance from a to b is the word norm of $a^{-1}b$ in Γ.
3. The diameter of X equals the maximum of the word norms in Γ of elements of G.

Proof
(1) This is equivalent to part (2) of Prop. 1.29.

(2) Let $(g_0, g_1, \ldots, g_n)$ be a walk of length n in X from a to b. (So $a = g_0$ and $b = g_n$.) Let $\gamma_j = g_{j-1}^{-1} g_j$ for $j = 1, \ldots, n$. That is, γ_j is an element of Γ, which gives us an edge from g_{j-1} to g_j. Then $(\gamma_1, \ldots, \gamma_n)$ is a word of length n in Γ that evaluates to $a^{-1}b$. Reversing this procedure, we see that conversely, every word of length n in Γ which evaluates to $a^{-1}b$ corresponds to a walk of length n in X from a to b. Thus, the distance from a to b equals the minimal length of all walks in X from a to b, which equals the minimal length of all words in Γ that evaluate to $a^{-1}b$, which equals the word norm of $a^{-1}b$ in Γ.

(3) If $g \in G$, then by (2) the distance from the identity element e to g is the word norm of g. Hence $\text{diam}(X)$ is at least the maximum of the word lengths of elements of G. Because by (2) every distance is a word norm, we have equality.

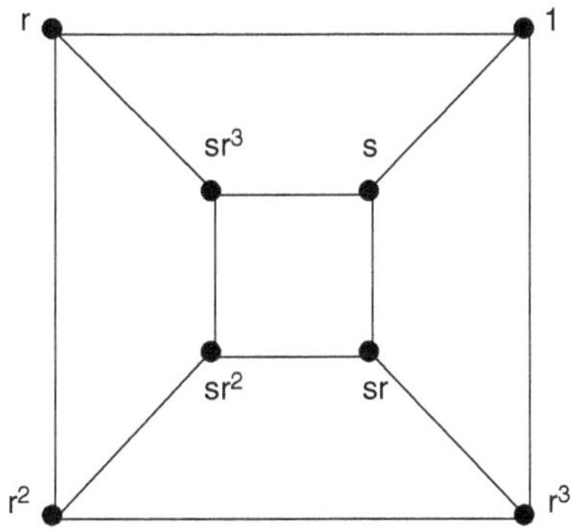

Figure 4.3 Cay(D_4, $\{s, r, r^3\}$)

Remark 4.18
We can now explain the etymology of the phrase "word norm." For in a vector space, we think of the norm of a vector as its distance from the origin. Analogously, by part (2) of Prop. 4.17, the word norm of a group element b equals the distance from e to b.

Example 4.19
Let G be the dihedral group D_4. (See Notations and conventions regarding the dihedral group.) Let $\Gamma = \{s, r, r^3\}$. Let $X = \text{Cay}(G, \Gamma)$. The graph X is shown in Figure 4.3. The word norm of sr^2 in Γ is 3, because a word of minimal length in Γ that evaluates to sr^2 is (s, r, r). This word corresponds to the path along the vertices $1, s, sr, sr^2$. (Note that there are other paths of length 3 in X from 1 to sr^2; these correspond to other words of length 3 in Γ that evaluate to sr^2.) The reader can verify that no element of G has word norm in Γ more than 3. This corresponds to the fact that $\text{diam}(X) = 3$.

Example 4.20
Generalizing the previous example, one can show that if $X_n = \text{Cay}(D_n, \{s, r, r^{-1}\})$, then $\text{diam}(X_n) \geq \frac{n-1}{2}$. Therefore (X_n) does not have logarithmic diameter and hence is not an expander family. In Exercise 5, we ask the reader to fill in the details in this argument.

Example 4.20 may lead us to wonder whether, using some other generating sets, the dihedral groups do in fact have logarithmic diameter—see Example 4.37 for the answer to this question.

In Examples 1.78 and 1.91, we showed that a certain sequence of Cayley graphs on symmetric groups (i.e., bubble-sort graphs) did not yield an expander family. In the following example, we use diameter estimates to furnish a third proof of this fact. Later, in Example 8.21, we see yet another proof.

Example 4.21
Recall from Example 1.78 that S_n is the symmetric group on n letters, $\sigma = (1, 2, \ldots, n) \in S_n$, $\tau = (1, 2) \in S_n$, $\Gamma = \{\sigma, \sigma^{-1}, \tau\}$, and $X_n = \text{Cay}(S_n, \Gamma)$.

We now show that (X_n) does not have logarithmic diameter and hence is not an expander family. (See Note 2.)

Suppose that $\gamma \in S_n$ and $i < j < k$ are integers between 1 and n. We say that the 3-tuple (i, j, k) is in *good position* with respect to $\gamma \in S_n$ if $\gamma(i) < \gamma(j) < \gamma(k)$ or $\gamma(j) < \gamma(k) < \gamma(i)$ or $\gamma(k) < \gamma(i) < \gamma(j)$. Otherwise we say that (i, j, k) is in *bad position* with respect to γ. For example, $(8, 2, 1)$ is in good position with respect to the permutation $\gamma = (5, 1, 2, 7, 8)(3, 6)$, because $2 = \gamma(1) < 5 = \gamma(8) < 7 = \gamma(2)$.

Notice that (i, j, k) is in good position with respect to γ if and only if (i, j, k) is in good position with respect to $\sigma\gamma$. Also notice that if (i, j, k) is in bad position with respect to γ, then (i, j, k) is in good position with respect to $\tau\gamma$ iff $\{\gamma(i), \gamma(j), \gamma(k)\} = \{1, 2, m\}$ for some m. Applying τ to γ can at most change $n - 2$ bad triples into good triples. (Count the number of m's.)

Now consider the permutation $\gamma : \{1, 2, \ldots, n\} \to \{1, 2, \ldots, n\}$ defined by $\gamma(a) = n - a + 1$. In row notation,

$$\gamma = \begin{pmatrix} 1 & 2 & \ldots & n \\ n & n-1 & \ldots & 1 \end{pmatrix}.$$

We now find a lower bound for the word norm of γ. Given any triple (i, j, k) with $i < j < k$, we have that $\gamma(k) < \gamma(j) < \gamma(i)$. Hence every triple is in bad position with respect to γ. In total, there are $\binom{n}{3}$ triples. Because every triple is in good position with respect to the identity element of S_n, any word in Γ that evaluates to γ must contain τ at least $\binom{n}{3}/(n-2)$ times. Thus, by Prop. 4.17, we have that $\operatorname{diam}(X_n) \geq \binom{n}{3}/(n-2) = \frac{1}{6}(n^2 - n)$.

Suppose that (X_n) is an expander family. Then by Corollary 4.8, we would have that $\operatorname{diam}(X_n) = O(\log |S_n|) = O(n \log(n))$, since

$$\log(n!) \leq \log(n^n) = n \log(n).$$

But

$$\lim_{n\to\infty} \frac{\operatorname{diam}(X_n)}{n \log(n)} \geq \lim_{n\to\infty} \frac{\frac{1}{6}(n^2 - n)}{n \log(n)} = \infty.$$

Hence, by Lemma B.2, $\operatorname{diam}(X_n) \neq O(\log |S_n|)$.

Example 4.21 may lead us to wonder whether, using some other generating sets, the symmetric groups do in fact have logarithmic diameter—see Note 3 for the answer to this question.

3. ABELIAN GROUPS NEVER YIELD EXPANDER FAMILIES: A COMBINATORIAL PROOF

Anyone who has taken a course in abstract algebra should be familiar with various common families of finite groups: dihedral groups, symmetric groups, alternating groups, and so on. If one has a family of groups, then via the Cayley graph construction one can easily produce a family of regular graphs. The easiest finite groups to work with are the cyclic groups, or more generally the finite abelian groups, that is, products of cyclic groups. If one is attempting to construct an expander family, it would be natural to first consider sequences of Cayley graphs on abelian groups.

In this section, we show that abelian groups are in fact useless for this purpose—a sequence of finite abelian groups never yields an expander family. We prove this by showing that no sequence of abelian groups has logarithmic diameter. To do so, we establish a few technical lemmas that we'll need.

Suppose we want to count the number of solutions to the equation $a_1 + a_2 + a_3 = 4$, where each a_i is a non-negative integer. Imagine six empty spaces. We choose four of these spaces to have dots $\cdot$ and the remaining spaces to be represented by vertical bars. The number of dots corresponds to the value assigned to the appropriate a_i, and the vertical bars indicate how to divide up the assignments of these values. For example, we have the following correspondences:

$$\begin{array}{cc} \cdot \mid \cdot \mid \cdot\ \cdot & \cdot \mid \mid \cdot\ \cdot\ \cdot \\ a_1 = 1, a_2 = 1, a_3 = 2 & a_1 = 1, a_2 = 0, a_3 = 3 \end{array}$$

The reader should pause to verify that this correspondence, between non-negative integer solutions to $a_1 + a_2 + a_3 = 4$ on the one hand and choices of four spaces out of six on the other hand, is bijective. Thus, there are $\binom{6}{4} = 15$ solutions to the equation. Generalizing this reasoning proves the following lemma.

Lemma 4.22
The number of solutions to the equation $a_1 + \cdots + a_n = k$, where the a_i are non-negative integers, is $\displaystyle\binom{n+k-1}{k}$.

Lemma 4.23
If $a, b \in \mathbb{N}$ with $b \leq a$, then

$$\binom{a}{b} \leq (a - b + 1)^b.$$

Proof
First observe that if $0 < q \leq p$, then $\frac{p+1}{q+1} \leq \frac{p}{q}$. Hence

$$\frac{a}{b} \leq \frac{a-1}{b-1} \leq \cdots \leq \frac{a-b+2}{2} \leq \frac{a-b+1}{1}.$$

So

$$\binom{a}{b} = \left(\frac{a}{b}\right)\left(\frac{a-1}{b-1}\right)\cdots\left(\frac{a-b+2}{2}\right)\left(\frac{a-b+1}{1}\right) \leq (a-b+1)^b. \quad ⚜$$

Remark 4.24
Much sharper bounds for $\binom{a}{b}$ are possible, but Lemma 4.23 will be sufficient for our purposes.

Proposition 4.25
No sequence of finite abelian groups has logarithmic diameter.

Proof
Let G be a finite abelian group; let $\Gamma \Subset G$; let $d = |\Gamma|$; let $\gamma_1, \ldots, \gamma_d$ be the elements of Γ. Let $X = \text{Cay}(G, \Gamma)$, and let $k = \text{diam}(X)$. If Γ does not generate G, then $k = \infty$. Otherwise, by Prop. 4.17, every element can be expressed as a word in Γ of length $\leq k$. Since G is abelian, we can rearrange the elements in the word, bringing the γ_1's to the front, then the γ_2's, and so on. That is, every element of G is of the form

$$e^{a_0}\gamma_1^{a_1}\cdots\gamma_d^{a_d},$$

where e is the identity element of G and $\sum_{i=0}^{d} a_i = k$, each a_i being a non-negative integer. (We introduced e so that the sum of the exponents a_i is fixed.) By Lemma 4.22, the number of distinct elements of this form is bounded above by $\binom{k+d}{k}$. Therefore, by Lemma 4.23, we have $|X| \leq \binom{k+d}{k} = \binom{k+d}{d} \leq (k+1)^d$, so $\text{diam}(X) \geq |X|^{1/d} - 1$.

For any sequence (X_n) of d-regular Cayley graphs on abelian groups, then, we have $\text{diam}(X_n) \geq \left|X_n\right|^{1/d} - 1$. But $|X_n|^{1/d} - 1$ is essentially a root function of $|X_n|$, and root functions grow faster than logarithmic functions. So $\text{diam}(X_n) \neq O(\log|X_n|)$. (Note: To make this last bit of reasoning more precise, use Lemma B.2.)

Corollary 4.26
No sequence of abelian groups yields an expander family.

Proof
Combine Prop. 4.25 and Corollary 4.8.

Together, Corollary 4.26 and the Quotients Nonexpansion Principle (Prop. 2.20) imply that if a sequence (G_n) of finite groups admits an unbounded sequence of abelian groups as quotients, then (G_n) does not yield an expander family. The following example illustrates this phenomenon.

Example 4.27
We continue Example 2.21. Recall that (G_n) admits $\mathbb{Z}_{p_n} \times \mathbb{Z}_{p_n}$ as a sequence of quotients. It now follows from Corollary 4.26 and the Quotients Nonexpansion Principle (Prop. 2.20) that (G_n) does not yield an expander family. In fact, we can state more strongly that (G_n) does not have logarithmic diameter, a fact we show in Example 4.38.

Observe that G_n is nonabelian, because $\begin{pmatrix} 1 & 1 & 0 \\ 0 & 1 & 0 \\ 0 & 0 & 1 \end{pmatrix}$ does not commute with $\begin{pmatrix} 1 & 1 & 1 \\ 0 & 1 & 1 \\ 0 & 0 & 1 \end{pmatrix}$. So we could not have applied Corollary 4.26 directly to (G_n).

In Section 5, we systematically combine Corollary 4.26 with the Subgroups and Quotients Nonexpansion Principles to find a large class of sequences of groups that never yield expander families.

4. DIAMETERS OF SUBGROUPS AND QUOTIENTS

In Chapter 2, we saw that if K is a subgroup or quotient of a group G, then the isoperimetric constant of a Cayley graph on G is bounded by the isoperimetric constant of a certain related Cayley graph on K (Lemmas 2.17 and 2.41). Moreover, we saw that similar statements hold for the second largest eigenvalues (Prop. 2.26 and Lemma 2.47). In this section, we prove the corresponding results for diameters. This section makes heavy use of notations and definitions from Chapter 2.

Definition 4.28 Let X, Y be graphs. Define the graph $C(X \times Y)$, called the *composite graph of X and Y*, as follows. The vertex set of $C(X \times Y)$ is $X \times Y$. The set of edges between a vertex (x_1, y_1) in $C(X \times Y)$ and a vertex (x_2, y_2) in $C(X \times Y)$ is the set of pairs (e_1, e_2) such that e_1 is an edge in X between x_1 and x_2, and e_2 is an edge in Y between y_1 and y_2.

Example 4.29
Figures 4.4 and 4.5 show two directed graphs X and Y and the composite directed graph $C(X \times Y)$. To simplify notation we wrote uv for the vertex (u, v) in Figure 4.5.

Figure 4.4 X (left) and Y (right)

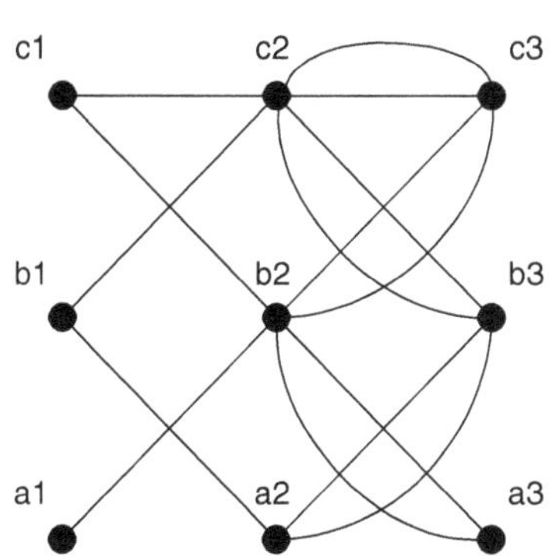

Figure 4.5 $C(X \times Y)$

Definition 4.30 Let X be a graph with vertex set V and edge multiset E. Suppose that X' is a graph with vertex set V and edge set E', where $E' \subset E$. Then we say that X' is a *spanning subgraph* of X.

Remark 4.31
The term *spanning* refers to the fact that X' uses *every* vertex of X. We will not need the more general notion of a *subgraph*, whose vertex set may be a proper subset of V.

Lemma 4.32
Recall Def. 2.1. Let G, H, Γ, T be as in Def. 2.30. Then $\text{Cay}(G, \Gamma)$ is isomorphic to a spanning subgraph of $C(\text{Cay}(H, \hat{\Gamma}) \times \text{Cos}(H\backslash G, \Gamma))$.

Proof
First we identify vertices of $\text{Cay}(G, \Gamma)$ with vertices of $C(\text{Cay}(H, \hat{\Gamma}) \times \text{Cos}(H\backslash G, \Gamma))$. Define $\phi : G \to H \times (H\backslash G)$ by $\phi(g) = (g(\overline{g})^{-1}, Hg)$. Then ϕ is onto, since $\phi(h\overline{a}) = (h, Ha)$. Also, ϕ is one-to-one, because if $(g_1(\overline{g_1})^{-1}, Hg_1) = (g_2(\overline{g_2})^{-1}, Hg_2)$, then $Hg_1 = Hg_2$, which implies that $\overline{g_1} = \overline{g_2}$, and because we know that $g_1(\overline{g_1})^{-1} = g_2(\overline{g_2})^{-1}$, this shows that $g_1 = g_2$. Since ϕ is bijective, we may therefore identify the vertex g of $\text{Cay}(G, \Gamma)$ with the vertex $(g(\overline{g})^{-1}, Hg)$ of the composite graph.

Let $\gamma \in \Gamma$ and $g \in G$. Then γ induces an edge in $\text{Cay}(G, \Gamma)$ from g to $g\gamma$. The corresponding edge in $C(\text{Cay}(H, \hat{\Gamma}) \times \text{Cos}(H\backslash G, \Gamma))$ comes from the pair (e_1, e_2), where e_1 is the edge in $\text{Cay}(H, \hat{\Gamma})$ from $g(\overline{g})^{-1}$ to $g\gamma(\overline{g\gamma})^{-1}$ induced by the Schreier generator $\overline{g}\gamma(\overline{\overline{g}\gamma})^{-1}$, and e_2 is the edge in $\text{Cos}(H\backslash G, \Gamma)$ induced by γ. (Since $\overline{g}\gamma$ and $g\gamma$ are both in $Hg\gamma$, by Lemma 2.34, $\overline{\overline{g}\gamma} = \overline{g\gamma}$.)

Lemma 4.33
Suppose X is a spanning subgraph of a finite graph Y. Then $\text{diam}(X) \geq \text{diam}(Y)$.

Proof
Let V be the common vertex set of X and Y, and let $p, q \in V$. Any walk in X from p to q is also a walk in Y from p to q. So the distance in Y from p to q is no more than the distance in X from p to q. The result follows.

Lemma 4.34
Let X, Y be finite graphs. Then $\text{diam}(C(X \times Y)) \geq \text{diam}(X)$ and $\text{diam}(C(X \times Y)) \geq \text{diam}(Y)$.

Proof
Let $x_1, x_2 \in X$ and $y_1, y_2 \in Y$. A walk of length ℓ in $C(X \times Y)$ from (x_1, y_1) to (x_2, y_2) projects down to a walk of length ℓ in X from x_1 to x_2. So the distance in X from x_1 to x_2 is no more than the distance in $C(X \times Y)$ from (x_1, y_1) to (x_2, y_2). Therefore, $\text{diam}(C(X \times Y)) \geq \text{diam}(X)$. One proves the other inequality similarly.

Proposition 4.35
Let G, H, Γ, T as in Def. 2.30. Then

$$\operatorname{diam}(\operatorname{Cay}(G, \Gamma)) \geq \operatorname{diam}(\operatorname{Cay}(H, \hat{\Gamma})), \text{ and}$$

$$\operatorname{diam}(\operatorname{Cay}(G, \Gamma)) \geq \operatorname{diam}(\operatorname{Cos}(H \backslash G, \Gamma)).$$

Proof
This follows from Lemmas 4.32, 4.33, and 4.34. ◬

Using Prop. 4.35, we can now formulate a diameter version of the Subgroups Nonexpansion Principle by replacing "yields an expander family" with "has logarithmic diameter."

Proposition 4.36
Let (G_n) be a sequence of finite groups. Suppose that (G_n) admits (H_n) as a bounded-index sequence of subgroups. If (H_n) does not have logarithmic diameter, then (G_n) does not have logarithmic diameter.

Proof
Suppose $(\operatorname{Cay}(G_n, \Gamma_n))$ has logarithmic diameter for some sets Γ_n such that $|\Gamma_n|$ is constant and $\Gamma_n \Subset G_n$ for all n. Let T_n be a set of transversals for H_n in G_n. Let M such that $[G_n : H_n] \leq M$ for all n. Let $\Lambda_n = \hat{\Gamma}_n \cup \{(M - [G_n : H_n]) |\Gamma_n| \cdot e_n\}$. (So Λ_n is essentially the set of Schreier generators, but with enough copies of the identity thrown in so that $|\Lambda_n| = M \cdot |\Gamma_n|$ for all n.) By Prop. 4.35, we have that

$$\begin{aligned}
\operatorname{diam}(\operatorname{Cay}(H_n, \Lambda_n)) &= \operatorname{diam}(\operatorname{Cay}(H_n, \hat{\Gamma}_n)) \\
&\leq \operatorname{diam}(\operatorname{Cay}(G_n, \Gamma_n)) \\
&\leq C \log |G_n| \\
&\leq C \log |H_n| + C \log M \\
&\leq 2C \log |H_n|
\end{aligned}$$

for some constant C and for sufficiently large n. But (H_n) does not have logarithmic diameter, so this is a contradiction. ◬

Example 4.37
Recall from Notations and conventions the dihedral groups D_n. Let $H_n = \langle r \rangle \cong \mathbb{Z}_n$. Note that $[D_n : H_n] = 2$ for all n. By Prop. 4.25, we know that (H_n) does not have logarithmic diameter. So by Prop. 4.36, we see that (D_n) does not have logarithmic diameter.

Example 4.38
We continue Example 4.27. We show that (G_n) does not have logarithmic diameter. Temporarily assume, to the contrary, that there exist a natural

number d and subsets $\Gamma_n \Subset G_n$ with $|\Gamma_n| = d$ for all n such that $\text{diam}(\text{Cay}(G_n, \Gamma_n)) = O(\log |G_n|)$. The sequence (G_n) does not admit a bounded-index sequence of proper subgroups, so Prop. 4.36 is of little help. Instead, use the homomorphism ϕ from Example 2.21 to identify $\mathbb{Z}_{p^n} \times \mathbb{Z}_{p^n}$ with a quotient of G_n. By Prop. 4.35, we have that

$$\begin{aligned}\text{diam}(\text{Cay}(\mathbb{Z}_{p^n} \times \mathbb{Z}_{p^n}, \overline{\Gamma_n})) &\leq \text{diam}(\text{Cay}(G_n, \Gamma_n)) \\ &= O(\log p_n^3) \\ &= O(\log p_n^2) \\ &= O(\log |\mathbb{Z}_{p^n} \times \mathbb{Z}_{p^n}|),\end{aligned}$$

which contradicts Prop. 4.25.

5. SOLVABLE GROUPS WITH BOUNDED DERIVED LENGTH

The main result of this section is Theorem 4.47, wherein we prove that if (G_n) is a sequence of solvable groups with bounded derived length, then (G_n) does not yield an expander family.

For the sake of readers not familiar with solvable groups, we very briefly provide several basic facts about them.

Definition 4.39 Let G be a group. An element in G of the form $a^{-1}b^{-1}ab$ for some $a, b \in G$ is called a *commutator*. Define G' to be the subgroup of G generated by the set of all commutators in G. We say that G' is the *commutator subgroup of* G.

Lemma 4.40
Let G be a group. Then:

1. $G' \lhd G$, and
2. If N is a normal subgroup of G, then G/N is abelian iff $G' < N$.

Proof
This is a standard exercise in elementary group theory, and we leave it as an exercise for the reader. Ⓐ

Remark 4.41
Lemma 4.40 tells us that the commutator subgroup is the smallest normal subgroup whose associated quotient is abelian.

Definition 4.42 Let G be a group. We recursively define a sequence of subgroups of G, as follows:

$$G^{(0)} = G, \text{ and } G^{(k+1)} = [G^{(k)}]'.$$

The group $G^{(k)}$ is called the *kth derived subgroup of* G.

Definition 4.43 Let G be a group. We say that G is *solvable with derived length* 0 if G is the trivial group. We say that G is *solvable with derived length* $k+1$ if $G^{(k)} \neq 1$ but $G^{(k+1)} = 1$.

Remark 4.44
Let G be a nontrivial group. Note that G is abelian iff G is solvable with derived length 1.

Example 4.45
Recall from Notations and conventions the dihedral group D_n. If $n = 1$ or $n = 2$, then D_n is abelian, so D_n is solvable with derived length 1. Now assume that $n \geq 3$. Observe that $s^{-1}r^{-1}sr = r^2$, so $r^2 \in D_n'$. Let $H = \langle r^2 \rangle$ be the subgroup generated by r^2. So $H < D_n'$. If n is even, define $\phi : D_n \to \mathbb{Z}_2 \times \mathbb{Z}_2$ by $\phi(r^j s^k) = (j, k)$; if n is odd, define $\phi : D_n \to \mathbb{Z}_2$ by $\phi(r^j s^k) = k$. The reader can verify that in both cases, ϕ is a well-defined surjective homomorphism with kernel H. Therefore $H \lhd D_n$ and D_n/H is abelian. By Lemma 4.40, we have that $D_n' < H$. Thus $D_n' = H$. Since H is abelian, by Remark 4.44, we have $D_n^{(2)} = H' = 1$. Therefore D_n is solvable with derived length 2.

Remark 4.46
Roughly speaking, to say that a finite group is solvable means that it is "built up out of abelian pieces." The derived length is the minimum number of required pieces.

Theorem 4.47
Let (G_n) be a sequence of finite nontrivial groups such that $|G_n| \to \infty$. Let k be a positive integer. Suppose that for all n, we have that G_n is solvable with derived length $\leq k$. Then (G_n) does not yield an expander family.

Proof
We prove the theorem by induction on k. In the base case ($k = 1$), we have by Remark 4.44 that G_n is abelian for all n, so the theorem holds by Corollary 4.26. Now we assume that the theorem is true for k and prove that it holds for $k+1$.

Case 1: The sequence (G_n') has bounded index in (G_n).

For all n, let ℓ_n be the derived length of G_n. Note that G_n' is solvable with derived length $\ell_n - 1 \leq k$. By the inductive hypothesis, (G_n') does not yield an expander family. Therefore by the Subgroups Non-expansion Principle (Prop. 2.46), we have that (G_n) does not yield an expander family.

Case 2: The sequence $(|G_n/G_n'|)$ is unbounded.

By Lemma 4.40, we know that G_n/G_n' is abelian. So by Corollary 4.26, it follows that (G_n/G_n') does not yield an expander family. Therefore, by the Quotients Nonexpansion Principle (Prop. 2.20), we have that (G_n) does not yield an expander family. ∎

Example 4.48
It follows immediately from Example 4.45 and Theorem 4.47 that the sequence (D_n) of dihedral groups does not yield an expander family.

Example 4.49
Taking notation as in Example 2.21, the reader can verify that $K_n = G'_n$. Using this fact, one can show that G_n is solvable with derived length 2. Theorem 4.47 recovers for us, once again, the fact that the sequence (G_n) of 3×3 unipotent groups does not yield an expander family.

6. SEMIDIRECT PRODUCTS AND WREATH PRODUCTS

In Section 7, we construct a sequence of finite solvable groups of derived length 2 and show that this sequence has logarithmic diameter—even though, by Theorem 4.47, it cannot yield an expander family. This construction makes use of an algebraic operation called the *wreath product*, which is a special case of the *semidirect product*. For the sake of readers unfamiliar with these topics, we provide a brief description of them.

We denote by $\text{Aut}(G)$ the automorphism group of a group G. (Recall that $\text{Aut}(G) = \{f : G \to G \mid f \text{ is an isomorphism}\}$ and that $\text{Aut}(G)$ is a group under function composition.)

Definition 4.50 Let G, K be groups. Let $\theta : K \to \text{Aut}(G)$ be a homomorphism. Define a binary operation $\star$ on $G \times K$ by

$$(g_1, k_1) \star (g_2, k_2) = (g_1[\theta(k_1)](g_2), k_1k_2).$$

The set $G \times K$, equipped with the operation $\star$, is called the *semidirect product of G and K with respect to* θ and is denoted $G \rtimes_\theta K$.

Remark 4.51
To prevent our computations from becoming overly cluttered with excess notation, we frequently omit the θ and simply write $G \rtimes K$ instead. Often we write gk instead of (g, k) and $g_1k_1g_2k_2$ instead of $(g_1, k_1) \star (g_2, k_2)$.

In the same vein, we often view G as a subgroup of $G \rtimes K$ by identifying $g \in G$ with $(g, 1)$. We are justified in doing so because the map $g \mapsto (g, 1)$ is an injective homomorphism. We similarly view K as a subgroup of $G \rtimes K$ by identifying $k \in K$ with $(1, k)$.

If θ is understood, then we denote $[\theta(k)](g)$ by ${}^k g$. Note that ${}^{k_1k_2}g = {}^{k_1}({}^{k_2}g)$, because θ is a homomorphism, and that ${}^k(g_1g_2) = ({}^kg_1)({}^kg_2)$, because $\theta(k)$ is a homomorphism. These two facts, which are easy to remember because of their formal resemblance to the laws of exponents, can often help speed up computations.

Proposition 4.52
Let G, K, θ be as in Def. 4.50. Then $G \rtimes K$ is a group.

Proof
Let e_G and e_K be the identity elements of G and K, respectively. Then e_Ge_K is the identity element of $G \rtimes K$. Also, $({}^{k^{-1}}g^{-1})k^{-1}$ is the inverse element of gk. We leave the proof of associativity as an exercise (see Exercise 3). ⊛

Example 4.53
In this example, we show that the dihedral groups D_n occur naturally as semidirect products. Define $\phi : \mathbb{Z}_n \to \mathbb{Z}_2$ by $\phi(a) = -a$. Then $\phi \in \operatorname{Aut}(\mathbb{Z}_n)$. Moreover, ϕ has order 2, because $\phi \circ \phi$ is the identity map ι on $\mathbb{Z}_n$. It follows that the map $\theta : \mathbb{Z}_2 \to \operatorname{Aut}(\mathbb{Z}_n)$ given by

$$\theta(b) = \begin{cases} \iota & \text{if } b = 0 \\ \phi & \text{if } b = 1 \end{cases}$$

is a well-defined homomorphism. Construct the semidirect product $\mathbb{Z}_n \rtimes \mathbb{Z}_2$ of $\mathbb{Z}_n$ and $\mathbb{Z}_2$ with respect to θ.

Recall the dihedral group $D_n = \langle r, s \mid r^n = s^2 = 1, rs = s^{-1}r \rangle$. Let $x = (0, 1) \in \mathbb{Z}_n \rtimes \mathbb{Z}_2$, and let $y = (1, 0) \in \mathbb{Z}_n \rtimes \mathbb{Z}_2$. Then

$$x^2 = (0, 1) \star (0, 1) = (0 + [\theta(1)](0), 1 + 1) = (0 + \phi(0), 0) = (0, 0).$$

So x has order 2. Similarly, one can compute that $y^k = (k, 0)$ for all k, so y has order n and $y^{-1} = (-1, 0)$. Moreover,

$$xy = (0,1) \star (1,0) = (0 + [\theta(1)](1), 1+0) = (0 + \phi(1), 1) = (-1, 1), \text{ and}$$
$$y^{-1}x = (-1,0) \star (0,1) = (-1 + [\theta(0)](0), 0+1) = (-1 + \iota(0), 1) = (-1, 1),$$

so $xy = y^{-1}x$. In other words, x and y behave exactly like s and r. Thus, we find that $\mathbb{Z}_n \rtimes \mathbb{Z}_2 \cong D_n$, with the isomorphism given by $(a, b) \mapsto r^a s^b$.

Lemma 4.54
$G \lhd G \rtimes K$, and $(G \rtimes K)/G \approx K$.

Proof
The proof is straightforward; we leave it as an exercise to the reader.

We now define a special semidirect product that will be of particular interest to us, namely, the *wreath product*. Let I be a finite set, and let G and K be groups. Let $G^I = \bigoplus_{i \in I} G$ be the direct product of several copies of G, one for each element of I. Elements of G^I are $|I|$-tuples $(g_i)_{i \in I}$, where $g_i \in G$ for all i. Let θ be an action of K on I. (In other words, θ is a homomorphism from K to S^I, where S^I is the symmetric group on I, that is, the group of all permutations of I.) Then θ induces a homomorphism from K to $\operatorname{Aut}(G^I)$, which we also denote by θ, defined by $\theta((g_i)_{i \in I}) = (g_{\theta(i)})_{i \in I}$. With this notation, then, we make the following definition.

Definition 4.55 The wreath product of G and K with respect to θ is denoted $G \wr_\theta K$ and is defined by

$$G \wr_\theta K := G^I \rtimes_\theta K.$$

As with semidirect products, we frequently omit the subscript θ when it is understood from context.

Example 4.56
Let $G = \mathbb{Z}_2$. Let $I = \mathbb{Z}_3$. So elements of G^I are triples (g_0, g_1, g_2), where each g_i is either a 0 or a 1. Let $K = \mathbb{Z}_3$. Define an action θ of K on I by $[\theta(a)](b) = a + b$. Then θ induces a homomorphism from K to $\text{Aut}(G^I)$. For example, we have ${}^2(1, 1, 0) = (1, 0, 1)$. (What happened was that all of the entries in $(1, 0, 1)$ "shifted forward" by 2, cycling around when necessary.) Using this action, construct the wreath product $G \wr K$. Elements of $G \wr K$ are ordered pairs, where the first entry is an element of G^I, and the second entry is an element of K. For example, $x = ((0, 1, 0), 2) \in G \wr K$ and $y = ((1, 1, 0), 1) \in G \wr K$. To give an example of a computation in $G \wr K$, we have

$$\begin{aligned} xy &= ((0, 1, 0), 2) \star ((1, 1, 0), 1) \\ &= ((0, 1, 0) + {}^2(1, 1, 0), 2 + 1) \\ &= ((0, 1, 0) + (1, 0, 1), 2 + 1) \\ &= ((1, 1, 1), 0). \end{aligned}$$

7. CUBE-CONNECTED CYCLE GRAPHS

As the authors were learning the material in this chapter and in Chapter 2, we had the following questions.

1. Proposition 4.36 shows that the Subgroups Nonexpansion Principle remains true if we replace "yields an expander family" with "has logarithmic diameter" in the statement of the theorem. Is the same true of the Quotients Nonexpansion Principle?
2. If a sequence of finite groups has logarithmic diameter, does it necessarily yield an expander family?
3. Can an unbounded sequence of solvable groups with bounded derived length have logarithmic diameter?

The answers to these questions are no, no, and yes, respectively—in each case, the opposite of what we had initially predicted. Moreover, a single example suffices to answer all three. We now discuss this example, the family of cube-connected cycle graphs. At the end of this section, we summarize the relevance of this family to questions (1), (2), (3) in Remark 4.68.

Throughout this section, we make the following notational conventions. Consider a fixed positive integer n. Let $\mathbf{e}_i$ denote the element of $\mathbb{Z}_2^n = \mathbb{Z}_2 \times \cdots \times \mathbb{Z}_2$ with a 1 in the ith coordinate and zeroes everywhere else. Let $\mathbf{0} = (0, \ldots, 0)$ denote the identity element of $\mathbb{Z}_2^n$.

Definition 4.57 Define an action θ of $\mathbb{Z}_n$ on $I = \mathbb{Z}_n$ by $[\theta(a)](b) = a + b$. Via this action construct the wreath product $G_n = \mathbb{Z}_2 \wr \mathbb{Z}_n$. Let

$$\Gamma_n = \{(\mathbf{e}_n, 0), \gamma, \gamma^{-1}\} \subset G_n,$$

where $\gamma = (\mathbf{0}, 1)$ and $\gamma^{-1} = (\mathbf{0}, -1)$. We define the *cube-connected cycle graph* CCC_n to be the Cayley graph $\text{Cay}(G_n, \Gamma_n)$.

Remark 4.58
Let θ be as in Definition 4.57. Then $G_n = \mathbb{Z}_2 \wr \mathbb{Z}_n = \mathbb{Z}_2^n \rtimes \mathbb{Z}_n$ where the action of $\mathbb{Z}_n$ on $\mathbb{Z}_2^n$ is given by

$$^1(a_1, a_2, \ldots, a_n) = (a_n, a_1, \cdots, a_{n-2}, a_{n-1}).$$

Because 1 generates $\mathbb{Z}_n$, this defines the group action for any element of $\mathbb{Z}_n$. For if $k = 1 + 1 + \cdots + 1 + 1 \in \mathbb{Z}_n$, then

$$\begin{aligned} ^k(a_1, a_2, \ldots, a_n) &= {}^1({}^1(\cdots{}^1({}^1(a_1, a_2, \ldots, a_n))\cdots)) \\ &= (a_{n-k+1}, a_{n-k+2}, \ldots, a_{n-k-1}, a_{n-k}). \end{aligned}$$

Example 4.59
Let $n = 3$. Then $(\mathbf{e}_1, 0) = (100, 0)$, $(\mathbf{e}_2, 0) = (010, 0)$, $(\mathbf{e}_3, 0) = (001, 0)$, $\gamma = (000, 1)$, and $\gamma^{-1} = (000, -1)$. Here we have abused notation by writing the elements of $\mathbb{Z}_2^3$ as binary strings instead of tuples. We continue with this abuse for the remainder of the section.

Consider the element $(100, 1)$ of G_3. Note that

$$\begin{aligned} (100, 1)(000, 1) &= (100 + 000, 1 + 1) = (100, 2), \\ (100, 1)(000, -1) &= (100 + 000, 1 - 1) = (100, 0), \text{ and} \\ (100, 1)(001, 0) &= (100 + 100, 1 + 0) = (000, 1). \end{aligned}$$

Hence $(100, 1)$ is adjacent to the vertices $(100, 2)$, $(100, 0)$, and $(000, 1)$ in CCC_3.

Let us give another example of a computation in CCC_3 that will arise in the proof of Proposition 4.64. Note that

$$\begin{aligned} \gamma(\mathbf{e}_3, 0)\gamma^{-1} &= (000, 1)(001, 0)(000, -1) = (000, 1)(001, -1) \\ &= (000 + 100, 1 - 1) = (100, 0) = (\mathbf{e}_1, 0). \end{aligned}$$

The reader should verify that $\gamma^2(\mathbf{e}_3, 0)\gamma^{-2} = \mathbf{e}_2$. (Hint: First establish that in $G \rtimes K$, if $g \in G$ and $k \in K$, then $kgk^{-1} = {}^kg$.)

Remark 4.60
Note that Γ_n is symmetric, so CCC_n is an honest graph, not merely a directed graph.

Remark 4.61
By Proposition 1.29, we have that CCC_n is 3-regular.

Remark 4.62
An element of $\mathbb{Z}_2^n$ can be thought of as a string of n binary digits. An **n-dimensional hypercube** is the graph whose vertices are the elements of $\mathbb{Z}_2^n$, where two vertices are adjacent, via an edge of multiplicity one, if they differ in exactly one digit (a.k.a. one bit), and they are nonadjacent otherwise.

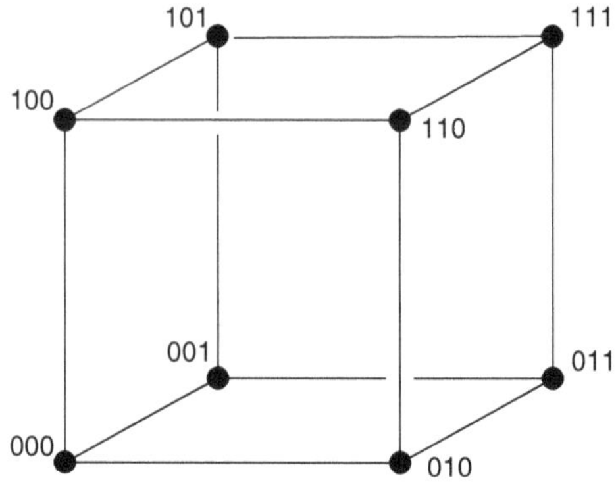

Figure 4.6 3-dimensional hypercube

The 3-dimensional hypercube can be arranged so as to resemble an ordinary cube, as in Figure 4.6.

Remark 4.63
We can visualize CCC_n as an n-dimensional hypercube, where each vertex has been replaced by an n-cycle. The vertices of the n-cycle are elements of the set $\mathbb{Z}_n = \{1, \cdots, n\}$. Each element j in the cycle that replaced vertex $b_1 b_2 \cdots b_n$ is adjacent to the elements $j-1$ and $j+1$ in the same cycle, as well as the element j in the cycle that replaced the vertex that differs from $b_1 b_2 \cdots b_n$ only in the jth digit. Figure 4.7 shows CCC_3. Hence the name "cube-connected cycles" graph.

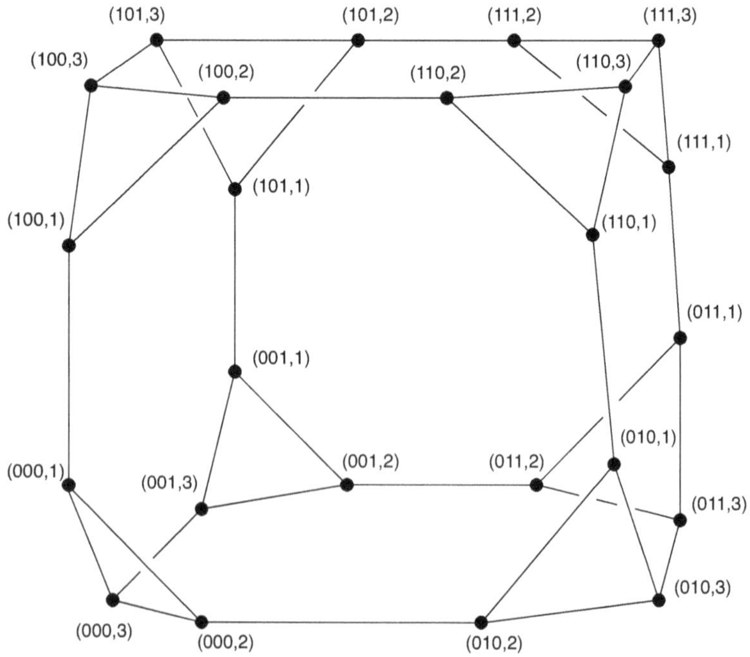

Figure 4.7 CCC_3

Proposition 4.64
For all n, we have $\operatorname{diam}(CCC_n) \leq 4n$.

Proof
First note that an arbitrary element of G_n is of the form $(\mathbf{e}_{j_1}\mathbf{e}_{j_2}\cdots\mathbf{e}_{j_k}, a)$ for some positive integers $j_1, \cdots, j_k$ with $1 \leq j_1 < j_2 < \cdots < j_k \leq n$ and $k \leq n$ and $1 \leq a \leq n$. By Proposition 4.17, it suffices to show that the word norm of $(\mathbf{e}_{j_1}\mathbf{e}_{j_2}\cdots\mathbf{e}_{j_k}, a)$ in Γ_n is less than or equal to $4n$.

Let $\mathbf{e} = (\mathbf{e}_n, 0)$. Note that for all positive integers c, we have

$$\begin{aligned}\gamma^c\mathbf{e}(\gamma^{-1})^c &= (\mathbf{0}, c)(\mathbf{e}_n, 0)(\mathbf{0}, -c)\\ &= ({}^c\mathbf{e}_n, c-c) = (\mathbf{e}_c, 0).\end{aligned}$$

This implies that

$$(\mathbf{e}_{j_1}\mathbf{e}_{j_2}\cdots\mathbf{e}_{j_k}, a) = \gamma^{j_1}\mathbf{e}\gamma^{j_2-j_1}\mathbf{e}\gamma^{j_3-j_2}\cdots\gamma^{j_k-j_{k-1}}\mathbf{e}(\gamma^{-1})^{j_k}\gamma^a. \tag{14}$$

In the right-hand side of Equation 14, we have that γ appears

$$j_1 + (j_2 - j_1) + \cdots + (j_k - j_{k-1}) + a = j_k + a$$

times; γ^{-1} appears j_k times; and $\mathbf{e} = (\mathbf{e}_n, 0)$ appears k times. So the word norm of $(\mathbf{e}_{j_1}\mathbf{e}_{j_2}\cdots\mathbf{e}_{j_k}, a)$ is less than or equal to $2j_k + a + k \leq 4n$. ◈

Corollary 4.65
The sequence (CCC_n) has logarithmic diameter.

Proof
We have that $|G_n| = |\mathbb{Z}_2^n|\,|\mathbb{Z}_n| = n2^n$, so $\log|CCC_n| = \log n + n\log 2$. Let $C = 4/\log 2$. By Proposition 4.64, for any n we have

$$\operatorname{diam}(CCC_n) \leq 4n \leq C(\log n + n\log 2) = C\log|CCC_n|.$$ ◈

Lemma 4.66
The group G_n is solvable with derived length 2.

Proof
First, G_n is not abelian, because $\gamma e_n \neq e_n\gamma$. So G_n does not have derived length 1. By Lemma 4.54, we have $\mathbb{Z}_2^n \lhd G_n$ and $G_n/\mathbb{Z}_2^n \cong \mathbb{Z}_n$. By Lemma 4.40, we have that $G_n' < \mathbb{Z}_2^n$, so G_n' is abelian, so $G_n^{(2)} = 1$. ◈

Corollary 4.67
The sequence (CCC_n) is not an expander family.

Proof
This follows from Lemma 4.66 and Theorem 4.47. ◈

Remark 4.68

We now answer the questions that we asked in the beginning of this section. Corollary 4.65, Lemma 4.66, and Corollary 4.67 answer question (2) with a no and question (3) with a yes. In the proof of Lemma 4.66 we noted that $\mathbb{Z}_2^n \lhd G_n$ and $G_n/\mathbb{Z}_2^n \cong \mathbb{Z}_n$. By Proposition 4.25, we have that $(\mathbb{Z}_n)$ does not have logarithmic diameter. Corollary 4.65 gives that (G_n) does have logarithmic diameter. Hence, we may answer question (1) with a no.

NOTES

1. The survey article [14] provides a wealth of results about diameters of finite groups, both general statements as well as many estimates for specific families of finite groups.
2. The line of reasoning in Example 4.21 comes from [87, p. 103].
3. In [15], Babai, Kantor, and Lubotzky show that the sequence (S_n) of symmetric groups has logarithmic diameter. In [76], Kassabov proves more strongly that (S_n) yields an expander family; this had been an open problem for many years. Kassabov's proof is highly intricate and has been described as a tour de force.
4. Our proof of Prop. 4.26 follows along the lines of a similar theorem in [8].
5. In [1], Abért and Babai explicitly construct an infinite family $\mathcal{G}$ of finite groups G_n, each generated by just two elements, so that $\mathcal{G}$ has uniform logarithmic diameter.
6. One can combine Props. 1.84 and 4.6 to obtain an upper bound for the diameter of a graph X in terms of $\lambda_1(X)$. In [39], however, Chung obtains a sharper estimate for a finite d-regular graph X with n vertices, namely:
$$\text{diam}(X) \leq \lceil \log(n-1)/\log(d/|\lambda_1(X)|) \rceil.$$
7. The use of the cube-connected cycles graph CCC_n as an interconnection pattern of processing elements goes back to [113]. Akers and Krishnamurthy [3] seem to have first realized CCC_n as a Cayley graph.
8. For any finite group G, nilpotent subgroup N of G with index r and nilpotency class c, and positive integer d, define $AB(G,N,d) = |N|^{(drc)^{-c}/2}$. Define $AB(G,d)$ to be the maximum, over all nilpotent subgroups N of G, of $AB(G,N,d)$. Let G be a finite group; let $\Gamma \Subset G$; let $d = |\Gamma|$; and let $X = \text{Cay}(G,\Gamma)$. Annexstein and Baumslag [13] show that
$$\text{diam}(X) \geq AB(G,d).$$
(The Annexstein-Baumslag bound is in the spirit of a celebrated theorem of Gromov [68], which states in part that balls in a Cayley graph on an infinite discrete group G grow polynomially iff G contains a nilpotent subgroup of finite index.) It follows that if $\log(AB(G_n,d))/|G_n|$ is unbounded for all d, then (G_n) does not have logarithmic diameter. (Our Prop. 4.25 is a special case of this statement.)
9. In [76], Kassabov discusses the following "difficult problem," also discussed in [88]: given a sequence (G_n) of finite groups, does (G_n) yield an expander family? He notes, "Currently there is no theory which can give a satisfactory answer to

this question. The answer is known only in a few special cases: If the family of finite groups comes from a finitely generated infinite group with property T (or its weaker versions) then the answer is YES. Also if all groups in the family are 'almost' abelian, then the answer is NO (see [90]) and this is essentially the only case where a negative answer ... is known." In this note, we tentatively offer some conjectures as to what a complete answer might look like.

Highly Speculative Conjecture 1: A sequence (G_n) of finite groups has logarithmic diameter iff $\log(AB(G_n, d))/|G_n|$ is bounded for some d. (See Note 8.)

Let $\mathscr{S}_0$ be set of all sequences of finite groups that do not have logarithmic diameter. (If Highly Speculative Conjecture 1 is correct, we have a characterization of all elements of $\mathscr{S}_0$.) Recursively define two sequences of sets as follows. Let $\mathscr{Q}_j$ be the set of all sequences of finite groups that admit some element of $\mathscr{S}_{j-1}$ as an unbounded sequence of quotients. Let $\mathscr{S}_j$ be the set of all sequences of finite groups that admit some element of $\mathscr{Q}_j$ as a bounded-index sequence of subgroups. Let

$$\mathscr{N} = \bigcup_{j=0}^{\infty} \mathscr{S}_j.$$

Inductively applying the Quotients and Subgroups Nonexpansion Principles, we see that no sequence in $\mathscr{N}$ yields an expander family.

Highly Speculative Conjecture 2: A sequence (G_n) of finite groups yields an expander family iff $(G_n) \notin \mathscr{N}$.

See Research Project Idea (2) for a possible simplification of this statement.

EXERCISES

1. Show that $\operatorname{diam}(\operatorname{Cay}(\mathbb{Z}_{4n}, \{1, -1, 2n\})) = n$.
2. For every integer $n \geq 2$, we define a graph X_n as follows. The vertex set of X_n is

$$\{v_0\} \cup \left(\bigcup_{i=0}^{n} (\mathbb{Z}_3 \times (\mathbb{Z}_2)^i) \right).$$

   That is, a vertex of X_n is either v_0 or else an $(i+1)$-tuple, where $0 \leq i \leq n$, so that the first entry is a 0, a 1, or a 2, and every other entry is either a 0 or a 1. We define adjacency in X_n as follows. (Note that there will be no multiple edges.) The vertex v_0 is adjacent to (0), (1), and (2). If $1 \leq i \leq n-1$, then an $(i+1)$-tuple $(a_0, a_1, \ldots, a_i)$ is adjacent to $(a_0, a_1, \ldots, a_{i-1})$, $(a_0, a_1, \ldots, a_i, 0)$, and $(a_0, a_1, \ldots, a_i, 1)$. Finally, an $(n+1)$-tuple $(a_0, a_1, \ldots, a_n)$ is adjacent to $(a_0, a_1, \ldots, a_{n-1})$, $(a_0 + 1, a_1, \ldots, a_n)$, and $(a_0 + 2, a_1, \ldots, a_n)$. Note that each X_n is 3-regular. (It may be helpful to draw, say, X_2 and X_3.) Prove that (X_n) has logarithmic diameter but is not an expander family.
3. Let G, H, θ as in Def. 4.50. Prove that $\star$ is an associative operation on $G \rtimes H$.
4. Prove Lemma 4.54.

5. Fill in the details in Example 4.20.
6. Fill in the details in Example 4.45.
7. Verify that the map ϕ in Example 4.27 is a surjective homomorphism.
8. If G is a group, and $N < G$, then we say that N is *characteristic in* G if $\phi(N) = N$ for all automorphisms ϕ of G. Prove the following.

   (a) Let G be a group. Then G' is a characteristic subgroup of G.
   (b) If N is a characteristic subgroup of G, then N is normal in G.
   (c) Suppose H is a characteristic subgroup of K, and K is a characteristic subgroup of G. Then H is a characteristic subgroup of G. (We remark that this shows one reason being characteristic is superior to being normal; for normality is not transitive, but being characteristic is.)
   (d) Let G be a group. Then $G^{(k)}$ is normal in G for all non-negative integers k.
   (e) Let G be a solvable group with derived length $\ell \geq 1$. Then $G/G^{(\ell-1)}$ is solvable with derived length $\ell - 1$.
   (f) Now use the previous part of this exercise, together with Props. 4.25 and 4.36, to provide an alternate proof of Thm. 4.47. (Hint: Divide into two cases, according to whether $(G_n^{(\ell_n-1)})$ has bounded index in (G_n), where ℓ_n is the derived length of G_n.)

9. For any field F, define the group of $k \times k$ unipotents with entries in F to be the set of $n \times n$ upper triangular matrices with 1s along the main diagonal. Under matrix multiplication, this set becomes a group. Fix a positive integer k. Let G_n be the group of $k \times k$ unipotents with entries in $\mathbb{Z}_{p_n}$, where p_n is the nth prime number. Prove that (G_n) does not yield an expander family.
10. Prove Lemma 4.40.
11. Let CCC_n be as in Definition 4.57. Prove that

$$h(CCC_n) \leq \begin{cases} \frac{4}{n} & \text{if } n \text{ is even} \\ \\ \frac{4}{n-1} & \text{if } n \text{ is odd.} \end{cases}$$

   (Hint: Take the "bottom half" of the cycle at each vertex.)
12. Prove that if G is a finite abelian group and X is a d-regular Cayley graph on G and $n = |G|$, then

$$h(X) \leq d\left(n^{\left(\frac{2}{n^{1/d}-1}\right)} - 1\right).$$

   Note that the right-hand side goes to 0 as n goes to infinity, in accordance with Corollary 4.26.

STUDENT RESEARCH PROJECT IDEAS

1. The example of the cube-connected cycle graphs shows us that a sequence of semidirect products of two abelian groups can have logarithmic diameter. Investigate conditions under which this can and cannot happen. One tractable family of such groups may be those of the form $\mathbb{Z}_p \rtimes \mathbb{Z}_q$, where p and q are primes and q divides $p - 1$. Theorem 6.1 in [14] may be relevant.

2. Let $\mathscr{S}'_0$ be set of all sequences of finite abelian groups. Recursively define two sequences of sets as follows. Let $\mathscr{Q}_j$ be the set of all sequences of finite groups that admit some element of $\mathscr{S}_{j-1}$ as an unbounded sequence of quotients. Let $\mathscr{S}_j$ be the set of all sequences of finite groups that admit some element of $\mathscr{Q}_j$ as a bounded-index sequence of subgroups. Let

$$\mathscr{N}' = \bigcup_{j=0}^{\infty} \mathscr{S}'_j.$$

Let $\mathscr{N}$ be as in Note 9. Is it true that $\mathscr{N} = \mathscr{N}'$?

5

Zig-Zag Products

In previous chapters, we have seen that there are significant obstacles to constructing an expander family. The most obvious attempts (e.g., Cayley graphs on abelian groups) do not work. So how *does* one construct an expander family?

The first explicit construction was given in 1973 by Margulis [92]. Over the next 30 or so years, many other expander families were constructed. Prior to 2002, though, the proofs that the spectral gaps in question were in fact bounded away from zero relied on algebraic techniques often using "heavy machinery" from analytic number theory, algebraic geometry, and the representation theory of finite simple groups of Lie type. In a 2002 article in *Annals of Mathematics,* Reingold, Vadhan, and Wigderson [116] significantly simplified matters by producing a straightforward combinatorial method for constructing an expander family; the proof that their spectral gaps are bounded away from zero requires only standard elementary techniques from linear algebra, such as the Rayleigh-Ritz theorem and the Cauchy-Schwarz inequality. The key to their construction is the "zig-zag product," a certain method for taking two graphs X and Y and creating a larger graph whose spectral gap is controlled by the spectra of X and Y. In Section 4, we present the Reingold-Vadhan-Wigderson expander family and prove that its spectral gaps are bounded away from zero.

We saw in Section 7 of Chapter 4 that the cube-connected cycle graphs CCC_n "almost" form an expander family, insofar as the sequence at least has logarithmic diameter. We constructed CCC_n by replacing each vertex of an n-dimensional hypercube with a "cloud" of vertices, corresponding to an n-cycle. The idea behind the zig-zag product of two graphs X and Y follows roughly along those lines: replace each vertex of X with a cloud corresponding to Y. Section 1 provides the details of the definition. In Section 2, we discuss the adjacency operator of the zig-zag product. In Section 3, we show that if X and Y have good expansion (as measured by λ), then so does their zig-zag product. Under certain circumstances, the zig-zag product of two Cayley graphs on two groups G_1 and G_2 equals a Cayley graph on the semidirect product $G_1 \rtimes G_2$. We detail this connection in Section 5.

1. DEFINITION OF THE ZIG-ZAG PRODUCT

Definition 5.1 Let X be a graph, let e be an edge in X, and let v be a vertex incident to e. If e is a loop, then define $e(v) = v$; otherwise, define $e(v) = w$, where w is the other vertex incident to e. Note that $e(v)$ is undefined if v is not an endpoint of e.

Definition 5.2 Let X be a d_X-regular graph, and let Y be a d_Y-regular graph such that $d_X = |Y|$.

Let V_X, V_Y be the vertex sets of X and Y, respectively. Let E_X, E_Y be the edge multisets of X and Y, respectively, with multiple edges treated as distinct elements.

For each vertex $v \in V_X$, let $E_v = \{e \in E_X \mid v \text{ is an endpoint of } e\}$, and let $L_v : V_Y \to E_v$ be a bijection. (The condition $d_X = |Y|$ guarantees that L_v exists.) We call L_v the *labeling at* v. Let L be the set $\{L_v \mid v \in V_Y\}$. We call L the *labeling from* Y *to* X.

We define the *zig-zag product* $X \textcircled{z}_L Y$ *with labeling* L as follows. The vertex set of $X \textcircled{z}_L Y$ is the Cartesian product $X \times Y$. If (x_1, y_1) and (x_2, y_2) are two vertices in $X \textcircled{z}_L Y$, then the multiplicity of the edge between them equals the number of ordered pairs $(z_1, z_2) \in E_Y \times E_Y$ such that y_1 is an endpoint of z_1; y_2 is an endpoint of z_2; and $L_{x_1}(z_1(y_1)) = L_{x_2}(z_2(y_2))$.

Remark 5.3
To avoid notational clutter, we sometimes drop the L and simply write $X \textcircled{z} Y$, if it is either clear or else irrelevant what labeling is being used.

We anticipate that most readers will be desperate for an example at this point.

Example 5.4
Consider the graphs X and Y shown in Figures 5.1 and 5.2. We have $V_X = \{a, b\}$ and $V_Y = \{1, 2, 3\}$. Note that X is 3-regular, and $|Y| = 3$. So once we choose a labeling L, we can define a zig-zag product of X and Y.

Take $E_X = \{e_1, e_2, e_3, e_4\}$, where e_1 is the loop at a; e_2 is the loop at b; e_3 is the "top" edge between a and b; and e_4 is the "bottom" edge between a and b. Then $E_a = \{e_1, e_3, e_4\}$ and $E_b = \{e_2, e_3, e_4\}$.

Define L_a and L_b by $L_a(3) = e_1, L_b(3) = e_2, L_a(1) = L_b(1) = e_3, L_a(2) = L_b(2) = e_4$. We depict this labeling $L = \{L_a, L_b\}$ by labeling the edges of X near each vertex of X, as in Figure 5.1.

Note that $r(1) = 2$ and $u(3) = 2$, by Def. 5.1. Then $L_a(r(1)) = e_4 = L_b(u(3))$, so (r, u) is an ordered pair in $E_Y \times E_Y$ such that 1 is an endpoint

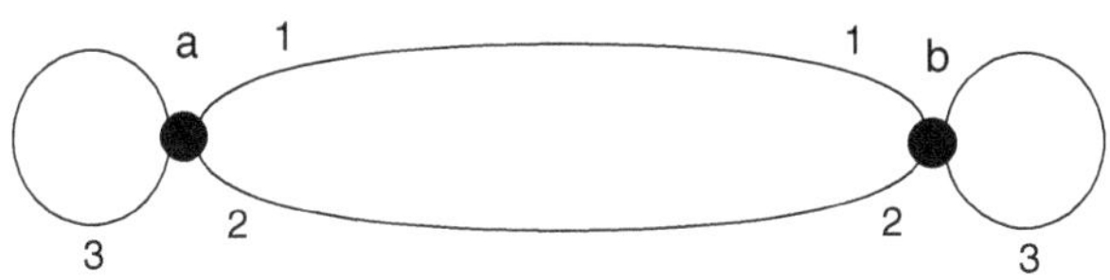

Figure 5.1 X

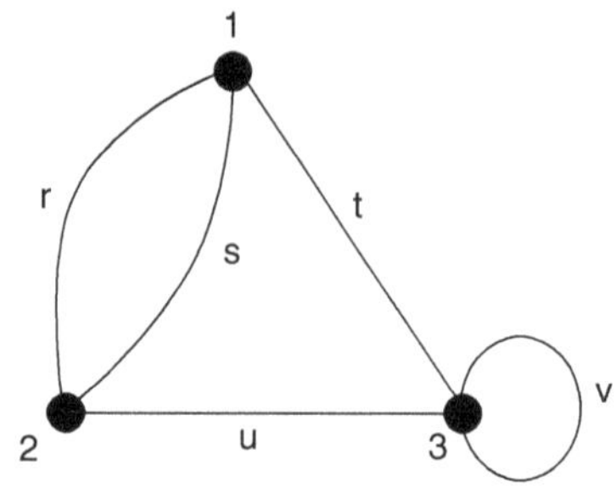

Figure 5.2 Y

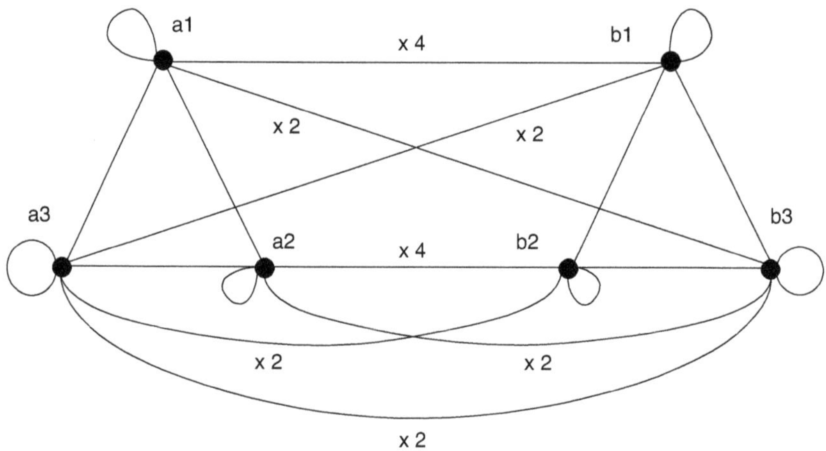

Figure 5.3 $Xⓩ_L Y$

of r; 3 is an endpoint of u; $L_a(r(1)) = L_b(u(3))$. In fact, there are two such ordered pairs, namely, (r, u) and (s, u). Hence, in $Xⓩ_L Y$, there is an edge of multiplicity two between $(a, 1)$ and $(b, 3)$. See Figure 5.3 for a picture of $Xⓩ_L Y$. Note that we are employing the sometimes convenient convention of writing the ordered pair (v, w) as a string vw. We do so for the remainder of the chapter.

We make frequent use of the following remark, which spells out the process for finding edges in zig-zag products.

Remark 5.5
Let (x_1, y_1) be a vertex in $Xⓩ_L Y$. To find vertices (x_2, y_2) adjacent to (x_1, y_1), we consider the following three-step process.

Step 1 (zig): Choose an edge z_1 in Y incident to y_1 and "move" to the vertex $(x_1, z_1(y_1))$.

Step 2: Let x_2 be the other endpoint of the edge $e = L_{x_1}(z_1(y_1))$. Let y' be the label of e at x_2. (In other words, $y' = L_{x_2}^{-1}(e)$.) Move to (x_2, y').

Step 3 (zag): Choose an edge z_2 in Y incident to y', let $y_2 = z_2(y')$, and move to (x_2, y_2).

(It is tempting to call Step 2 the "hyphen" step, because it comes between the zig and the zag.)

It follows from the definition of the zig-zag product that (x_1, y_1) and (x_2, y_2) are adjacent in $X \textcircled{z}_L Y$ via an edge corresponding to the pair (z_1, z_2) constructed in Steps 1–3. In fact, every edge incident to (x_1, y_1) corresponds to a pair (z_1, z_2) constructed according to this three-step process.

Definition 5.6 Let X be a d_X-regular graph, let Y be a d_Y-regular graph with vertex set V_Y such that $d_X = |Y|$. Let v be a vertex in X. Define the *v-cloud* in $X \textcircled{z} Y$ to be $\{(v, y) \mid y \in V_Y\}$.

Remark 5.7
It is useful to think of the vertex set of $X \textcircled{z} Y$ as arising by taking each vertex v of X and replacing it with the v-cloud.

Example 5.8
Continuing Example 5.4, we have that the vertex set of $X \textcircled{z}_L Y$ consists of the a-cloud $\{a1, a2, a3\}$ and the b-cloud $\{b1, b2, b3\}$, as shown in Figure 5.3.

Using the three-step process from Remark 5.5, we find all edges in $X \textcircled{z}_L Y$ incident to $a1$. At the zig step, we might choose $z_1 = r$, whereupon we would move to $a2$. At Step 2, then, we wind up at $b2$. At the zag step, we might choose $z_2 = u$, finally ending at $b3$. Proceeding in a similar manner, we can find all edges incident to $a1$, as follows:

$a1$ has a loop of multiplicity 1, via the zig $z_1 = t$ and the zag $z_2 = t$.
$a1$ has a single edge to $a2$, via the zig $z_1 = t$ and the zag $z_2 = u$.
$a1$ has a single edge to $a3$, via the zig $z_1 = t$ and the zag $z_2 = v$.
$a1$ has an edge of multiplicity 2 to $b3$, via the pairs (r, u) and (s, u).
$a1$ has an edge of multiplicity 4 to $b1$, via the pairs (r, r), (r, s), (s, r), and (s, s).

We can similarly compute that the multiplicities of all edges in $X \textcircled{z}_L Y$ are as shown in Figure 5.3.

Example 5.9
Recall that we are employing the sometimes convenient convention of writing the ordered pair (v, w) as a string vw. Figures 5.4 and 5.5 show two graphs X and Y and a labeling M. Note that X and Y are the same two graphs from

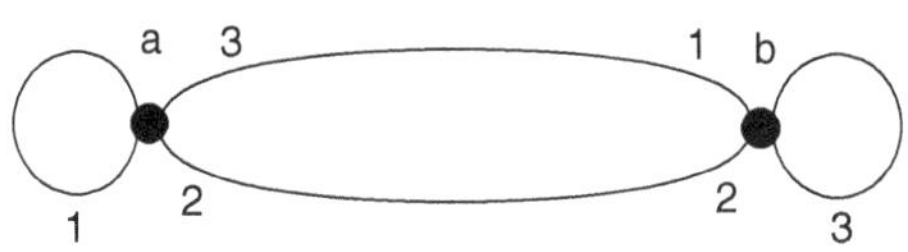

Figure 5.4 X

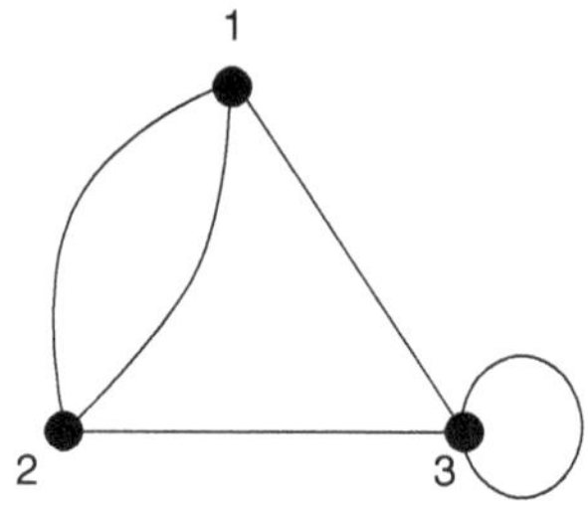

Figure 5.5 Y

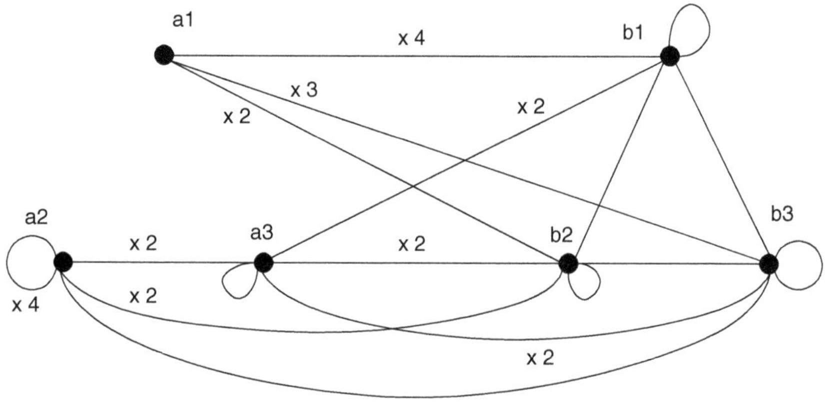

Figure 5.6 $X \textcircled{Z}_M Y$

Example 5.4; all we have done is change the labeling. The zig-zag product $X \textcircled{Z}_M Y$ is shown in Figure 5.6. Note that $X \textcircled{Z}_M Y$ does not have a loop at $a1$, whereas $X \textcircled{Z}_L Y$ from Example 5.4 has a loop at every vertex. Hence $X \textcircled{Z}_L Y$ and $X \textcircled{Z}_M Y$ are not isomorphic graphs. The moral of this story is that zig-zag products may indeed depend on the choice of labeling. (See Exercise 5 for an example that illustrates that whether the zig-zag product is connected may depend on the choice of labeling.)

Proposition 5.10
Let X, Y, L be as in Def. 5.2. Then $X \textcircled{Z}_L Y$ is a d_Y^2-regular graph.

Proof
Begin at the vertex (x_1, y_1) of $X \textcircled{Z}_L Y$. Consider the three-step process in Remark 5.5. At the zig step, there are d_Y choices of edges z_1 incident in Y to y_1. Independent of that choice, at the zag step there are d_Y choices of edges z_2 in Y incident to $L_{e(x_1)}^{-1}(e)$, where $e = L_{x_1}(z_1(y_1))$. Hence there are d_Y^2 pairs (z_1, z_2) that yield edges in $X \textcircled{Z}_L Y$ incident to (x_1, y_1). ∎

Remark 5.11
We can generalize the zig-zag construction a bit as follows. Let Y be a finite graph. Let $X_1, \ldots, X_n$ be a collection of graphs, each d_Y-regular, each with the

same vertex set. Let X be the union of the graphs $X_1, \ldots, X_n$. (That is, the multiplicity of the edge between two vertices in X equals the sum from $j = 1$ to n of the multiplicity of the edge in X_j between those two vertices.) Taking a separate labeling L_j for each graph X_j, we may form a zig-zag product $X \circledZ Y$ by taking the union of the graphs $X_j \circledZ Y$.

2. ADJACENCY MATRICES AND ZIG-ZAG PRODUCTS

In this section, we express the adjacency operator of $X \circledZ Y$ in terms of the adjacency operators of X and Y. This expression will, in Section 3, allow us to find a lower bound for the spectral gap of $X \circledZ Y$ in terms of the spectra of X and Y.

Let X be a d_X-regular graph, and let Y be a d_Y-regular graph such that $d_X = |Y|$. Let L be a labeling, as in Def. 5.2. We define two graphs Z and H as follows. Let $V = V_X \times V_Y$. Take V to be the vertex set both of Z and of H.

In Z, we define the multiplicity of the edge between (x_1, y_1) and (x_2, y_2) to be equal to the multiplicity of the edge between y_1 and y_2 in Y if $x_1 = x_2$, and we define this multiplicity to be 0 if $x_1 \neq x_2$.

In H, we define the multiplicity of the edge between (x_1, y_1) and (x_2, y_2) to be 1 if $L_{x_1}(y_1) = L_{x_2}(y_2)$ and 0 otherwise.

(Because Z and H depend on X, Y, and L, we should properly write $Z(X, Y, L)$ and $H(X, Y, L)$. To avoid such abominable notation, we assume thoughout this section and Section 3 that X, Y, and L are fixed.)

Intuitively, Z defines both the zig step and the zag step, and H defines the hyphen step. More precisely, we have the following proposition.

Proposition 5.12
$X \circledZ Y = Z \cdot H \cdot Z.$

Proof
This follows directly from Def. 5.2 and Def. 1.95. ◬

Example 5.13
Figure 5.7 shows two graphs X and Y, with a labeling from Y to X. Figures 5.8 and 5.9 show the corresponding graphs Z and H. The zig step from Remark 5.5

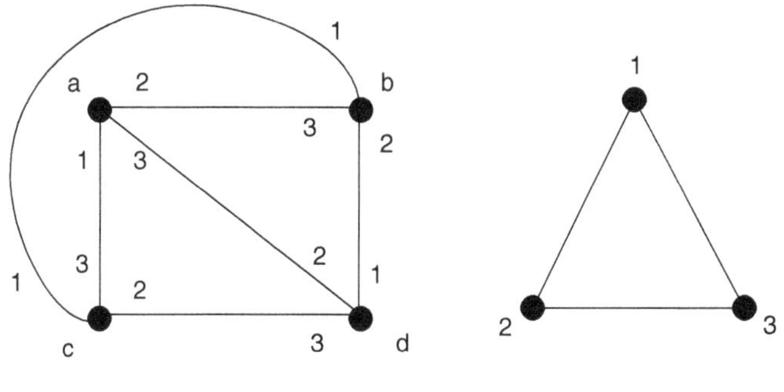

Figure 5.7 A graph X (left) and a graph Y (right), with a labeling from Y to X

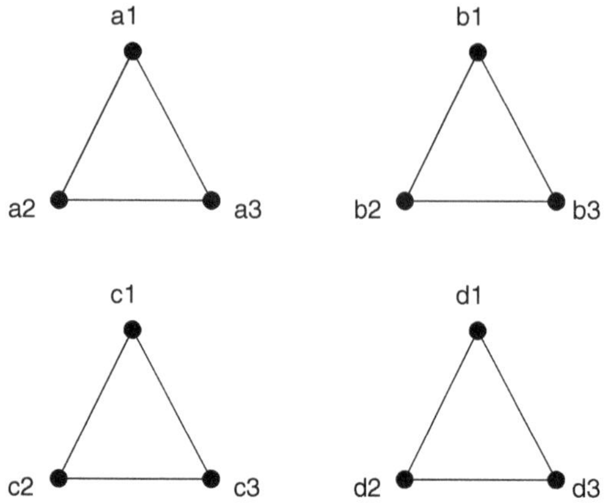

Figure 5.8 The "zig" graph Z

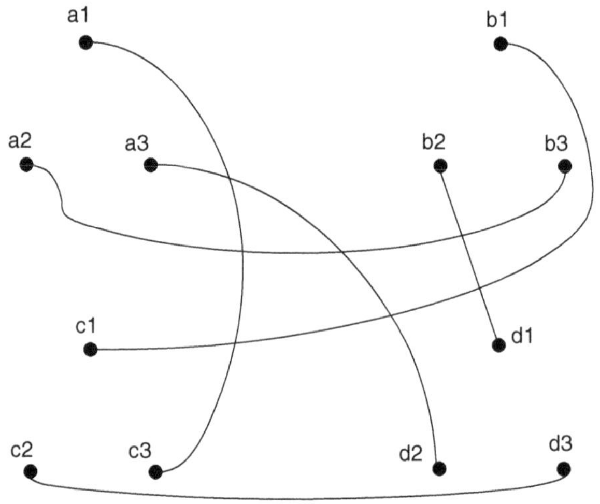

Figure 5.9 The "hyphen" graph H

corresponds to taking a step along an edge in Z. The hyphen step corresponds to taking a step along an edge in H. The zag step is again from Z. So $X \textcircled{z} Y$ is $Z \cdot H \cdot Z$. Figure 5.10 shows $X \textcircled{z} Y$.

We now fix an ordering $x_1, \ldots, x_n$ of the vertices of X and an ordering $y_1, \ldots, y_{d_X}$ of the vertices of Y. Order the vertices of $X \textcircled{z} Y$ lexicographically—that is, the ordering $x_1y_1, \ldots, x_1y_{d_X}, x_2y_1, \ldots, x_2y_{d_X}, \ldots, x_ny_1, \ldots, x_ny_{d_X}$.

Let $\tilde{B}$, $\tilde{A}$, and $\tilde{M}$ be the adjacency matrices of Z, H, and $X \textcircled{z} Y$, respectively, in terms of this ordering.

Corollary 5.14
$\tilde{M} = \tilde{B}\tilde{A}\tilde{B}$.

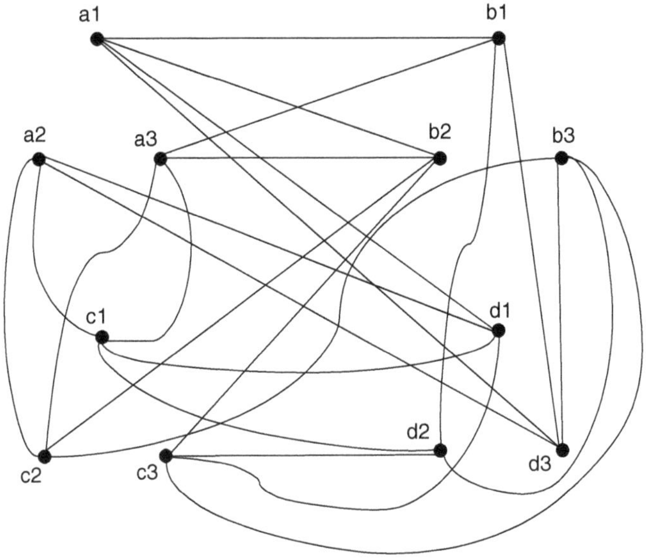

Figure 5.10 $X \textcircled{z} Y$

Proof
This follows from Prop. 1.97 and Prop. 5.12. ⊛

Remark 5.15
In the informal language of coverings, $X \textcircled{z} Y$ is an "upstairs" graph, whereas Y is a "downstairs" graph. It is common to use tildes for objects that live upstairs. Hence $\tilde{B}$ gets a tilde, whereas B does not. To help keep track of where things live, we frequently (perhaps ad molestiam) use tildes for upstairs objects.

Let A be the adjacency matrix of X, and let B be the adjacency matrix of Y, with respect to our fixed orderings. Edges in Z come from edges in Y, so we expect a relationship between B and $\tilde{B}$. Similarly, edges in H come from edges in X, so we expect a relationship between A and $\tilde{A}$. In Props. 5.16 and 5.19, we establish these relationships.

Proposition 5.16
$\tilde{B}$ equals the $|X| \cdot |Y| \times |X| \cdot |Y|$ block diagonal matrix

$$\begin{pmatrix} B & 0 & \dots & 0 \\ 0 & B & \dots & 0 \\ \vdots & \vdots & \ddots & \vdots \\ 0 & 0 & \dots & B \end{pmatrix}.$$

Proof
This follows directly from the definition of Z. ⊛

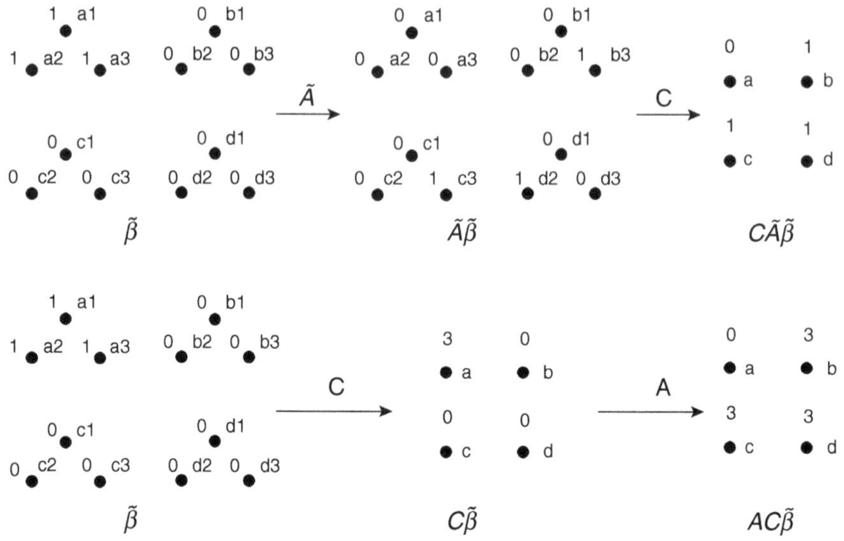

Figure 5.11 $C\tilde{A}\tilde{\beta}$ vs. $AC\tilde{\beta}$

Because $X \textcircled{Z} Y$, Z, and H each have the same vertex set, the spaces $L^2(X \textcircled{Z} Y)$, $L^2(Z)$, and $L^2(H)$ are identical. Define an operator $C : L^2(X \textcircled{Z} Y) \to L^2(X)$ by

$$Cf(x) = \sum_{y \in V_Y} f(x, y).$$

In other words, C sums over clouds.

Recall our convention from Section 1.3 that we may choose to regard A (respectively, $\tilde{A}$) not as a matrix but as a linear operator $A : L^2(X) \to L^2(X)$ (respectively, $\tilde{A} : L^2(X \textcircled{Z} Y) \to L^2(X \textcircled{Z} Y)$).

Definition 5.17 We say that $\tilde{\beta} \in L^2(X \textcircled{Z} Y)$ *is constant on clouds* if $\tilde{\beta}(x, y_1) = \tilde{\beta}(x, y_2)$ for all $x \in V_X, y_1, y_2 \in V_Y$.

Example 5.18
We continue Example 5.13. Consider the vector $\tilde{\beta}$ that equals 1 on the a-cloud and 0 outside it. In Figure 5.11, we compute $C\tilde{A}\tilde{\beta}$ and $AC\tilde{\beta}$. Note that $C\tilde{A}\tilde{\beta}$ first "spreads out" the 1's before summing over clouds, whereas $AC\tilde{\beta}$ first sums the 1's into a 3, then "spreads out" the 3. Hence $AC\tilde{\beta} = 3C\tilde{A}\tilde{\beta}$.

Proposition 5.19
If $\tilde{\beta} \in L^2(X \times Y)$ is constant on clouds, then $AC\tilde{\beta} = d_X C\tilde{A}\tilde{\beta}$.

Proof
Let $v \in V_X$, and let $\tilde{e}_v$ be the vector in $L^2(X \textcircled{Z} Y)$ defined by $\tilde{e}_v(x, y) = 1$ if $v = x$ and $\tilde{e}_v(x, y) = 0$ if $v \neq x$. We show that $AC\tilde{e}_v = d_X C\tilde{A}\tilde{e}_v$ for all $v \in X$. This will suffice, because the vectors $\tilde{e}_v$ form a basis for the subspace of vectors in $L^2(X \textcircled{Z} Y)$ that are constant on clouds.

Let $w \in V_X$. Then $(C\tilde{A}\tilde{e}_v)(w)$ equals the number of edges in X between v and w. Recall Remark 1.33. We have $C\tilde{e}_v = |Y|\delta_v = d_X\delta_v$, so $AC\tilde{e}_v = d_X A\delta_v$. So $(AC\tilde{e}_v)(w)$ equals d_X times the number of edges in X between v and w. Hence $AC\tilde{e}_v = d_X C\tilde{A}\tilde{e}_v$. ∎

3. EIGENVALUES OF ZIG-ZAG PRODUCTS

The main theorem of this section (Theorem 5.26) provides an upper bound on $\lambda(X \textcircled{z} Y)$ in terms of $\lambda(X)$ and $\lambda(Y)$. (Recall Def. 3.3.) This bound is the main tool we use to show that the graphs constructed in Section 4 indeed form an expander family.

The idea of the proof of Theorem 5.26 is to take any nonzero vector $\tilde{\alpha}$ in $L^2_0(X \textcircled{z} Y)$ and decompose it into one part that is constant on clouds and another part that sums to 0 on clouds. Using this decomposition, along with Corollary 5.14, we estimate the Rayleigh quotient $\frac{|\langle \tilde{M}\tilde{\alpha}, \tilde{\alpha}\rangle|}{\langle \tilde{\alpha}, \tilde{\alpha}\rangle}$ by a sum of three terms, each of which contains a single $\tilde{A}$ or $\tilde{B}$. We then use Props. 5.19 and 5.16 to relate these terms to A and B, which in turn can be estimated in terms of $\lambda(X)$ and $\lambda(Y)$.

We continue to use notations from Section 2 (such as A, B, etc.) throughout this section. Before proceeding, we first need to introduce some additional notation. For any vertex $v \in V_X$, define $\tilde{f}_v \in L^2(X \textcircled{z} Y)$ by $\tilde{f}_v(x, y) = |Y|^{-1/2}$ if $v = x$ and $\tilde{f}_v(x, y) = 0$ if $v \neq x$. So $\tilde{f}_v$ is a unit vector that is constant on the v-cloud and 0 outside the v-cloud. For any $\tilde{\alpha} \in L^2_0(X \textcircled{z} Y)$ and any vertex $v \in V_X$, define $\tilde{\alpha}_v^{\parallel} = \langle \tilde{\alpha}, \tilde{f}_v\rangle \tilde{f}_v$. Define $\tilde{\alpha}^{\parallel} = \sum_{v \in V_X} \tilde{\alpha}_v^{\parallel}$. Define $\tilde{\alpha}^{\perp} = \tilde{\alpha} - \tilde{\alpha}^{\parallel}$. Intuitively, $\tilde{\alpha}^{\parallel}$ is the part of $\tilde{\alpha}$ that is constant on clouds, and $\tilde{\alpha}^{\perp}$ is the part that sums to 0 on clouds.

We denote by $\mathbf{0}$ the zero vector in $L^2(X \textcircled{z} Y)$.

Lemma 5.20
Let $\tilde{\alpha} \in L^2(X \textcircled{z} Y)$. Then $C\tilde{\alpha}^{\perp} = \mathbf{0}$.

Proof
For any $v \in V_X$, we have

$$\begin{aligned}
(C\tilde{\alpha}^{\perp})(v) &= \sum_{y \in V_Y} \tilde{\alpha}^{\perp}(v, y) \\
&= |Y|^{1/2} \langle \tilde{\alpha}^{\perp}, \tilde{f}_v\rangle \\
&= |Y|^{1/2} \langle \tilde{\alpha} - \langle \tilde{\alpha}, \tilde{f}_v\rangle \tilde{f}_v, \tilde{f}_v\rangle \\
&= |Y|^{1/2} (\langle \tilde{\alpha}, \tilde{f}_v\rangle - \langle \tilde{\alpha}, \tilde{f}_v\rangle \langle \tilde{f}_v, \tilde{f}_v\rangle) \\
&= 0.
\end{aligned}$$

∎

Lemma 5.21
If $\tilde{\beta} \in L^2(X \textcircled{z} Y)$, then $\left\|\tilde{A}\tilde{\beta}\right\| = \left\|\tilde{\beta}\right\|$.

Proof
H has degree 1. Hence $\tilde{A}\tilde{\beta}$ is the same as $\tilde{\beta}$, but with entries permuted. ∎

The next lemma is a key step toward our main theorem. We begin with an arbitrary vector $\tilde{\alpha}$ in $L_0^2(X \textcircled{z} Y)$, and then find an upper bound for $\left|\langle \tilde{M}\tilde{\alpha}, \tilde{\alpha}\rangle\right|$. What makes this upper bound useful is that its various pieces involve either $\tilde{\alpha}^{\parallel}$ or $\tilde{\alpha}^{\perp}$, seperately.

Lemma 5.22
For any $\tilde{\alpha} \in L_0^2(X \textcircled{z} Y)$, we have

$$|\langle \tilde{M}\tilde{\alpha}, \tilde{\alpha}\rangle| \leq d_Y^2 |\langle \tilde{A}\tilde{\alpha}^{\parallel}, \tilde{\alpha}^{\parallel}\rangle| + 2d_Y \left\|\tilde{\alpha}^{\parallel}\right\| \cdot \left\|\tilde{B}\tilde{\alpha}^{\perp}\right\| + \left\|\tilde{B}\tilde{\alpha}^{\perp}\right\|^2.$$

Proof
First note that if $\tilde{\beta}$ is constant on clouds, then by Prop. 5.16, we have that $\tilde{B}\tilde{\beta} = d_Y \tilde{\beta}$. Now,

$$\begin{aligned}
\langle \tilde{M}\tilde{\alpha}, \tilde{\alpha}\rangle &= \langle \tilde{B}\tilde{A}\tilde{B}\tilde{\alpha}, \tilde{\alpha}\rangle \\
&= \langle \tilde{A}\tilde{B}\tilde{\alpha}, \tilde{B}\tilde{\alpha}\rangle \qquad \text{(by Lemma A.31)} \\
&= \langle \tilde{A}\tilde{B}(\tilde{\alpha}^{\parallel} + \tilde{\alpha}^{\perp}), \tilde{B}(\tilde{\alpha}^{\parallel} + \tilde{\alpha}^{\perp})\rangle \\
&= \langle \tilde{A}\tilde{B}\tilde{\alpha}^{\parallel}, \tilde{B}\tilde{\alpha}^{\parallel}\rangle + \langle \tilde{A}\tilde{B}\tilde{\alpha}^{\parallel}, \tilde{B}\tilde{\alpha}^{\perp}\rangle + \langle \tilde{A}\tilde{B}\tilde{\alpha}^{\perp}, \tilde{B}\tilde{\alpha}^{\parallel}\rangle + \langle \tilde{A}\tilde{B}\tilde{\alpha}^{\perp}, \tilde{B}\tilde{\alpha}^{\perp}\rangle \\
&= d_Y^2 \langle \tilde{A}\tilde{\alpha}^{\parallel}, \tilde{\alpha}^{\parallel}\rangle + d_Y \langle \tilde{A}\tilde{\alpha}^{\parallel}, \tilde{B}\tilde{\alpha}^{\perp}\rangle + d_Y \langle \tilde{A}\tilde{B}\tilde{\alpha}^{\perp}, \tilde{\alpha}^{\parallel}\rangle + \langle \tilde{A}\tilde{B}\tilde{\alpha}^{\perp}, \tilde{B}\tilde{\alpha}^{\perp}\rangle.
\end{aligned}$$

Therefore, by Cauchy-Schwarz (Prop. A.20), the triangle inequality, and Lemma 5.21, we have

$$\begin{aligned}
|\langle \tilde{M}\tilde{\alpha}, \tilde{\alpha}\rangle| &\leq d_Y^2 |\langle \tilde{A}\tilde{\alpha}^{\parallel}, \tilde{\alpha}^{\parallel}\rangle| + d_Y \left\|\tilde{A}\tilde{\alpha}^{\parallel}\right\| \cdot \left\|\tilde{B}\tilde{\alpha}^{\perp}\right\| \\
&\quad + d_Y \left\|\tilde{A}\tilde{B}\tilde{\alpha}^{\perp}\right\| \cdot \left\|\tilde{\alpha}^{\parallel}\right\| + \left\|\tilde{A}\tilde{B}\tilde{\alpha}^{\perp}\right\| \cdot \left\|\tilde{B}\tilde{\alpha}^{\perp}\right\| \\
&= d_Y^2 |\langle \tilde{A}\tilde{\alpha}^{\parallel}, \tilde{\alpha}^{\parallel}\rangle| + 2d_Y \left\|\tilde{\alpha}^{\parallel}\right\| \cdot \left\|\tilde{B}\tilde{\alpha}^{\perp}\right\| + \left\|\tilde{B}\tilde{\alpha}^{\perp}\right\|^2.
\end{aligned}$$

Lemma 5.23
If Y is nonbipartite, then for any $\tilde{\alpha} \in L_0^2(X \textcircled{z} Y)$, we have

$$\left\|\tilde{B}\tilde{\alpha}^{\perp}\right\| \leq \lambda(Y) \left\|\tilde{\alpha}^{\perp}\right\|.$$

Proof
For any $v \in V_X$, define $\alpha_v^{\perp} \in L^2(Y)$ by $\alpha_v^{\perp}(y) = \tilde{\alpha}^{\perp}(v, y)$. Note that $\tilde{\alpha}^{\perp} = \left(\alpha_{v_1}^{\perp} \mid \alpha_{v_2}^{\perp} \mid \ldots \mid \alpha_{v_n}^{\perp}\right)^t$. Also note that $\alpha_v^{\perp} \in L_0^2(Y)$, by Lemma 5.20. Then $\tilde{B}\tilde{\alpha}^{\perp} = \left(B\alpha_{v_1}^{\perp} \mid B\alpha_{v_2}^{\perp} \mid \ldots \mid B\alpha_{v_n}^{\perp}\right)^t$ by Prop. 5.16.

By decomposing each $\alpha_{v_j}^{\perp}$ into eigenvectors of B as in Theorem A.53 and the proof of Prop. 3.13, we find that $\left\|B\alpha_{v_j}^{\perp}\right\| \leq \lambda(Y) \left\|\alpha_{v_j}^{\perp}\right\|$. (Here we are using the fact that Y is nonbipartite.) Hence, we have

$$\begin{aligned}\left\|\tilde{B}\tilde{\alpha}^{\perp}\right\|^2 &= \left\|B\alpha_{v_1}^{\perp}\right\|^2 + \left\|B\alpha_{v_2}^{\perp}\right\|^2 + \cdots + \left\|B\alpha_{v_n}^{\perp}\right\|^2 \\ &\leq \lambda(Y)^2\left(\left\|\alpha_{v_1}^{\perp}\right\|^2 + \left\|\alpha_{v_2}^{\perp}\right\|^2 + \cdots + \left\|\alpha_{v_n}^{\perp}\right\|^2\right) \\ &= \lambda(Y)^2\left\|\tilde{\alpha}^{\perp}\right\|^2.\end{aligned}$$

Lemma 5.24
Let $\tilde{\beta}_1, \tilde{\beta}_2 \in L^2(X ⓩ Y)$ such that $\tilde{\beta}_2$ is constant on clouds. Then $\langle C\tilde{\beta}_1, C\tilde{\beta}_2\rangle = d_X\langle\tilde{\beta}_1, \tilde{\beta}_2\rangle$.

We leave the proof of Lemma 5.24 as an exercise for the reader.

Lemma 5.25
If X is nonbipartite, then for any $\tilde{\alpha} \in L^2_0(X ⓩ Y)$, we have

$$|\langle\tilde{A}\tilde{\alpha}^{\|}, \tilde{\alpha}^{\|}\rangle| \leq \frac{\lambda(X)}{d_X}\cdot\langle\tilde{\alpha}^{\|}, \tilde{\alpha}^{\|}\rangle.$$

Proof
First note that $C\tilde{\alpha} = C\tilde{\alpha}^{\|}$, by Lemma 5.20. By Lemma 5.24 and Prop. 5.19, we have

$$\langle\tilde{A}\tilde{\alpha}^{\|}, \tilde{\alpha}^{\|}\rangle = \frac{\langle C\tilde{A}\tilde{\alpha}^{\|}, C\tilde{\alpha}^{\|}\rangle}{d_X} = \frac{\langle AC\tilde{\alpha}^{\|}, C\tilde{\alpha}^{\|}\rangle}{d_X^2} = \frac{\langle AC\tilde{\alpha}, C\tilde{\alpha}\rangle}{d_X^2}.$$

Because $\tilde{\alpha} \in L^2_0(X ⓩ Y)$, we have that $C\tilde{\alpha} \in L^2_0(X)$. Therefore, by Prop. 3.13 and again by Lemma 5.24, we have

$$|\langle\tilde{A}\tilde{\alpha}^{\|}, \tilde{\alpha}^{\|}\rangle| \leq \frac{\lambda(X)\langle C\tilde{\alpha}, C\tilde{\alpha}\rangle}{d_X^2} = \frac{\lambda(X)\langle C\tilde{\alpha}^{\|}, C\tilde{\alpha}^{\|}\rangle}{d_X^2} = \frac{\lambda(X)\langle\tilde{\alpha}^{\|}, \tilde{\alpha}^{\|}\rangle}{d_X}.$$

Theorem 5.26
Let X be a d_X-regular nonbipartite graph, and let Y be a d_Y-regular nonbipartite graph such that $d_X = |Y|$. Then for any choice of labeling from X to Y, we have

$$\lambda(X ⓩ Y) \leq \frac{d_Y^2\lambda(X)}{d_X} + d_Y\lambda(Y) + \lambda(Y)^2.$$

Proof

Let $\tilde{\alpha}$ be a nonzero vector in $L_0^2(X \textcircled{z} Y)$. Then by Lemmas 5.22, 5.23, and 5.25, we have

$$|\langle \tilde{M}\tilde{\alpha}, \tilde{\alpha}\rangle| \le \frac{d_Y^2}{d_X}\lambda(X)\left\|\tilde{\alpha}^{\parallel}\right\|^2 + 2d_Y\lambda(Y)\left\|\tilde{\alpha}^{\parallel}\right\| \cdot \left\|\tilde{\alpha}^{\perp}\right\| + \lambda(Y)^2\left\|\tilde{\alpha}^{\perp}\right\|^2. \tag{15}$$

Let $p = \dfrac{\|\tilde{\alpha}^{\parallel}\|}{\|\tilde{\alpha}\|}$ and $q = \dfrac{\|\tilde{\alpha}^{\perp}\|}{\|\tilde{\alpha}\|}$. Since $\langle \tilde{\alpha}^{\parallel}, \tilde{\alpha}^{\perp}\rangle = 0$, we have $p^2 + q^2 = 1$. So, $p \le 1$ and $q \le 1$. Also, $0 \le (p-q)^2 = 1 - 2pq$, so $2pq \le 1$. Therefore, dividing both sides of Equation 15 by $\langle \tilde{\alpha}, \tilde{\alpha}\rangle$, we have

$$\frac{|\langle \tilde{M}\tilde{\alpha}, \tilde{\alpha}\rangle|}{\langle \tilde{\alpha}, \tilde{\alpha}\rangle} \le \frac{d_Y^2}{d_X}\lambda(X)p^2 + 2\lambda(Y)d_Y pq + \lambda(Y)^2 q^2 \le \frac{d_Y^2}{d_X}\lambda(X) + \lambda(Y)d_Y + \lambda(Y)^2.$$

Therefore, by the Rayleigh-Ritz theorem, we have

$$\lambda(X \textcircled{z} Y) \le \frac{d_Y^2\lambda(X)}{d_X} + d_Y\lambda(Y) + \lambda(Y)^2.$$

Remark 5.27

Theorem 5.26 can be stated a bit more cleanly in terms of *normalized* adjacency matrices. (See Remark 2.50.) Letting μ be the normalized version of λ, we have

$$\mu(X \textcircled{z} Y) \le \mu(X) + \mu(Y) + \mu(Y)^2.$$

4. AN ACTUAL EXPANDER FAMILY

In this section, we give an iterative construction of an expander family. To get it started, we first need a "base graph" W. We will see that such a graph must satisfy several conditions for the construction to be well defined and for Theorem 5.26 to guarantee that the spectral gaps are bounded away from zero. Specifically, we need a regular nonbipartite graph W such that $|W| = d_W^4$ and $\lambda(W) \le \frac{d_W}{5}$. We proceed to construct just such a graph.

Let p be a prime number greater than 35. Let $G = \mathbb{Z}_p^8$. In other words, G is the direct product of 8 copies of $\mathbb{Z}_p$. (You may well ask: Why 35? Why 8? The reasons for these seemingly arbitrary choices will become clear as we go along.)

Define $\gamma : \mathbb{Z}_p^2 \to G$ by $\gamma(x,y) = (x, xy, xy^2, xy^3, xy^4, xy^5, xy^6, xy^7)$. Let $\Gamma = \{\gamma(x,y) \mid (x,y) \in \mathbb{Z}_p^2\}$. (Here we regard Γ as a multiset, not a set.) Note that $\Gamma \Subset G$. Let

$$W = \mathrm{Cay}(G, \Gamma). \tag{16}$$

Let $\omega = e^{\frac{2\pi i}{p}}$.

Definition 5.28 We write an element $\mathbf{a} \in G$ as $\mathbf{a} = (a_0, a_1, a_2, a_3, a_4, a_5, a_6, a_7)$. For two elements $\mathbf{a}, \mathbf{b} \in G$, we define $\mathbf{a} \cdot \mathbf{b} = a_0b_0 + a_1b_1 + a_2b_2 + a_3b_3 + a_4b_4 + a_5b_5 + a_6b_6 + a_7b_7$. Denote the identity element of G by $\mathbf{0}$.

Lemma 5.29

$$\sum_{\mathbf{x}\in G} \omega^{\mathbf{c}\cdot\mathbf{x}} = \begin{cases} p^8 & \text{if } \mathbf{c} = \mathbf{0} \\ 0 & \text{if } \mathbf{c} \neq \mathbf{0}. \end{cases}$$

Proof
This is immediate if $\mathbf{c} = \mathbf{0}$, because $|G| = p^8$. If $\mathbf{c} \neq \mathbf{0}$, then $c_j \neq 0$ for some j. Without loss of generality, assume $c_0 \neq 0$. Then we have

$$\begin{aligned} \sum_{\mathbf{x}\in G} \omega^{\mathbf{c}\cdot\mathbf{x}} &= \sum_{\mathbf{x}\in G} \omega^{c_0x_0}\omega^{c_1x_1}\ldots\omega^{c_7x_7} \\ &= \sum_{x_0\in\mathbb{Z}_p}\sum_{x_1\in\mathbb{Z}_p}\cdots\sum_{x_7\in\mathbb{Z}_p} \omega^{c_0x_0}\omega^{c_1x_1}\ldots\omega^{c_7x_7} \\ &= \left(\sum_{x_0\in\mathbb{Z}_p} \omega^{c_0x_0}\right)\left(\sum_{x_1\in\mathbb{Z}_p}\cdots\sum_{x_7\in\mathbb{Z}_p} \omega^{c_1x_1}\ldots\omega^{c_7x_7}\right) = 0, \end{aligned}$$

where the last equality comes from Lemma 1.51. ◈

For any $\mathbf{a} \in G$, define $f_{\mathbf{a}} \in L^2(W)$ by $f_{\mathbf{a}}(\mathbf{x}) = \omega^{\mathbf{a}\cdot\mathbf{x}}$. Note that $f_{\mathbf{a}}$ is well defined, since $\omega^p = 1$.

Let A be the adjacency operator of W. Define $\lambda_{\mathbf{a}} = \sum_{x,y\in\mathbb{Z}_p} \omega^{\mathbf{a}\cdot\gamma(x,y)}$. We now show that the eigenvalues of W are precisely the numbers $\lambda_{\mathbf{a}}$.

Lemma 5.30
For all $\mathbf{a} \in G$, we have $Af_{\mathbf{a}} = \lambda_{\mathbf{a}} f_{\mathbf{a}}$.

Proof
For any $\mathbf{b} \in G$, we have

$$\begin{aligned} (Af_{\mathbf{a}})(\mathbf{b}) &= \sum_{\gamma(x,y)\in\Gamma} f_{\mathbf{a}}(\mathbf{b} + \gamma(x,y)) \\ &= \sum_{x,y\in\mathbb{Z}_p} \omega^{\mathbf{a}\cdot\mathbf{b}+\mathbf{a}\cdot\gamma(x,y)} \\ &= \lambda_{\mathbf{a}} f_{\mathbf{a}}(\mathbf{b}). \end{aligned}$$

◈

Lemma 5.31
$\{f_{\mathbf{a}} \mid \mathbf{a} \in G\}$ is a linearly independent set.

Proof
We will show that $\langle f_{\mathbf{a}}, f_{\mathbf{b}} \rangle = 0$ if $\mathbf{a} \neq \mathbf{b}$. By Lemma A.16, this will suffice. We compute that

$$\begin{aligned}\langle f_{\mathbf{a}}, f_{\mathbf{b}} \rangle &= \sum_{\mathbf{x} \in G} f_{\mathbf{a}}(\mathbf{x}) \overline{f_{\mathbf{b}}(\mathbf{x})} \\ &= \sum_{\mathbf{x} \in G} \omega^{\mathbf{a} \cdot \mathbf{x}} \overline{\omega^{\mathbf{b} \cdot \mathbf{x}}} \\ &= \sum_{\mathbf{x} \in G} \omega^{\mathbf{a} \cdot \mathbf{x}} \omega^{-\mathbf{b} \cdot \mathbf{x}} \\ &= \sum_{\mathbf{x} \in G} \omega^{(\mathbf{a} - \mathbf{b}) \cdot \mathbf{x}} = 0,\end{aligned}$$

where the last equality is by Lemma 5.29. ◈

Lemma 5.32
Every eigenvalue of W equals $\lambda_{\mathbf{a}}$ for some $\mathbf{a} \in G$.

Proof
By Lemma 5.30, every $f_{\mathbf{a}}$ is an eigenvector of W with eigenvalue $\lambda_{\mathbf{a}}$. Using Lemma 5.31 and the fact that the number of vectors $f_{\mathbf{a}}$ is precisely $|W|$, we see that the vectors $f_{\mathbf{a}}$ form an orthonormal eigenbasis. ◈

Remark 5.33
In Chapter 7, we develop a general technique for finding eigenvalues of Cayley graphs. Lemma 5.32 is a special case of Exercise 4 from Chapter 7.

The following lemma allows us to estimate $\lambda(W)$. For its proof, we assume the reader is familiar with (or at least, we ask the reader to accept) the result from elementary field theory that a nonzero polynomial of degree d with coefficients in $\mathbb{Z}_p$ has no more than d roots in $\mathbb{Z}_p$.

Lemma 5.34
For all $\mathbf{a} \neq \mathbf{0}$, we have $0 \leq \lambda_{\mathbf{a}} \leq 7p$.

Proof
We have

$$\begin{aligned}\lambda_{\mathbf{a}} &= \sum_{\gamma(x,y) \in \Gamma} \omega^{\mathbf{a} \cdot \gamma(x,y)} \\ &= \sum_{x,y \in \mathbb{Z}_p} \omega^{a_0 x + a_1 xy + \cdots + a_7 xy^7} \\ &= \sum_{y \in \mathbb{Z}_p} \sum_{x \in \mathbb{Z}_p} \omega^{(a_0 + a_1 y + \cdots + a_7 y^7) x}.\end{aligned}$$

Define the polynomial $g(t) = a_0 + a_1 t + \cdots + a_7 t^7$. For any fixed $y \in \mathbb{Z}_p$, by Lemma 1.51 we have

$$\sum_{x \in \mathbb{Z}_p} \omega^{g(y)x} = \begin{cases} p & \text{if } g(y) = 0 \\ 0 & \text{if } g(y) \neq 0. \end{cases}$$

Hence $\lambda_{\mathbf{a}} = np$, where n is the number of roots of g in $\mathbb{Z}_p$. Since $\mathbf{a} \neq \mathbf{0}$, we have that g is a nonzero polynomial of degree less than or equal to 7, and so $n \leq 7$. Ⓐ

The following lemma enumerates the properties of W we need.

Lemma 5.35
Let W be as Equation 16. Then

1. $|W| = d_W^4$, and
2. W is nonbipartite, and
3. $\lambda(W) < \frac{d_W}{5} = \frac{p^2}{5}$.

Proof

1. $|W| = |G| = p^8$, and $d_W = |\Gamma| = p^2$.
2. By Lemma 5.32 and Lemma 5.34, we know that $-p^2$ is not an eigenvalue of W.
3. By Lemma 5.32, Lemma 5.34, and the fact that $p > 35$, we have

$$\lambda(W) \leq 7p < \frac{p^2}{5}.$$ Ⓐ

At long last, we can define an actual expander family.

Definition 5.36 Let W be as in Equation 16. Recursively define a sequence (W_n) as follows.

$$W_1 = W^2, \text{ and } W_{n+1} = W_n^2 \textcircled{z} W,$$

where $W_n^2 \textcircled{z} W$ is formed using any choice of labeling from W to W_n^2.

Remark 5.37
It is straightforward to show by induction, using Prop. 5.10, that $\deg(W_n^2) = p^8 = |W|$ for all positive integers n, so it is legal to form the zig-zag product $W_n^2 \textcircled{z} W$.

Theorem 5.38
Let (W_n) be as in Def. 5.36. Then (W_n) is an expander family.

Proof
First, note that by Prop. 5.10, we have that $\deg(W_n) = p^4$ for all n. Moreover, by induction we have $|W_n| = p^{8n}$, so $|W_n| \to \infty$.

Now we show by induction that $\lambda(W_n) < 2p^4/5$ for all n. This will imply that the spectral gap of W_n is at least $3p^4/5$ for all n, so by Corollary 1.87, this suffices to show that (W_n) is an expander family.

Also, W_n^2 is nonbipartite, because the square of any graph with an edge will necessarily contain a loop.

By Prop. 1.100, Lemma 5.35, and the fact that $p > 35$, we have

$$\lambda(W^2) < \frac{p^4}{25} \leq \frac{2p^4}{5}.$$

This establishes the base case. For the inductive step, assume $\lambda(W_n) < 2p^4/5$. By Lemma 5.35, we know W is nonbipartite. By Lemma 5.35, Prop. 1.100, and Theorem 5.26, we have

$$\begin{aligned}\lambda(W_n^2 \textcircled{z} W) &\leq \frac{p^4\lambda(W_n^2)}{p^8} + p^2\lambda(W) + \lambda(W)^2 \\ &< \frac{4p^4}{25} + p^2 \cdot \frac{p^2}{5} + \frac{p^4}{25} = \frac{2p^4}{5}.\end{aligned}$$

5. ZIG-ZAG PRODUCTS AND SEMIDIRECT PRODUCTS

Under certain circumstances, the zig-zag product of a Cayley graph on a group G and a Cayley graph on a group K equals a Cayley graph on the semidirect product $G \rtimes K$. (Recall the definition of semidirect product from Chapter 4.) In this section, we spell out the precise correspondence between zig-zag products of graphs and semidirect products of groups.

Let G, K be finite groups. Let $\theta : K \to \text{Aut}(G)$ be a homomorphism, and denote $[\theta(k)](g)$ by ${}^k g$. Let $\gamma \in G$ such that $\gamma^2 = e_G$, where e_G is the identity element of G. Let $\Gamma = \{{}^k\gamma \ : \ k \in K\}$. (In other words, Γ is the orbit of γ under the action of K.) Note that each element of Γ equals its own inverse; hence, Γ is a symmetric. Let Λ be a symmetric subset of K. Construct the Cayley graphs $X = \text{Cay}(G, \Gamma)$ and $Y = \text{Cay}(K, \Lambda)$. For each $g \in G$—that is, for each vertex in $\text{Cay}(G, \Gamma)$—let E_g be as in Def. 5.2. Define $L_g : K \to E_g$ by taking $L_g(k)$ to be the edge between g and $g({}^k\gamma)$. Let L be the labeling defined by the maps L_g.

Proposition 5.39
$\text{Cay}(G, \Gamma) \textcircled{z}_L \text{Cay}(H, \Lambda) = \text{Cay}(G \rtimes_\theta H, \Upsilon)$, where Υ is the multiset $\{\sigma_1\gamma\sigma_2 \mid (\sigma_1, \sigma_2) \in \Lambda \times \Lambda\}$.

Proof
Let Z, H be as in Section 2. We first claim that the zig graph Z equals the Cayley graph $\text{Cay}(G \rtimes K, \Lambda)$. Each has $G \times K$ as its vertex set. Moreover, in both the multiplicity of the edge between (g_1, k_1) and (g_2, k_2) equals the multiplicity of $k_1^{-1}k_2$ as an element of Λ.

Next, we claim that the hyphen graph H equals the Cayley graph $\text{Cay}(G \rtimes K, \{\gamma\})$. Each has $G \times K$ as its vertex set. Note that $\gamma^2 = e_G$ implies that $\{\gamma\}$

is symmetric. Both H and $\mathrm{Cay}(G \rtimes K, \{\gamma\})$ are 1-regular. Two vertices (g_1, k_1) and (g_2, k_2) are adjacent in H

iff $L_{g_1}(k_1) = L_{g_2}(k_2)$
iff the edge in $\mathrm{Cay}(G, \Gamma)$ between g_1 and $g_1({}^{k_1}\gamma)$ equals the edge between g_2 and $g_2({}^{k_2}\gamma)$
iff $g_1 = g_2({}^{k_2}\gamma)$ and $g_2 = g_1({}^{k_1}\gamma)$
iff $k_1 = k_2$ and $g_1 = g_2({}^{k_2}\gamma)$ (since the *multiset* Γ is indexed by K, so if k_1, k_2 are distinct, then we regard ${}^{k_1}\gamma$ and ${}^{k_2}\gamma$ as distinct)
iff $(g_1, k_1) = (g_2, k_2) \cdot \gamma$ in $G \rtimes K$
iff (g_1, k_1) and (g_2, k_2) are adjacent in $\mathrm{Cay}(G \rtimes K, \{\gamma\})$.

The result follows from Prop. 5.12 and Prop. 1.102. ◬

Example 5.40
Let $G = \mathbb{Z}_2 \times \mathbb{Z}_2$. Let $\gamma = (1, 0) \in G$. Note that γ has order 2. Let $K = \mathbb{Z}_3 = \{\overline{0}, \overline{1}, \overline{2}\}$. (To distinguish elements of G from elements of K we use overscores for the latter.) Define $\phi : G \to G$ by $\phi(a, b) = (b, a + b)$. Note that $\phi \in \mathrm{Aut}(G)$. Also note that ϕ has order 3 in $\mathrm{Aut}(G)$. Hence there is a unique, well-defined homomorphism $\theta : K \to \mathrm{Aut}(G)$ with $\theta(\overline{1}) = \phi$. We compute ${}^{\overline{0}}\gamma = (1, 0)$, ${}^{\overline{1}}\gamma = (0, 1)$, and $\gamma^{\overline{2}} = (1, 1)$. Let Γ be the orbit of γ under the action of K—that is, $\Gamma = \{(1, 0), (0, 1), (1, 1)\}$. Let $\Lambda = \{\overline{1}, \overline{2}\} \subset K$.

The conditions of Prop. 5.39 are now met. Figures 5.12 and 5.13 show the graphs $X = \mathrm{Cay}(G, \Gamma)$ and $Y = \mathrm{Cay}(H, \Lambda)$. Labelings at each vertex of X are given by L. Edge labelings at each vertex of Y are given by the corresponding elements of Λ. Let $\Upsilon = \{\overline{1} \cdot (1, 0) \cdot \overline{1}, \overline{1} \cdot (1, 0) \cdot \overline{2}, \overline{2} \cdot (1, 0) \cdot \overline{1}, \overline{2} \cdot (1, 0) \cdot \overline{2}\}$.

We begin at the vertex $(0, 0)\overline{0}$ of $X \textcircled{z}_L Y$. For the zig, choose the edge of Y labeled $\overline{1}$ at $\overline{0}$. So we move to $(0, 0)\overline{1} = [(0, 0)\overline{0}] \cdot \overline{1}$. At step 2, we are required to move to $(0, 1)\overline{1} = (0, 0)(0, 1)\overline{1} = [(0, 0)\overline{1}] \cdot (1, 0)$. For the zag, choose the edge of Y labeled $\overline{1}$ at $\overline{1}$. So we move to $(0, 1)\overline{2} = [(0, 1)\overline{1}] \cdot \overline{1}$.

So the vertex $(0, 0)\overline{0}$ is adjacent, via the pair $(\overline{1}, \overline{1})$, to the vertex $(0, 1)\overline{2}$. Alternatively, regarding $X \textcircled{z}_L Y$ as the Cayley graph $\mathrm{Cay}(G \rtimes K, \Upsilon)$, we have that $(0, 0)\overline{0}$ is adjacent to $(0, 1)\overline{2}$ via the element $\overline{1} \cdot (1, 0) \cdot \overline{1} \in \Upsilon$, since $(0, 1)\overline{2} = [(0, 0)\overline{0}] \cdot (\overline{1} \cdot (1, 0) \cdot \overline{1})$.

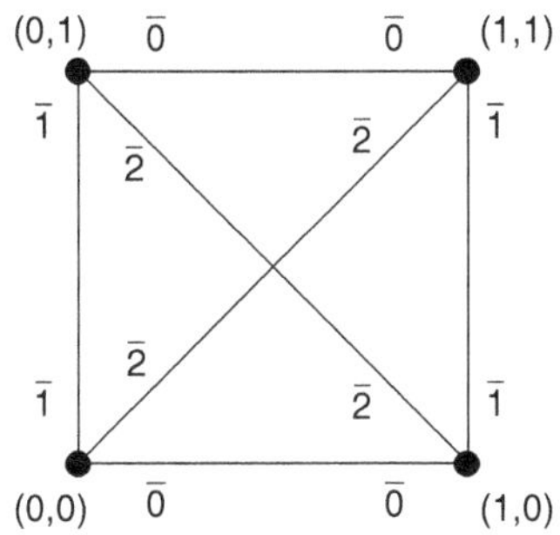

Figure 5.12 X

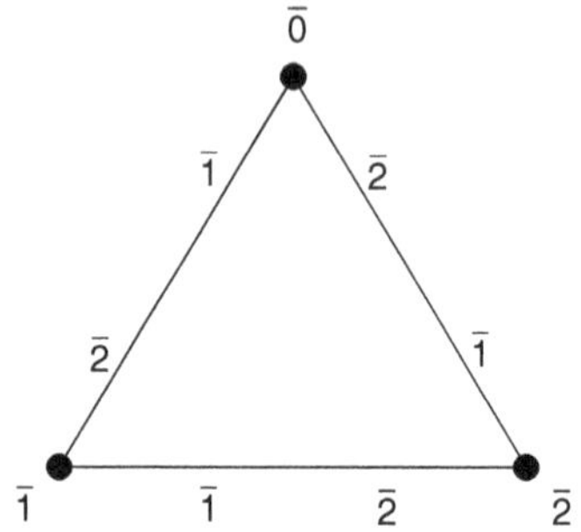

Figure 5.13 Y

Remark 5.41
We have dealt only with the cases of orbits of elements of order 1 or 2. Prop. 5.39 holds more generally, though, for orbits of arbitrary elements. To do so, one must first define zig-zag products of directed graphs. This is not particularly difficult—we chose not to do so only for the sake of keeping the presentation concise—and we invite the reader to work through this more general construction.

Using Remark 5.11, we find that Prop. 5.39 then continues to hold when the generating set is an arbitrary union of orbits.

NOTES

1. In [116], Reingold, Vadhan, and Wigderson prove Theorem 5.26 but also prove a stronger upper bound for $\lambda(X \textcircled{z} Y)$. They define two other related graph products, namely, the "modified zig-zag product" and the "(balanced) replacement product." In addition, they discuss the "explicitness level" (i.e., computation time) of their constructions.
2. Prop. 5.39 is proved in [6]. The papers [120] and [96] exploit the connection between zig-zag products and semidirect products to produce iterative constructions of expander families (though in the second paper, not of constant degree). In [90], Lubotzky and Weiss pose the following question: if a sequence of finite groups (G_n) yields an expander family, then is $(\mathrm{Cay}(G_n, \Gamma_n))$ an expander family for *any* constant-degree sequence of generating sets (Γ_n)? A counterexample, also exploiting Prop. 5.39, is discussed in [6].
3. In [9], Alon, Schwartz, and Shapira give another iterative construction of expander families, using the "(balanced) replacement product" of [116]. What is noteworthy is that their proof is purely combinatorial, dealing directly with the isoperimetric constant instead of with eigenvalues.

EXERCISES

1. Let Y be a regular graph. Find a graph X such that $X \textcircled{z} Y = Y^2$.
2. Let n be a positive integer. Recall the complete graph K_{n+1} and the cycle graph C_n from Examples 1.52 and 1.53. Give the vertices of C_n the labels $1, \ldots, n$ in the obvious way. For any vertex v of K_{n+1}, define the labeling $L_v : E_v \to \{1, \ldots, n\}$

at v to be the function that maps the edge between v and $v + b$ to b. Let X_n be the zig-zag product $K_{n+1} \textcircled{z} C_n$ with respect to these labeling. Prove that (X_n) is not an expander family.

3. Let X be a regular graph with degree d, and let Y be a connected nonbipartite regular graph with d vertices. Let p, q be two points in the same cloud in $X \textcircled{z} Y$. Show that there exists a walk in $X \textcircled{z} Y$ from p to q. (Hint: Use Exercise (6) of Chapter 1.)
4. Prove that if in Exercise 3 we furthermore assume that X is connected, then $X \textcircled{z} Y$ is connected.
5. This exercise demonstrates that whether a zig-zag product is connected may depend on the choice of labeling. Recall that K_5 is a complete graph with five vertices, and C_4 is a cycle graph with four vertices.

   (a) Find a choice of labeling from C_4 to K_5 such that $K_5 \textcircled{z} C_4$ is not connected. (Hint: Regard K_5 and C_4 as Cayley graphs on $\mathbb{Z}_5$ and $\mathbb{Z}_4$, respectively, and then use Prop. 5.39.)
   (b) Find a choice of labeling from C_4 to K_5 such that $K_5 \textcircled{z} C_4$ is connected.

6. Draw $\text{Cay}(G, \Gamma) \textcircled{z}_L \text{Cay}(H, \Lambda)$ from Example 5.40 in two different ways. First use Def. 5.2. Then draw $\text{Cay}(G \rtimes_\theta H, \Upsilon)$, with Υ as in Prop. 5.39. By the time you have completed this exercise, you should feel that Prop. 5.39 is obvious.
7. Prove Lemma 5.24.

STUDENT RESEARCH PROJECT IDEAS

1. Develop some recognition algorithms for zig-zag products. That is, given a graph Z, how can one determine whether Z is of the form $X \textcircled{z} Y$?
2. Theorem 5.26 provides some information about $\lambda(X \textcircled{z} Y)$ in terms of $\lambda(X)$ and $\lambda(Y)$. Investigate the effect of zig-zag products on some other standard graph invariant, such as the chromatic number, diameter, girth, and so on.
3. Prop. 5.39 gives a condition under which the zig-zag product of two Cayley graphs is again a Cayley graph. Are there other such conditions? For example, are the graphs (W_n) of Def. 5.36 Cayley graphs?
4. See Exercise 2. Do there exist labelings from C_n to K_{n+1} such that $(K_{n+1} \textcircled{z} C_n)$ has logarithmic diameter? ... is an expander family?

PART THREE

Representation-Theoretic Techniques

6

Representations of Finite Groups

A linear representation of a finite group G provides a way to view the group elements as linear transformations. We can express the adjacency matrix A of a Cayley graph in terms of a certain representation called the regular representation. This observation gives us access to a whole slew of theorems and techniques from the very well-developed theory of representations. In particular, we can decompose the large matrix A into little pieces, each of which corresponds to an irreducible representation of G. The irreducible representations of many families of finite groups (e.g., abelian groups, dihedral groups, symmetric groups, simple groups of Lie type) have been extensively studied. In this way, we can study the action of A on $L^2(G)$ and, in some cases, approximate the eigenvalues of A.

Throughout this chapter we use the notations and theorems from linear algebra given in Appendix A. The reader may wish to skim this appendix to review this material.

1. REPRESENTATIONS OF FINITE GROUPS

Groups act on a variety of sets. When the set has the added structure of a vector space, and the action of the group preserves this structure (i.e., the group elements act as linear transformations of the vector space to itself), then one can derive properties of the group from this action. A representation of a finite group G is a homomorphism from the group G into the group of invertible linear transformations $GL(V)$ of a vector space V to itself. In this section, we establish the basic terminology to talk about representations.

Definition 6.1 Let V be a vector space. The *general linear group of* V is

$$GL(V) = \{L : V \to V \mid L \text{ is a bijective linear transformation}\}.$$

Remark 6.2
Note that $GL(V)$ is a group where the group operation is composition of functions. The identity element of $GL(V)$ is the function $I : V \to V$ such that $I(v) = v$ for all $v \in V$. Given $L \in GL(V)$, the inverse of L is the function L^{-1}.

Example 6.3
Let $V = \mathbb{C}, L_1 : \mathbb{C} \to \mathbb{C}$ be given by $L_1(z) = iz$, and $L_2 : \mathbb{C} \to \mathbb{C}$ be given by $L_2(z) = -iz$. Then $L_1, L_2 \in GL(\mathbb{C})$. Furthermore, $(L_1 \circ L_2)(z) = L_1(-iz) = i(-iz) = z$. Similarly, $(L_2 \circ L_1)(z) = z$. Hence, $L_1^{-1} = L_2$.

Definition 6.4 Let G be a finite group. A *representation* of G is a group homomorphism $\rho : G \to GL(V)$ where V is a finite-dimensional vector space over $\mathbb{C}$. We define the *degree* of ρ to be the dimension of V as a vector space over $\mathbb{C}$.

Remark 6.5
If $\rho : G \to GL(V)$ is a representation of G, then $\rho(gh) = \rho(g) \circ \rho(h)$ for all $g, h \in G$. If e_G is the identity element of G, then $\rho(e_G) = I$ where I is the identity map given in Remark 6.2.

Remark 6.6
In some texts, the type of representation defined above is called a *linear representation* of G.

Remark 6.7
Technically, a representation of a group consists of a pair (V, ρ) where $\rho : G \to GL(V)$ is a group homomorphism. However, we will be sloppy and call ρ a representation of G, or even V if the map ρ is understood.

Remark 6.8
Let e_G be the identity of G. A representation $\rho : G \to GL(V)$ induces an action of G on V given by $g \cdot \mathbf{v} = \rho(g)(\mathbf{v})$ satisfying

1. $g \cdot (\alpha\mathbf{v} + \beta\mathbf{w}) = \alpha(g \cdot \mathbf{v}) + \beta(g \cdot \mathbf{w})$
2. $(gh) \cdot \mathbf{v} = g \cdot (h \cdot \mathbf{v})$
3. $e_G \cdot \mathbf{v} = \mathbf{v}$

for all $\mathbf{v}, \mathbf{w} \in V$, $g, h \in G$, and $\alpha, \beta \in \mathbb{C}$. Part (1) follows because for each $g \in G$, we have that $\rho(g)$ is a linear transformation of V. Parts (2) and (3) follow because ρ is a homomorphism (see Remark 6.5). Conversely, note that any action of G on V satisfying (1), (2), and (3) provides a representation of G.

In the future, we may drop the ρ and just write $g \cdot \mathbf{v}$ for $\rho(g)(\mathbf{v})$ if the representation is understood. We may also drop the dot from $g \cdot \mathbf{v}$ and just write $g\mathbf{v}$.

Definition 6.9 Let $\rho : G \to GL(V)$ be a representation of a group G. If in addition V has an inner product $\langle \cdot, \cdot \rangle$ such that

$$\left\langle \rho(g)\mathbf{v}, \rho(g)\mathbf{w} \right\rangle = \langle \mathbf{v}, \mathbf{w} \rangle$$

for all $g \in G$ and $\mathbf{v}, \mathbf{w} \in V$, then we say that ρ is a *unitary* representation with respect to $\langle \cdot, \cdot \rangle$. If we think of ρ as defining an action of G on V as in Remark 6.8, then we say that $\langle \cdot, \cdot \rangle$ is *G-invariant* if $\left\langle g \cdot \mathbf{v}, g \cdot \mathbf{w} \right\rangle = \langle \mathbf{v}, \mathbf{w} \rangle$ for all $g \in G$ and $\mathbf{v}, \mathbf{w} \in V$.

Example 6.10
Let G be a finite group. Let $\phi : G \to GL(\mathbb{C})$ be given by $\phi(g)\alpha = \alpha$ for all $g \in G$ and $\alpha \in \mathbb{C}$. We call ϕ the *trivial representation* of G. Since the dimension of $\mathbb{C}$ as a vector space over the complex numbers is 1, the degree of the trivial representation is equal to 1. Recall the standard inner product on $\mathbb{C}$ given by $\langle \alpha, \beta \rangle_2 = \alpha\overline{\beta}$ for $\alpha, \beta \in \mathbb{C}$. Since

$$\langle \phi(g)\alpha, \phi(g)\beta \rangle_2 = \langle \alpha, \beta \rangle_2$$

for any $\alpha, \beta \in \mathbb{C}$ and $g \in G$, we see that ϕ is a unitary representation with respect to $\langle \cdot, \cdot \rangle_2$.

Remark 6.11
Alternatively, we could have first defined the *trivial action* of G on $\mathbb{C}$ by $g \cdot \alpha = \alpha$ for all $g \in G$ and $\alpha \in \mathbb{C}$. As outlined, the trivial action gives rise to the trivial representation and vice versa.

Example 6.12
Let $a \in \mathbb{Z}$ and $n \geq 2$ be a positive integer. Suppose that $k_1 \equiv k_2 (\mathrm{mod}\, n)$. Then $k_1 = k_2 + nl$ for some integer l. Hence

$$\exp\left(\frac{2\pi i a k_1}{n}\right) = \exp\left(\frac{2\pi i a k_2}{n}\right) \exp(2\pi i a l) = \exp\left(\frac{2\pi i a k_2}{n}\right). \qquad (17)$$

Let $\rho_a : \mathbb{Z}_n \to GL(\mathbb{C})$ where $\rho_a(k)z = e^{\frac{2\pi i a k}{n}} z$. Then ρ_a is well defined by Equation 17. It is easy to see that ρ_a is a representation of $\mathbb{Z}_n$ of degree 1. Note that $\rho_a = \rho_b$ if and only if $a \equiv b (\mathrm{mod}\, n)$. Hence the formula gives us n representations of $\mathbb{Z}_n$. These are $\rho_0, \rho_1, \ldots, \rho_{n-1}$. Later we see that these are all of the representations of $\mathbb{Z}_n$ of degree 1.

If $\alpha, \beta \in \mathbb{C}$, then

$$\langle \rho_a(k)\alpha, \rho_a(k)\beta \rangle_2 = e^{\frac{2\pi i a k}{n}} \alpha e^{\frac{-2\pi i a k}{n}} \overline{\beta} = \alpha\overline{\beta} = \langle \alpha, \beta \rangle_2 .$$

Hence, ρ_a is a unitary representation with respect to the standard inner product on $\mathbb{C}$.

Suppose that $\theta \in \mathbb{R}$. Then $e^{i\theta}z$ is obtained by rotating z counterclockwise by the angle θ on the circle of radius $|z|$ centered at 0. See Figure 6.1. Hence, $\rho_a(k)$ rotates complex numbers about 0 by the angle $2\pi\, ak/n$.

Consider the case $n = 4$. Then $\rho_1(k)z = e^{k\pi i/2}z$. Hence $\rho_1(k)$ rotates the complex number z by $\pi k/2$ degrees. For example, $\rho_1(1)(1 + i) = -1 + i$.

Example 6.13
Let G be a finite group. Recall that

$$L^2(G) = \{f : G \to \mathbb{C}\}.$$

The complex vector space $L^2(G)$ can be made into a representation of G via the *right regular representation*, $R : G \to GL(L^2(G))$, defined by $(R(\gamma)f)(g) = f(g\gamma)$ for all $f \in L^2(G)$ and $\gamma, g \in G$.

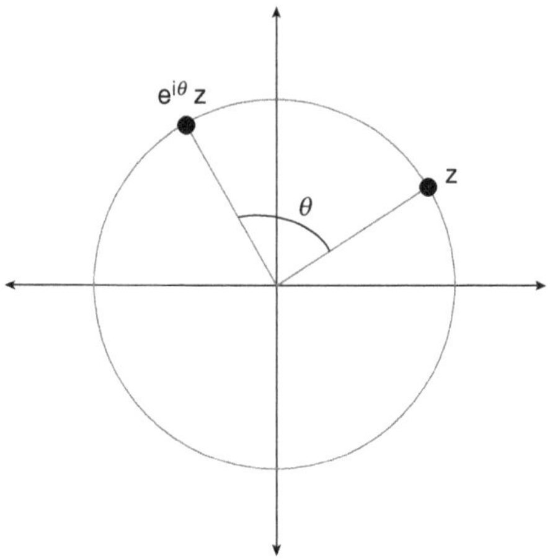

Figure 6.1

As an example, consider the function $f \in L^2(\mathbb{Z}_4)$ given by

$$f(n) = \begin{cases} 1 & \text{if } n = 0 \\ 0 & \text{if } n = 1 \\ 1 - 15i & \text{if } n = 2 \\ -2/3 & \text{if } n = 3 \end{cases}.$$

Then $(R(1)f)(n) = f(n+1)$. Hence,

$$(R(1)f)(n) = \begin{cases} 0 & \text{if } n = 0 \\ 1 - 15i & \text{if } n = 1 \\ -2/3 & \text{if } n = 2 \\ 1 & \text{if } n = 3 \end{cases}.$$

The reader should compute the functions $R(0)f$, $R(2)f$, and $R(3)f$.

We return to the general case of an arbitrary group G. Let us show that R is a homomorphism. Let $\gamma, \gamma', g \in G$ and $f \in L^2(G)$. Then

$$(R(\gamma\gamma')f)(g) = f(g\gamma\gamma')$$

while

$$(R(\gamma)(R(\gamma')f))(g) = (R(\gamma')f)(g\gamma) = f(g\gamma\gamma').$$

Therefore, $R(\gamma\gamma') = R(\gamma)R(\gamma')$.

If $f, h \in L^2(G)$ and $\gamma \in G$, then

$$\begin{aligned}\langle R(\gamma)f, R(\gamma)h\rangle_2 &= \sum_{g\in G}(R(\gamma)f)(g)\overline{(R(\gamma)h)(g)}\\ &= \sum_{g\in G} f(g\gamma)\overline{h(g\gamma)}\\ &= \sum_{x\in G} f(x)\overline{h(x)}\\ &= \langle f, h\rangle_2 .\end{aligned}$$

Therefore, R is unitary with respect to the standard inner product on $L^2(G)$.

Recall from Remark 1.33 that the dimension of $L^2(G)$ is $|G|$. Thus, the degree of R is $|G|$.

Definition 6.14 The *general linear group*, denoted by $GL(n, \mathbb{C})$, is the group of all invertible $n \times n$ matrices over the complex numbers. The group operation is multiplication of matrices.

Definition 6.15 Let G be a finite group. A *matrix representation* of G is a group homomorphism $\pi : G \to GL(n, \mathbb{C})$. The *degree* of π is n.

Definition 6.16 Recall the definition of a unitary matrix given in Definition A.32. We say that $\pi : G \to GL(n, \mathbb{C})$ is a *unitary matrix representation* of G if $\pi(g)$ is a unitary matrix for all $g \in G$.

Example 6.17
Let $a \in \mathbb{Z}$ and $n \geq 2$ be a positive integer. Consider the matrix representation $\phi_a : \mathbb{Z}_n \to GL(1, \mathbb{C})$ given by $\phi_a(k) = \left(e^{\frac{2\pi iak}{n}}\right)$. By Equation 17 ϕ_a is well defined. Note that $\phi_a = \phi_b$ iff $a \equiv b(\mathrm{mod}\, n)$.

Because $\exp\left(\frac{2\pi iak}{n}\right)\exp\left(\frac{-2\pi iak}{n}\right) = 1$, we have that $\phi_a(k)$ is a unitary matrix for all $a \in \mathbb{Z}$ and $k \in \mathbb{Z}_n$. Hence, ϕ_a is a unitary representation of $\mathbb{Z}_n$ of degree 1.

As an example, let $n = 4$. Then

$$\phi_0(n) = \begin{cases} (1) & \text{if } n = 0 \\ (1) & \text{if } n = 1 \\ (1) & \text{if } n = 2 \\ (1) & \text{if } n = 3 \end{cases}, \qquad \phi_1(n) = \begin{cases} (1) & \text{if } n = 0 \\ (i) & \text{if } n = 1 \\ (-1) & \text{if } n = 2 \\ (-i) & \text{if } n = 3 \end{cases},$$

$$\phi_2(n) = \begin{cases} (1) & \text{if } n = 0 \\ (-1) & \text{if } n = 1 \\ (1) & \text{if } n = 2 \\ (-1) & \text{if } n = 3 \end{cases}, \quad \text{and} \quad \phi_4(n) = \begin{cases} (1) & \text{if } n = 0 \\ (-i) & \text{if } n = 1 \\ (-1) & \text{if } n = 2 \\ (i) & \text{if } n = 3 \end{cases}.$$

Example 6.18
Consider the symmetric group

$$S_3 = \{(), (1,2), (2,3), (1,3), (1,2,3), (1,3,2)\}.$$

Let $\xi = e^{2\pi i/3}$. Define a unitary representation π of S_3 as follows.

$$\pi(()) = \begin{pmatrix} 1 & 0 \\ 0 & 1 \end{pmatrix}, \qquad \pi((1,2)) = \begin{pmatrix} 0 & \xi^2 \\ \xi & 0 \end{pmatrix}, \qquad \pi((1,3)) = \begin{pmatrix} 0 & \xi \\ \xi^2 & 0 \end{pmatrix},$$

$$\pi((2,3)) = \begin{pmatrix} 0 & 1 \\ 1 & 0 \end{pmatrix}, \qquad \pi((1,2,3)) = \begin{pmatrix} \xi^2 & 0 \\ 0 & \xi \end{pmatrix}, \qquad \pi((1,3,2)) = \begin{pmatrix} \xi & 0 \\ 0 & \xi^2 \end{pmatrix}.$$

The reader may wish to verify that π is in fact a representation of S_3 (or wait until Example 6.41). To do this, check that $\pi(\sigma_1\sigma_2) = \pi(\sigma_1)\pi(\sigma_2)$ for all $\sigma_1, \sigma_2 \in S_3$. Note that π has degree 2.

In Example 6.41 we see that this matrix representation comes from the permutation representation of S_3.

Example 6.19
Recall that a transposition of S_n is an element of the form (a, b). Let $\sigma \in S_n$. Recall that σ is called an even (respectively odd) permutation if it can be written as a product of an even (respectively odd) number of transpositions.

Consider the homomorphism $\text{sgn} : S_n \to \{-1, 1\}$ where

$$\text{sgn}(\sigma) = \begin{cases} 1 & \text{if } \sigma \text{ is an even permutation} \\ -1 & \text{if } \sigma \text{ is an odd permutation} \end{cases}.$$

The unitary matrix representation $\pi : S_n \to GL(1, \mathbb{C})$ given by $\pi(\sigma) = (\text{sgn}(\sigma))$ is called the *alternating representation* of S_n. Note that π has degree 1.

For example, consider the alternating representation π when $n = 3$. We see that $() = (1,2)(1,2)$, $(1,2,3) = (1,3)(1,2)$, and $(1,3,2) = (1,2)(1,3)$ are even, whereas $(1,2)$, $(1,3)$, and $(2,3)$ are odd. Hence, $\pi(()) = \pi((1,2,3)) = \pi((1,3,2)) = (1)$ and $\pi((1,2)) = \pi((1,3)) = \pi((2,3)) = (-1)$.

Recall the definition of an ordered basis and the definition of a linear transformation with respect to a basis. These are given in Appendix A.

Remark 6.20
Let G be a group. One can go back and forth between the two different definitions of representations as follows.

Let $\pi : G \to GL(n, \mathbb{C})$ be a matrix representation of G of degree n. Then the map $\rho : G \to GL(\mathbb{C}^n)$ defined by $\rho(g)(\mathbf{v}) = \pi(g)\mathbf{v}$ is a representation of G of degree n.

Let $\rho : G \to GL(V)$ be a representation of degree n. Fix an ordered basis β for V. We can define a matrix representation $\pi : G \to GL(n, \mathbb{C})$

by $\pi(g) = [\rho(g)]_\beta$. The degree of π is n. Note that π is a homomorphism by Proposition A.47.

Suppose we chose a different basis, say β'. Then, by Proposition A.44, there exists an invertible matrix Q such that $[\rho(g)]_{\beta'} = Q^{-1}[\rho(g)]_\beta Q$. That is, $[\rho(g)]_\beta$ and $[\rho(g)]_{\beta'}$ are similar matrices.

Definition 6.21 Let $\rho : G \to GL(V)$ and $\phi : G \to GL(W)$ be two representations of G. If $\theta : V \to W$ is a map such that $\theta(\rho(g)\mathbf{v}) = \phi(g)\theta(\mathbf{v})$ for all $g \in G$ and $\mathbf{v} \in V$, then we say that θ is *G-invariant*. A *G-homomorphism* is a G-invariant linear transformation. If in addition θ is a vector space isomorphism, then we say that V and W are *equivalent* and write $V \cong W$. If V and W have associated maps ρ and ϕ, respectively, we also write $\rho \cong \phi$.

Remark 6.22

Roughly speaking, to say that θ is G-invariant means that θ commutes with the action of G. In other words, it doesn't matter whether we first apply θ and then apply the G-action or if we first apply the G-action and then apply θ. That's the meaning of the equation $\theta(\rho(g)\mathbf{v}) = \phi(g)\theta(\mathbf{v})$.

Example 6.23

Let $a \in \mathbb{Z}$ and $n \geq 2$ be a positive integer. Recall Example 6.12. Let $\rho_a : \mathbb{Z}_n \to GL(\mathbb{C})$ where $\rho_a(m)z = e^{2\pi iam/n}z$.

Recall Definition 6.13. The regular representation R acts on the vector space $L^2(\mathbb{Z}_n)$. Consider the function $\chi_a \in L^2(\mathbb{Z}_n)$ defined by $\chi_a(m) = e^{2\pi iam/n}$. Let

$$W_a = \text{span}\{\chi_a\} = \{\alpha\chi_a \mid \alpha \in \mathbb{C}\}.$$

The reader should verify that W_a is a one-dimensional subspace of $L^2(\mathbb{Z}_n)$ in the sense of vector spaces. Let $\alpha\chi_a \in W_a$. Then

$$(R(m)(\alpha\chi_a))(k) = \alpha\chi_a(k+m) = \chi_a(m)\alpha\chi_a(k).$$

Hence $R(m)\alpha\chi_a = (\chi_a(m)\alpha)\chi_a \in W_a$. Define $\rho : G \to GL(W_a)$ by $\rho(m) = R(m)|_{W_a}$. Using terminology we'll introduce later, we say that ρ is a "subrepresentation" of R.

We now show that $\rho \cong \rho_a$. Let $\theta : \mathbb{C} \to W_a$ be given by $\theta(\alpha) = \alpha\chi_a$. Then θ is a one-to-one and onto linear transformation. Also

$$\begin{aligned}\theta(\rho_a(m)\alpha) &= \theta(e^{\frac{2\pi iam}{n}}\alpha) = e^{\frac{2\pi iam}{n}}\alpha\chi_\alpha = \chi_a(m)\alpha\chi_a \\ &= R(m)(\alpha\chi_a) = \rho(m)(\theta(\alpha)).\end{aligned}$$

Hence $\rho \cong \rho_a$.

Intuitively, here's what's going on. W_a is indexed by the complex number α, and $R(m)$ acts on W_a by multiplying by $\chi_a(m)$. $\mathbb{C}$ is also indexed the complex numbers and $\rho_a(m)$ acts on $\mathbb{C}$ by multiplying by $\chi_a(m)$. Thus ρ and ρ_a are essentially the same representation. They just act on different one-dimensional vector spaces.

Remark 6.24
Let $\rho : G \to GL(V)$ and $\phi : G \to GL(W)$ be two representations of G, and suppose that θ is a G-homomorphism from V to W. Let β be an ordered basis for V and β' be an ordered basis for W. Then

$$[\theta]_{\beta}^{\beta'}\ [\rho(g)]_{\beta}[\mathbf{v}]_{\beta} = [\phi(g)]_{\beta'}\ [\theta]_{\beta}^{\beta'}\ [\mathbf{v}]_{\beta}$$

for all $\mathbf{v} \in V$ and $g \in G$. Thus,

$$[\theta]_{\beta}^{\beta'}\ [\rho(g)]_{\beta} = [\phi(g)]_{\beta'}\ [\theta]_{\beta}^{\beta'}$$

for all $g \in G$. Moreover, if θ is an isomorphism, then

$$[\theta]_{\beta}^{\beta'}\ [\rho(g)]_{\beta}\left([\theta]_{\beta}^{\beta'}\right)^{-1} = [\phi(g)]_{\beta'}. \tag{18}$$

Equation 18, then, shows us what the matrix equivalent of Def. 6.21 should be.

Definition 6.25 Let G be a finite group and $\phi, \pi : G \to GL(n, \mathbb{C})$ be two matrix representations of G. We say that ϕ and π are *equivalent* if there is a matrix $X \in GL(n, \mathbb{C})$ such that $X\phi(g)X^{-1} = \pi(g)$ for all $g \in G$. If this is the case, then we write $\phi \cong \pi$.

Example 6.26
Consider the matrix representations π and ϕ of $\mathbb{Z}_4$ defined by

$$\pi(0) = \begin{pmatrix} 1 & 0 & 0 & 0 \\ 0 & 1 & 0 & 0 \\ 0 & 0 & 1 & 0 \\ 0 & 0 & 0 & 1 \end{pmatrix}, \qquad \pi(1) = \begin{pmatrix} 0 & 1 & 0 & 0 \\ 0 & 0 & 1 & 0 \\ 0 & 0 & 0 & 1 \\ 1 & 0 & 0 & 0 \end{pmatrix},$$

$$\pi(2) = \begin{pmatrix} 0 & 0 & 1 & 0 \\ 0 & 0 & 0 & 1 \\ 1 & 0 & 0 & 0 \\ 0 & 1 & 0 & 0 \end{pmatrix}, \qquad \pi(3) = \begin{pmatrix} 0 & 0 & 0 & 1 \\ 1 & 0 & 0 & 0 \\ 0 & 1 & 0 & 0 \\ 0 & 0 & 1 & 0 \end{pmatrix}$$

and

$$\phi(0) = \begin{pmatrix} 1 & 0 & 0 & 0 \\ 0 & 1 & 0 & 0 \\ 0 & 0 & 1 & 0 \\ 0 & 0 & 0 & 1 \end{pmatrix}, \qquad \phi(1) = \begin{pmatrix} 1 & 0 & 0 & 0 \\ 0 & i & 0 & 0 \\ 0 & 0 & -1 & 0 \\ 0 & 0 & 0 & -i \end{pmatrix},$$

$$\phi(2) = \begin{pmatrix} 1 & 0 & 0 & 0 \\ 0 & -1 & 0 & 0 \\ 0 & 0 & 1 & 0 \\ 0 & 0 & 0 & -1 \end{pmatrix}, \qquad \phi(3) = \begin{pmatrix} 1 & 0 & 0 & 0 \\ 0 & -i & 0 & 0 \\ 0 & 0 & -1 & 0 \\ 0 & 0 & 0 & i \end{pmatrix}.$$

The reader should verify that π and ϕ are representations of $\mathbb{Z}_4$. (Hint: Look at the next example.) Let

$$X = \begin{pmatrix} 1 & 1 & 1 & 1 \\ 1 & i & -1 & -i \\ 1 & -1 & 1 & -1 \\ 1 & -i & -1 & i \end{pmatrix}. \quad \text{Then} \quad X^{-1} = \frac{1}{4}\begin{pmatrix} 1 & 1 & 1 & 1 \\ 1 & -i & -1 & i \\ 1 & -1 & 1 & -1 \\ 1 & i & -1 & -i \end{pmatrix}.$$

The reader should verify that $X\phi(n)X^{-1} = \pi(n)$ for all $n \in \mathbb{Z}_4$. Hence $\pi \cong \phi$.

Example 6.27
Let π, ϕ, and X be as in Example 6.26. Let β be the ordered basis $[\delta_0, \delta_1, \delta_2, \delta_3]$ of $L^2(\mathbb{Z}_4)$ where $\delta_a(b) = 1$ if $a = b$, and $\delta_a(b) = 0$ if $a \neq b$. Recall from Remark 1.33 that β is called the standard basis for $L^2(\mathbb{Z}_4)$. The reader should verify that $\pi(n) = [R(n)]_\beta$ where R is the regular representation of $\mathbb{Z}_4$. As an example to guide the reader, we show that $\pi(1) = [R(1)]_\beta$. Note that $(R(1)\delta_0)(n) = \delta_0(n+1) = \delta_3(n)$. Thus $R(1)\delta_0 = \delta_3$. Similarly, $R(1)\delta_1 = \delta_3$, $R(1)\delta_2 = \delta_1$, and $R(1)\delta_3 = \delta_2$. Hence

$$[R(1)]_\beta = \begin{pmatrix} 0 & 1 & 0 & 0 \\ 0 & 0 & 1 & 0 \\ 0 & 0 & 0 & 1 \\ 1 & 0 & 0 & 0 \end{pmatrix} = \pi(1).$$

Let $\beta' = [\chi_0, \chi_1, \chi_2, \chi_3]$ where $\chi_a(k) = \exp(2\pi iak/4) = i^{ak}$. Using Proposition A.16 and the standard inner product on $L^2(\mathbb{Z}_4)$ given in Definition 1.31, it can be shown that β' is a basis for $L^2(\mathbb{Z}_4)$. The reader should verify that $\phi(n) = [R(n)]_{\beta'}$. For example, we show that $\phi(1) = [R(1)]_{\beta'}$. Note that $(R(1)\chi_0)(n) = \chi_0(n+1) = \chi_0(n)$. Thus, $R(1)\chi_0 = \chi_0$. Note also that

$$(R(1)\chi_1)(n) = \chi_1(n+1) = \begin{cases} i & \text{if } n = 0 \\ -1 & \text{if } n = 1 \\ -i & \text{if } n = 2 \\ 1 & \text{if } n = 3 \end{cases} = i\,\chi_1(n).$$

Thus $R(1)\chi_1 = i\chi_1$. Similarly, $R(1)\chi_2 = -\chi_2$ and $R(1)\chi_3 = -i\chi_3$. Hence

$$[R(1)]_{\beta'} = \begin{pmatrix} 1 & 0 & 0 & 0 \\ 0 & i & 0 & 0 \\ 0 & 0 & -1 & 0 \\ 0 & 0 & 0 & -i \end{pmatrix} = \phi(1).$$

Recall Definition A.42. The matrix X from Example 6.26 is the change-of-basis matrix from β' to β. That is, $X = [I]_{\beta'}^{\beta}$. Thus Proposition A.44 gives that

$$\phi(n) = [R(n)]_{\beta'} = X^{-1}[R(n)]_\beta X = X^{-1}\pi(n)X.$$

Remark 6.28
We show that two different equivalent matrix representations of G can be thought of as arising from the same representation of G. Suppose that $\phi : G \to GL(n, \mathbb{C})$ and $\pi : G \to GL(n, \mathbb{C})$ are two equivalent matrix representations of G. Then there exists a matrix $X \in GL(n, \mathbb{C})$ so that $X\phi(g)X^{-1} = \pi(g)$ for all $g \in G$. Suppose that

$$X = \left(\mathbf{v}_1 \middle| \mathbf{v}_2 \middle| \cdots \middle| \mathbf{v}_n \right),$$

where the vectors $\mathbf{v}_i$ are the columns of X. Because X is invertible, the vectors $\mathbf{v}_i$ form a basis for $\mathbb{C}^n$. Let $\beta = [\mathbf{e}_1, \ldots, \mathbf{e}_n]$ denote the standard ordered basis for $\mathbb{C}^n$ where $\mathbf{e}_i$ is the vector with a one in the ith spot and zeros everywhere else. Let $\beta' = [\mathbf{v}_1, \ldots, \mathbf{v}_n]$. Recall Definition A.42. Note that X is the change-of-basis matrix $[I]^{\beta}_{\beta'}$. Define the representation $\rho : G \to GL(\mathbb{C}^n)$ of G so that $[\rho(g)]_\beta = \pi(g)$ for all $g \in G$. Do this by defining ρ so that $[\rho(g)(\mathbf{e}_i)]_\beta$ is the ith column of $\pi(g)$ and then extending ρ to all of $\mathbb{C}^n$ using the formula

$$\rho(g)(c_1\mathbf{e}_1 + \cdots + c_n\mathbf{e}_n) = c_1\rho(g)(\mathbf{e}_1) + \cdots + c_n\rho(g)(\mathbf{e}_n).$$

Then $\phi(g) = X^{-1}[\rho(g)]_\beta X = [\rho(g)]_{\beta'}$ for all $g \in G$. Hence, ϕ and π arise from from the same representation ρ of G, but each comes about by picking a different basis for $\mathbb{C}^n$.

2. DECOMPOSING REPRESENTATIONS INTO IRREDUCIBLE REPRESENTATIONS

When working with mathematical objects, it is often useful to analyze them in terms of the basic building blocks from which they are composed, much the way a chemist might study how elements come together to form a molecule. The canonical example is the prime numbers, the basic building blocks for the natural numbers under multiplication. For representations, the basic building blocks are so-called irreps, or irreducible representations. The main result of this section is Maschke's theorem, which plays a role similar to that of the existence half of the Fundamental Theorem of Algebra: it states that every representation can be broken down into constituent pieces, each of which is an irrep. Later, in Section 3, we establish the (essential) uniqueness of this decomposition.

Definition 6.29 Let G be a finite group and $\rho : G \to GL(V)$ a representation of G. We say that a subspace W of V is a *G-invariant subspace,* or *subrepresentation of V,* if $\rho(g)(\mathbf{w}) \in W$ for all $g \in G$ and $\mathbf{w} \in W$.

The G-invariant subspaces $\{\mathbf{0}\}$ and V are called the *trivial* subrepresentations of V.

We say that V is *reducible* if it contains a nontrivial G-invariant subspace W. Otherwise, we say that V is *irreducible,* or that V is an *irrep.*

Example 6.30
Let $R : \mathbb{Z}_n \to GL(L^2(\mathbb{Z}_n))$ and W_a be as in Example 6.23. In Example 6.23, it was shown that W_a is a subrepresentation of $L^2(\mathbb{Z}_n)$. Thus the regular representation of $\mathbb{Z}_n$ is reducible.

Remark 6.31
Suppose that $\rho : G \to GL(V)$ is a representation of G and that W is a subrepresentation of V. Recall that $\rho(g)|_W$ denotes the restriction of $\rho(g)$ to the subspace W. Since W is a subrepresentation of V, we have that $\rho(g)|_W : W \to W$. Hence we may define a new representation $\rho_W : G \to GL(W)$ by $\rho_W(g) = \rho(g)|_W$. Note that the degree of ρ_W is equal to the dimension of W as a vector space.

Remark 6.32
Suppose that W and W' are subrepresentations of V and that $V = W \oplus W'$. Let $\mathbf{v} \in V$. Then $v = w + w'$ for some uniquely determined $w \in W$ and $w' \in W'$. Let ρ be the representation of G corresponding to V. Let ρ_W and $\rho_{W'}$ be as in Remark 6.31. Then

$$\rho(g)(\mathbf{v}) = \rho(g)(\mathbf{w}) + \rho(g)(\mathbf{w}') = \rho_W(g)(\mathbf{w}) + \rho_{W'}(g)(\mathbf{w}').$$

Definition 6.33 Under the conditions of Remark 6.32 we say that ρ is the *direct sum* of ρ_W and $\rho_{W'}$, and we write $\rho = \rho_W \oplus \rho_{W'}$.

If n is a positive integer and π is a representation of G, then we write $n\pi$ to mean $\pi \oplus \pi \oplus \cdots \oplus \pi$ where π is repeated n times.

Example 6.34
We define the permutation representation of the symmetric group S_n. Let

$$V_n = \{c_1\mathbf{x}_1 + \cdots + c_n\mathbf{x}_n \mid c_1, \cdots, c_n \in \mathbb{C}\}$$

where the $\mathbf{x}_i$ are thought of as formal vectors. Let $\mathbf{v} = c_1\mathbf{x}_1 + \cdots + c_n\mathbf{x}_n$, $\mathbf{w} = d_1\mathbf{x}_1 + \cdots + d_n\mathbf{x}_n$, and $\alpha \in \mathbb{C}$. Define addition of vectors by the formula

$$\mathbf{v} + \mathbf{w} = (c_1 + d_1)\mathbf{x}_1 + \cdots + (c_n + d_n)\mathbf{x}_n$$

and scalar multiplication by the formula

$$\alpha(c_1\mathbf{x}_1 + \cdots + c_n\mathbf{x}_n) = (\alpha c_1)\mathbf{x}_1 + \cdots + (\alpha c_n)\mathbf{x}_n.$$

Define the standard inner product on V_n as follows.

$$\left\langle c_1\mathbf{x}_1 + \cdots + c_n\mathbf{x}_n, d_1\mathbf{x}_1 + \cdots + d_n\mathbf{x}_n\right\rangle_2 = c_1\overline{d_1} + \cdots + c_n\overline{d_n}.$$

Note that V_n is isomorphic to $\mathbb{C}^n$ as an inner product space.

Define the representation $\rho : S_n \to GL(V_n)$ by the formula

$$\rho(\sigma)(c_1\mathbf{x}_1 + c_2\mathbf{x}_2 + \cdots + c_n\mathbf{x}_n) = c_1\mathbf{x}_{\sigma(1)} + c_2\mathbf{x}_{\sigma(2)} + \cdots + c_n\mathbf{x}_{\sigma(n)},$$

where $\sigma \in S_n$. As an example, consider $\sigma = (1, 2, 3) \in S_3$ and the vector $4\mathbf{x}_1 - 3\mathbf{x}_2 + 0\mathbf{x}_3 \in V_n$. Then $\rho(\sigma)(4\mathbf{x}_1 - 3\mathbf{x}_2 + 0\mathbf{x}_3) = 0\mathbf{x}_1 + 4\mathbf{x}_2 - 3\mathbf{x}_3$.

The representation ρ induces the action $\sigma \cdot \mathbf{v} = \rho(\sigma)(\mathbf{v})$ of S_n on V_n, where $\sigma \in S_n$ and $\mathbf{v} \in V_n$. Later, we sometimes write $\sigma \cdot (4\mathbf{x}_1 - 3\mathbf{x}_2 + 0\mathbf{x}_3) = 0\mathbf{x}_1 + 4\mathbf{x}_2 - 3\mathbf{x}_3$.

Consider the subspace

$$V_{per} = \{c_1\mathbf{x}_1 + \cdots + c_n\mathbf{x}_n \in V_n \mid c_1 + \cdots + c_n = 0\}$$

of V_n. Let $\sigma \in S_n$ and $\mathbf{v} = c_1\mathbf{x}_1 + \cdots + c_n\mathbf{x}_n \in V_{per}$. Because σ is a permutation of $\{1, 2, \ldots, n\}$, we have that summing over the coordinates of the vector $\rho(\sigma)(\mathbf{v})$ gives

$$\sum_{i=1}^{n} c_{\sigma(i)} = \sum_{j=1}^{n} c_j = 0.$$

Hence, $\rho(\sigma)(\mathbf{v})$ is an element of V_{per}. Therefore, V_{per} is an S_n-invariant subspace of V_n.

Note that the map $\rho_{per}(\sigma) = \rho(\sigma)|_{V_{per}}$ is a representation of S_n of degree $n - 1$. The representation ρ_{per} is called the *permutation representation* of S_n. We will frequently (and somewhat sloppily) call the vector space V_{per} the permutation representation.

If $\mathbf{v}_1 = c_1\mathbf{x}_1 + \cdots + c_n\mathbf{x}_n$ and $\mathbf{v}_2 = d_1\mathbf{x}_1 + \cdots + d_n\mathbf{x}_n$ are in V_n, then

$$\langle \sigma \cdot \mathbf{v}_1, \sigma \cdot \mathbf{v}_2 \rangle_2 = \sum_{i=1}^{n} c_{\sigma(i)}\overline{d_{\sigma(i)}} = \sum_{j=1}^{n} c_j\overline{d_j} = \langle \mathbf{v}_1, \mathbf{v}_2 \rangle_2 .$$

Thus, ρ and ρ_{per} are unitary with respect to the standard inner product on V_n and V_{per}. In Example 6.63 we will see that V_{per} is irreducible.

Remark 6.35
In some other books, the phrase "permutation representation" has a different meaning. Namely, it refers to a homomorphim $G \to S_n$.

Example 6.36
Let ρ, ρ_{per}, V_n, and V_{per} be as in Example 6.34. Consider the subspace

$$V_{const} = \{\alpha\mathbf{x}_1 + \cdots + \alpha\mathbf{x}_n \mid \alpha \in \mathbb{C}\}.$$

of V_n. Note that $\rho(\sigma)(\mathbf{v}) = \mathbf{v}$ for all $\mathbf{v} \in V_{const}$. Hence, V_{const} is an S_n-invariant subspace of V_n.

Under the standard inner product on V_n we have that $V_{const}^{\perp} = V_{per}$. Hence, by Proposition A.27 we have that $V_n = V_{const} \oplus V_{per}$. Let ρ_{const} denote the restriction of ρ to V_{const}. Then

$$\rho = \rho_{const} \oplus \rho_{per}.$$

Let $\theta : V_{const} \to \mathbb{C}$ be the linear transformation given by $\theta(\alpha\mathbf{x}_1 + \cdots + \alpha\mathbf{x}_n) = \alpha$. Note that θ is a bijection. Let ϕ be the trivial representation from

Example 6.10. Then θ is an S_n-homomorphism because

$$\begin{aligned}\theta(\rho(\sigma)(\alpha\mathbf{x}_1 + \cdots + \alpha\mathbf{x}_n)) &= \theta(\alpha\mathbf{x}_1 + \cdots + \alpha\mathbf{x}_n) = \alpha \\ &= \phi(\sigma)\theta(\alpha\mathbf{x}_1 + \cdots + \alpha\mathbf{x}_n)\end{aligned}$$

for all $\alpha \in \mathbb{C}$ and $\sigma \in S_n$. Therefore, V_{const} is equivalent to the trivial representation.

Definition 6.37 Let X, Y, and Z be matrices. We say that $X = Y \oplus Z$ if

$$X = \left(\begin{array}{c|c} Y & 0 \\ \hline 0 & Z \end{array}\right),$$

where the two 0's are the appropriately sized zero matrices. To simplify our notation, if X is a matrix, we sometimes write nX to mean $X \oplus X \oplus \cdots \oplus X$, where X is repeated n times.

Given two matrix representations π_1 and π_2 of a group G, define the *direct sum* of π_1 and π_2, denoted $\pi_1 \oplus \pi_2$, to be the matrix representation of G given by $(\pi_1 \oplus \pi_2)(g) = \pi_1(g) \oplus \pi_2(g)$. If n is a positive integer, then $n\pi_1$ means $\pi_1 \oplus \pi_1 \oplus \cdots \oplus \pi_1$ where π_1 is repeated n times.

Let $\phi : G \to GL(n, \mathbb{C})$ be a matrix representation for a finite group G. We say that ϕ is *reducible* if $\phi \cong \pi_1 \oplus \pi_2$ for some matrix representations π_1 and π_2 of G. Otherwise, we say that ϕ is *irreducible*. If ϕ is irreducible, then we sometimes call it an *irrep*.

Example 6.38

$$\begin{pmatrix} 1 & -1 \\ 0 & 5 \end{pmatrix} \oplus \begin{pmatrix} 3 & -3 & 2 \\ 1 & 1 & 1 \\ 5 & 3 & 1 \end{pmatrix} = \left(\begin{array}{cc|ccc} 1 & -1 & 0 & 0 & 0 \\ 0 & 5 & 0 & 0 & 0 \\ \hline 0 & 0 & 3 & -3 & 2 \\ 0 & 0 & 1 & 1 & 1 \\ 0 & 0 & 5 & 3 & 1 \end{array}\right).$$

Example 6.39
Let π and ϕ be as in Example 6.26, and ϕ_0, ϕ_1, ϕ_2, and ϕ_3 be as in Example 6.17. Then π and ϕ are reducible since

$$\begin{aligned}\pi(n) \cong \phi(n) &= \begin{pmatrix} \phi_0(n) & 0 & 0 & 0 \\ 0 & \phi_1(n) & 0 & 0 \\ 0 & 0 & \phi_2(n) & 0 \\ 0 & 0 & 0 & \phi_3(n) \end{pmatrix} \\ &= \phi_0(n) \oplus \phi_1(n) \oplus \phi_2(n) \oplus \phi_3(n).\end{aligned}$$

Remark 6.40
Invariant subspaces allow us to simplify a representation by picking an appropriate basis that puts the representation's matrices, with respect to this basis, into block diagonal form. More precisely, suppose $\rho : G \to GL(V)$

is a representation of a finite group G. Furthermore, suppose that $V = V_1 \oplus V_2 \oplus \cdots \oplus V_n$, where each V_i is a G-invariant subspace of V with respect to ρ. Let β_i be a basis for V_i for $i = 1, \ldots, n$, $\beta = \cup_{i=1}^{n} \beta_i$, and $\rho_i(g) = \rho(g)|_{V_i}$ be the restriction of ρ to V_i. Then

$$[\rho(g)]_\beta = \begin{pmatrix} [\rho_1(g)]_{\beta_1} & 0 & 0 & \cdots & 0 \\ 0 & [\rho_2(g)]_{\beta_2} & 0 & \cdots & 0 \\ 0 & 0 & [\rho_3(g)]_{\beta_3} & \cdots & 0 \\ \vdots & \vdots & \vdots & \ddots & \vdots \\ 0 & 0 & 0 & \cdots & [\rho_n(g)]_{\beta_n} \end{pmatrix}$$
$$= [\rho_1(g)]_{\beta_1} \oplus [\rho_2(g)]_{\beta_2} \oplus \cdots \oplus [\rho_n(g)]_{\beta_n}$$

for every $g \in G$.

Example 6.41
Let V_n, ρ, V_{per}, and ρ_{per} be as in Example 6.34. Let ρ_{const} and V_{const} be as in Example 6.36. Let $\xi = e^{2\pi i/3}$,

$$\mathbf{v}_1 = \mathbf{x}_1 + \mathbf{x}_2 + \mathbf{x}_3,$$
$$\mathbf{v}_2 = \mathbf{x}_1 + \xi \mathbf{x}_2 + \xi^2 \mathbf{x}_3,$$
$$\text{and } \mathbf{v}_3 = \mathbf{x}_1 + \xi^2 \mathbf{x}_2 + \xi \mathbf{x}_3.$$

One may use Proposition A.16 to show that $\beta_1 = [\mathbf{v}_1]$ is an ordered basis for V_{const}, $\beta_2 = [\mathbf{v}_2, \mathbf{v}_3]$ is an ordered basis for V_{per}, and $\beta = \beta_1 \cup \beta_2$ is an ordered basis for V_3. Recall from Example 6.36 that $V_3 = V_{const} \oplus V_{per}$.

Let us compute the matrix $[\rho((1,2))]_\beta$. Recall that we sometimes write $(1,2) \cdot \mathbf{v}$ for $\rho((1,2))(\mathbf{v})$. Note that

$$(1,2) \cdot \mathbf{v}_1 = 1\mathbf{v}_1 + 0\mathbf{v}_2 + 0\mathbf{v}_3$$
$$(1,2) \cdot \mathbf{v}_2 = 0\mathbf{v}_1 + 0\mathbf{v}_2 + \xi \mathbf{v}_3$$
$$(1,2) \cdot \mathbf{v}_3 = 0\mathbf{v}_1 + \xi^2 \mathbf{v}_2 + 0\mathbf{v}_3$$

Hence, $[\rho((1,2))]_\beta = \left(\begin{array}{c|cc} 1 & 0 & 0 \\ \hline 0 & 0 & \xi^2 \\ 0 & \xi & 0 \end{array}\right)$.

Similar computations give us the following matrices.

$$[\rho(())]_\beta = \left(\begin{array}{c|cc} 1 & 0 & 0 \\ \hline 0 & 1 & 0 \\ 0 & 0 & 1 \end{array}\right), \quad [\rho((1,2))]_\beta = \left(\begin{array}{c|cc} 1 & 0 & 0 \\ \hline 0 & 0 & \xi^2 \\ 0 & \xi & 0 \end{array}\right), \quad [\rho((1,3))]_\beta = \left(\begin{array}{c|cc} 1 & 0 & 0 \\ \hline 0 & 0 & \xi \\ 0 & \xi^2 & 0 \end{array}\right),$$

$$[\rho((2,3))]_\beta = \left(\begin{array}{c|cc} 1 & 0 & 0 \\ \hline 0 & 0 & 1 \\ 0 & 1 & 0 \end{array}\right), \quad [\rho((1,2,3))]_\beta = \left(\begin{array}{c|cc} 1 & 0 & 0 \\ \hline 0 & \xi^2 & 0 \\ 0 & 0 & \xi \end{array}\right), \quad [\rho((1,3,2))]_\beta = \left(\begin{array}{c|cc} 1 & 0 & 0 \\ \hline 0 & \xi & 0 \\ 0 & 0 & \xi^2 \end{array}\right).$$

Note that

$$[\rho(\sigma)]_\beta = [\rho_{const}(\sigma)]_\beta \oplus [\rho_{per}(\sigma)]_{\beta'}$$

for all $\sigma \in S_3$. In particular, note that the matrices in Example 6.18 are in fact the matrices $[\rho_{per}(\sigma)]_{\beta'}$.

Before we state the main theorem of this section, we need a lemma and a definition.

Lemma 6.42
Let $\rho : G \to GL(V)$ be a representation of a finite group G. Then there exists a G-invariant inner product on V.

Proof
Let $\beta = [\mathbf{v}_1, \mathbf{v}_2, \cdots, \mathbf{v}_n]$ be any ordered basis for V. Define an inner product on V as follows. Given two vectors

$$\mathbf{v} = a_1\mathbf{v}_1 + a_2\mathbf{v}_2 + \cdots + a_n\mathbf{v}_n \text{ and } \mathbf{w} = b_1\mathbf{v}_1 + b_2\mathbf{v}_2 + \cdots + b_n\mathbf{v}_n,$$

define

$$\langle \mathbf{v}, \mathbf{w} \rangle = a_1\overline{b_1} + a_2\overline{b_2} + \cdots + a_n\overline{b_n}.$$

We leave it as an exercise to show that this is an inner product on V. Unfortunately, ρ may not be unitary with respect to this inner product. We use $\langle \cdot, \cdot \rangle$ to define another inner product on V with respect to which ρ is unitary. For any $\mathbf{v}, \mathbf{w} \in V$ let

$$\langle \mathbf{v}, \mathbf{w} \rangle' = \sum_{g \in G} \langle \rho(g)(\mathbf{v}), \rho(g)(\mathbf{w}) \rangle.$$

We leave it as an exercise to show that ρ is unitary with respect to $\langle \cdot, \cdot \rangle'$. ◈

We conclude with the main result of this section: every unitary representation of a finite group can be completely decomposed into orthogonal irreducible representations.

Recall the definition of an orthogonal direct sum given in Definition A.24.

Theorem 6.43 (Maschke's theorem)
Let V be a unitary representation of a finite group G with a G-invariant inner product $\langle \cdot, \cdot \rangle$. Then there exist irreducible subrepresentations $V_1, \ldots, V_n$ of V such that V equals the orthogonal direct sum

$$V = V_1 \oplus V_2 \oplus \cdots \oplus V_n.$$

Proof
The basic idea of the proof is as follows. If V itself is not irreducible, then decompose it as the direct sum of a subrepresentation and its orthogonal complement. Then iterate this process. We fill in the details of this argument.

We prove the theorem by induction on the dimension of V. If $\dim(V) = 1$, then we are done, for any representation of degree 1 is irreducible. Suppose that $\dim(V) > 1$ and that the theorem holds for all representations of G of smaller dimension. If V is irreducible, then we are done. Otherwise, there is a nontrivial G-invariant subspace W of V. Recall (see Definition A.26) that

$$W^{\perp} = \{\mathbf{v} \in V \mid \langle \mathbf{v}, \mathbf{w} \rangle = 0 \text{ for all } \mathbf{w} \in W\}$$

is a subspace of V. By Proposition A.27, $V = W \oplus W^{\perp}$.

Let $g \in G$, $\mathbf{w} \in W$, and $\mathbf{v} \in W^{\perp}$. Recall Remark 6.8. Because $\langle \cdot, \cdot \rangle$ is G-invariant and $g^{-1} \cdot \mathbf{w} \in W$, we have that

$$\langle g \cdot \mathbf{v}, \mathbf{w} \rangle = \langle g^{-1} \cdot (g \cdot \mathbf{v}), g^{-1} \cdot \mathbf{w} \rangle = \langle \mathbf{v}, g^{-1} \cdot \mathbf{w} \rangle = 0.$$

Hence, $g \cdot \mathbf{v} \in W^{\perp}$. So, $W^{\perp}$ is a G-invariant subspace of V.

Now apply the induction hypothesis to W and $W^{\perp}$ to finish the proof of the proposition. ⊛

Maschke's theorem has the following matrix version.

Corollary 6.44
If $\rho : G \to GL(n, \mathbb{C})$ is a unitary matrix representation for a finite group G, then there is a unitary matrix $X \in GL(n, \mathbb{C})$ such that

$$X^{-1}\rho(g)X = \begin{pmatrix} \rho_1(g) & 0 & 0 & \cdots & 0 \\ 0 & \rho_2(g) & 0 & \cdots & 0 \\ 0 & 0 & \rho_3(g) & \cdots & 0 \\ \vdots & \vdots & \vdots & \ddots & \vdots \\ 0 & 0 & 0 & \cdots & \rho_n(g) \end{pmatrix}$$
$$= \rho_1(g) \oplus \rho_2(g) \oplus \cdots \oplus \rho_n(g)$$

for every $g \in G$, where each ρ_i is an irreducible matrix representation of G.

Proof
See Exercise 1. ⊛

Remark 6.45
Suppose $\rho : G \to GL(V)$ is a unitary representation with respect to some inner product $\langle \cdot, \cdot \rangle$. Let $\beta = [\mathbf{v}_1, \mathbf{v}_2, \ldots, \mathbf{v}_n]$ be an ordered orthonormal basis for V with respect to $\langle \cdot, \cdot \rangle$. The existence of such a basis is guaranteed by

Proposition A.18. Suppose that

$$[\rho(g)]_\beta = \left([\rho(g)\mathbf{v}_1]_\beta \,\middle|\, [\rho(g)\mathbf{v}_2]_\beta \,\middle|\, \cdots \,\middle|\, [\rho(g)\mathbf{v}_n]_\beta \right)$$

$$= \begin{pmatrix} c_{1,1} & c_{1,2} & \cdots & c_{1,n} \\ \vdots & \vdots & \ddots & \vdots \\ c_{n,1} & c_{n,2} & \cdots & c_{n,n} \end{pmatrix}.$$

Because β is an orthonormal basis for V, we have that

$$c_{1,i}\overline{c_{1,j}} + \cdots + c_{n,i}\overline{c_{n,j}} = \left\langle c_{1,i}\mathbf{v_1} + \cdots + c_{n,i}\mathbf{v_n}, c_{1,j}\mathbf{v_1} + \cdots + c_{n,j}\mathbf{v_n} \right\rangle$$

$$= \left\langle \rho(g)\mathbf{v_i}, \rho(g)\mathbf{v_j} \right\rangle = \left\langle \mathbf{v_i}, \mathbf{v_j} \right\rangle$$

$$= \begin{cases} 1 & \text{if } i \neq j \\ 0 & \text{if } i = j \end{cases}.$$

The columns of $[\rho(g)]_\beta$ form an orthonormal basis for $\mathbb{C}^n$. That is, $[\rho(g)]_\beta$ is a unitary matrix for all $g \in G$.

Suppose that $\pi : G \to GL(n, \mathbb{C})$ is a matrix representation of G. Then, by the previous paragraph and Lemma 6.42, there exists a change-of-basis matrix $T \in GL(n, \mathbb{C})$ such that the matrix representation $\phi(g) = T^{-1}\pi(g)T$ is unitary. Thus, any matrix representation of G is equivalent to a unitary matrix representation of G.

3. SCHUR'S LEMMA AND CHARACTERS OF REPRESENTATIONS

In this section, we introduce the character of a representation. We will show that two representations are equivalent if and only if they have the same character. One consequence is that the direct sum decomposition in Maschke's theorem is essentially unique.

Definition 6.46 Let G be a finite group. Let $\pi : G \to GL(n, \mathbb{C})$ be a matrix representation of G. The *character* of π, denoted χ_π, is defined to be $\chi_\pi(g) = \mathrm{tr}(\pi(g))$ for all $g \in G$, where "tr" denotes the trace of a matrix. Let $\rho : G \to GL(V)$ be a representation of G. The character of ρ, denoted χ_ρ, is $\chi(g) = \mathrm{tr}[\rho]_\beta$, where β is any basis for V. We frequently drop the subscript and just write χ when the representation is understood.

Suppose that χ is the character of a representation (or matrix representation) ρ on a finite group G. Let e_G be the identity element of G. The degree of χ is defined to be $\chi(e_G)$.

We say that a character χ is an *irreducible character* if there exists an irreducible representation ρ with $\chi = \chi_\rho$.

Remark 6.47

By Lemma A.61, similar matrices have the same trace. So the character of ρ is well defined, independent of the choice of β.

We will see in Corolary 6.65 that two representations are equivalent if and only if they have the same characters. Hence we could have defined an irreducible character as follows. Let ρ be a representation and χ_ρ be the character of ρ. We say that χ_ρ is irreducible if ρ is irreducible.

Remark 6.48

Suppose that χ is a character of some group G. Then $\chi \in L^2(G)$.

Let G be a group and $g, h \in G$. Recall that g and h are *conjugate* in G if $g = xhx^{-1}$ for some $x \in G$; and that the *conjugacy class* of g is

$$K_g = \{xgx^{-1} | x \in G\}.$$

Example 6.49

In an abelian group G, the conjugacy class of $g \in G$ is $K_g = \{g\}$.

Example 6.50

By Exercise 3, the conjugacy classes of S_3 are

$$K_{()} = \{()\},\ K_{(1,2)} = \{(1,2), (1,3), (2,3)\},\ \text{and}\ K_{(1,2,3)} = \{(1,2,3), (1,3,2)\}.$$

Lemma 6.51

Let G be a finite group, and let ϕ and π be representations (or matrix representations) of G. Furthermore, let χ and ψ be the characters of ϕ and π, respectively. Then

1. If g and h are conjugate in G, then $\chi(g) = \chi(h)$. That is, characters are constant on conjugacy classes.
2. If e_G is the identity of G, then $\chi(e_G)$ equals the degree of ϕ.
3. If $\phi \cong \pi$, then $\chi = \psi$.

Proof

We give the proof where π and ϕ are matrix representations of G, but the same proof works for representations of G by choosing bases for the corresponding vector spaces.

1. Suppose that $g = xhx^{-1}$ for some $x \in G$. Lemma A.61 implies that

$$\chi(g) = \operatorname{tr}(\phi(g)) = \operatorname{tr}(\phi(x)\phi(h)\phi(x)^{-1}) = \operatorname{tr}(\phi(h)) = \chi(h).$$

2. Since ϕ is a homomorphism, $\phi(e_G) = I$, where I is the $n \times n$ identity matrix and n is the degree of ϕ. Hence, $\chi(e_G) = \operatorname{tr}(I) = n$.
3. If $\phi \cong \pi$, then $\phi(g) = X\pi(g)X^{-1}$ for all $g \in G$, where X is some fixed invertible matrix. Thus,

$$\chi(g) = \operatorname{tr}(\phi(g)) = \operatorname{tr}(X\pi(g)X^{-1}) = \operatorname{tr}(\pi(g)) = \psi(g).$$

Ⓐ

We see in the next section that there are only finitely many irreps of a group (up to isomorphism). Therefore, by Lemma 6.51(3), there are only a finite number of irreducible characters of G. Lemma 6.51 allows us to make the *character table* of a group G. This is the table whose rows are indexed by the irreducible characters of G and whose columns are indexed by the conjugacy classes of G. The entries in the table are the values of the characters on the elements of the conjugacy classes of G.

Example 6.52
Let $\xi = e^{2\pi i/3}$ and let $\phi_a : \mathbb{Z}_3 \to \mathbb{C}$ be as in Example 6.17. Recall that $\phi_a(k) = (\xi^{ak})$ for $a = 0, 1, 2$. We will show in Proposition 7.11 that ϕ_0, ϕ_1, ϕ_2 are the only irreducible matrix representations of $\mathbb{Z}_3$. Let χ_0, χ_1, and χ_2 be the characters of ϕ_0, ϕ_1, and ϕ_2. Then the character table of $\mathbb{Z}_3$ is:

	$\{0\}$	$\{1\}$	$\{2\}$
χ_0	1	1	1
χ_1	1	ξ	ξ^2
χ_2	1	ξ^2	ξ

Example 6.53
Recall Example 6.41. Let $\xi = e^{2\pi i/3}$. Define three representations π_0, π_1, and π_2 on S_3 by taking $\pi_0(\sigma) = (1)$ to be the trivial representation, $\pi_1(\sigma) = (\mathrm{sgn}(\sigma))$ the alternating representation, and

$$\pi_2(()) = \begin{pmatrix} 1 & 0 \\ 0 & 1 \end{pmatrix}, \quad \pi_2((1,2)) = \begin{pmatrix} 0 & \xi^2 \\ \xi & 0 \end{pmatrix}, \quad \pi_2((1,3)) = \begin{pmatrix} 0 & \xi \\ \xi^2 & 0 \end{pmatrix},$$

$$\pi_2((2,3)) = \begin{pmatrix} 0 & 1 \\ 1 & 0 \end{pmatrix}, \quad \pi_2((1,2,3)) = \begin{pmatrix} \xi^2 & 0 \\ 0 & \xi \end{pmatrix}, \quad \pi_2((1,3,2)) = \begin{pmatrix} \xi & 0 \\ 0 & \xi^2 \end{pmatrix}.$$

It can be shown that up to equivalence, these are all of the irreducible representations of S_3 (see Example 6.61 and Exercise 4).

Let χ_i be the character of π_i for $i = 0, 1, 2$. Label the conjugacy classes of S_3 as in Example 6.50. The character table of S_3 is:

	$K_{()}$	$K_{(1,2)}$	$K_{(1,2,3)}$
χ_0	1	1	1
χ_1	1	-1	1
χ_2	2	0	-1

In calculating the traces, we used the fact that $1 + \xi + \xi^2 = \frac{\xi^3 - 1}{\xi - 1} = 0$.

Theorem 6.54 (Schur's lemma)
Let V and W be irreducible representations of G, and let $\theta : V \to W$ be a G-homomorphism. Then either θ is the zero map or θ is an isomorphism.

Proof
Recall Remark 6.8. Let $ker(\theta)$ be the kernel of θ and $im(\theta)$ the image of θ. We claim that $ker(\theta)$ is a subrepresentation of V, and $im(\theta)$ is a subrepresentation of W. Because $\theta : V \to W$ is a linear transformation, $ker(\theta)$ is a subspace of V and $im(\theta)$ is a subspace of W (see [56, p. 63]). Suppose that $\mathbf{v} \in ker(\theta)$ and $g \in G$. Then $\theta(g \cdot \mathbf{v}) = g \cdot \theta(\mathbf{v}) = g \cdot \mathbf{0} = \mathbf{0}$. Thus, $g \cdot \mathbf{v} \in ker(\theta)$. So $ker(\theta)$ is a subrepresentation of V. Now suppose that $\theta(\mathbf{v}) \in im(\theta)$ for some $\mathbf{v} \in V$. If $g \in G$, then $g \cdot \theta(\mathbf{v}) = \theta(g \cdot \mathbf{v}) \in im(\theta)$. Thus, $im(\theta)$ is a subrepresentation of W.

Now suppose $\theta : V \to W$ is a G-homomorphism. Since $ker(\theta)$ is a subrepresentation of V, and V is irreducible, we must have that either $ker(\theta) = V$ or $ker(\theta) = \{\mathbf{0}\}$. Because $im(\theta)$ is a subrepresentation of W and W is irreducible, we must have that either $im(\theta) = \{\mathbf{0}\}$ or $im(\theta) = W$. Putting these facts together, we either have that $ker(\theta) = V$ and $im(\theta) = \{\mathbf{0}\}$, or $ker(\theta) = \{\mathbf{0}\}$ and $im(\theta) = W$. Thus, θ is either the zero map or else an isomorphism. ⌂

The following is a matrix version of Schur's lemma.

Corollary 6.55
Let $\pi : G \to GL(n, \mathbb{C})$ and $\phi : G \to GL(m, \mathbb{C})$ be irreducible matrix representations of a finite group G. Suppose that Z is an $n \times m$ matrix and that $\pi(g)Z = Z\phi(g)$ for all $g \in G$. Then either Z is the zero matrix or Z is an invertible square matrix and $n = m$, in which case $\pi \cong \phi$.

Proof
See Exercise 5. ⌂

One of the most useful consequences of Schur's lemma is the following corollary, which states that a square matrix commutes with every element in the image of an irreducible matrix representation iff it equals a constant times the identity matrix.

Corollary 6.56
Let $\pi : G \to GL(n, \mathbb{C})$ be an irreducible matrix representation of G. A matrix $X \in GL(n, \mathbb{C})$ satisfies $X\pi(g) = \pi(g)X$ for all $g \in G$ if and only if $X = cI$, where $c \in \mathbb{C}$ and I is the $n \times n$ identity matrix.

Proof
Suppose that X satisfies $X\pi(g) = \pi(g)X$ for all $g \in G$. Let c be an eigenvalue of X. Then $\det(X - cI) = 0$ and so $X - cI$ is not invertible. Since $(X - cI)\pi(g) = \pi(g)(X - cI)$ for all $g \in G$, by Corollary 6.55, we have that $X - cI = 0$. Hence, $X = cI$.

Conversely, every matrix of the form $X = cI$ commutes with $\pi(g)$ for all $g \in G$. ⌂

Lemma 6.57
Let $\pi : G \to GL(n, \mathbb{C})$ and $\phi : G \to GL(m, \mathbb{C})$ be irreducible unitary matrix representations of a finite group G. Furthermore, suppose that their matrices are given by $\pi(g) = (\pi_{i,j}(g))_{1 \le i,j \le n}$ and $\phi(g) = (\phi_{i,j}(g))_{1 \le i,j \le m}$.

The coordinate functions $\pi_{i,j}$ and $\phi_{r,s}$ of π and ϕ are elements of $L^2(G)$, and they satisfy the following orthogonality relations with respect to the standard inner product on $L^2(G)$.

1. The coordinate functions of inequivalent, irreducible unitary representations are orthogonal. More precisely, if π and ϕ are inequivalent, then

$$\left\langle \pi_{i,j}, \phi_{r,s} \right\rangle_2 = 0$$

for all i, j, r, s.

2. The coordinate functions of an irreducible unitary representation are pairwise orthogonal. More precisely, we have that

$$\left\langle \pi_{i,j}, \pi_{r,s} \right\rangle_2 = \begin{cases} |G|/n & \text{if } i = r \text{ and } j = s. \\ 0 & \text{otherwise.} \end{cases}$$

Proof
(1) Suppose that π and ϕ are inequivalent. Given arbitrary j and s, let $E_{j,s}$ be the $n \times m$ matrix with a 1 in the j, s position and 0s everywhere else. Define the $n \times m$ matrix

$$Y = \sum_{g \in G} \pi(g)\, E_{j,s}\, \phi(g^{-1}) = \sum_{g \in G} \pi(g)\, E_{j,s}\, \overline{\phi(g)}^t, \tag{19}$$

where the second equality follows because ϕ is unitary and so $\phi(g)\overline{\phi(g)}^t = I$. For each $h \in G$, note that

$$\pi(h)Y = \sum_{g \in G} \pi(hg)E_{j,s}\phi(g^{-1}) = \sum_{x \in G} \pi(x)E_{j,s}\phi(x^{-1}h) = Y\phi(h), \tag{20}$$

where we substituted $x = hg$ to get the second equality. Since π and ϕ are inequivalent, by Corollary 6.55 and Equation 20, we must have that $Y = 0$.

A short calculation shows that

$$0 = \sum_{g \in G} \pi(g)\, E_{j,s}\, \overline{\phi(g)}^{\,t}$$

$$= \sum_{g \in G} \begin{pmatrix} \pi_{1,j}(g)\overline{\phi_{1,s}(g)} & \pi_{1,j}(g)\overline{\phi_{2,s}(g)} & \cdots & \pi_{1,j}(g)\overline{\phi_{m,s}(g)} \\ \pi_{2,j}(g)\overline{\phi_{1,s}(g)} & \pi_{2,j}(g)\overline{\phi_{2,s}(g)} & \cdots & \pi_{2,j}(g)\overline{\phi_{m,s}(g)} \\ \vdots & \vdots & \ddots & \vdots \\ \pi_{n,j}(g)\overline{\phi_{1,s}(g)} & \pi_{n,j}(g)\overline{\phi_{2,s}(g)} & \cdots & \pi_{n,j}(g)\overline{\phi_{m,s}(g)} \end{pmatrix}.$$

Hence, for any i, j, r, s, we have

$$\left\langle \pi_{i,j}, \phi_{r,s} \right\rangle_2 = \sum_{g \in G} \pi_{i,j}(g)\overline{\phi_{r,s}(g)} = 0.$$

(2) Let $E_{j,s}$ be defined as in the first part of this proof. Define

$$Y = \sum_{g \in G} \pi(g)\, E_{j,s}\, \pi(g^{-1}). \tag{21}$$

As before, for each $h \in G$, we have that

$$\pi(h)Y = \sum_{g \in G} \pi(hg)E_{j,s}\pi(g^{-1}) = \sum_{x \in G} \pi(x)E_{j,s}\pi(x^{-1}h) = Y\pi(h).$$

By Corollary 6.56 we must have that $Y = cI$ where $c \in \mathbb{C}$ and I is the $n \times n$ identity matrix. Taking the trace of Equation 21, and using Lemma A.61, we get

$$\begin{aligned} c &= \frac{\operatorname{tr}(Y)}{n} = \frac{1}{n}\sum_{g \in G} \operatorname{tr}(\pi(g)E_{j,s}\pi(g)^{-1}) = \frac{1}{n}\sum_{g \in G} \operatorname{tr}(E_{j,s}) \\ &= \begin{cases} \frac{|G|}{n} & \text{if } j = s. \\ 0 & \text{otherwise.} \end{cases} \end{aligned}$$

As in the first part of this proof, we have that

$$cI = Y = \sum_{g \in G} \begin{pmatrix} \pi_{1,j}(g)\overline{\pi_{1,s}(g)} & \pi_{1,j}(g)\overline{\pi_{2,s}(g)} & \cdots & \pi_{1,j}(g)\overline{\pi_{n,s}(g)} \\ \pi_{2,j}(g)\overline{\pi_{1,s}(g)} & \pi_{2,j}(g)\overline{\pi_{2,s}(g)} & \cdots & \pi_{2,j}(g)\overline{\pi_{n,s}(g)} \\ \vdots & \vdots & \ddots & \vdots \\ \pi_{n,j}(g)\overline{\pi_{1,s}(g)} & \pi_{n,j}(g)\overline{\pi_{2,s}(g)} & \cdots & \pi_{n,j}(g)\overline{\pi_{n,s}(g)} \end{pmatrix}.$$

Hence,

$$\begin{aligned} \left\langle \pi_{i,j}, \pi_{r,s} \right\rangle_2 &= \sum_{g \in G} \pi_{i,j}(g)\overline{\pi_{r,s}(g)} = Y_{i,r} = (cI)_{i,r} \\ &= \begin{cases} |G|\,/n & \text{if } i = r \text{ and } j = s. \\ 0 & \text{otherwise.} \end{cases} \end{aligned}$$

Theorem 6.58
Let π and ϕ be irreducible representations (or matrix representations) of G with corresponding characters χ and ψ, respectively. Then

$$\langle \chi, \psi \rangle_2 = \begin{cases} |G| & \text{if } \pi \cong \phi. \\ 0 & \text{if } \pi \ncong \phi. \end{cases}$$

Proof
We prove this theorem where $\pi : G \to GL(n, \mathbb{C})$ and $\phi : G \to GL(m, \mathbb{C})$ are matrix representations. The same proof works for representations by

picking bases for the corresponding vector spaces. By Remark 6.45, we may assume that π and ϕ are unitary.

Suppose that $\pi \cong \phi$. By Lemma 6.51 we have that $\chi = \psi$. Hence, by Lemma 6.57

$$\begin{aligned}\langle \chi, \psi \rangle_2 &= \langle \chi, \chi \rangle_2 \\ &= \sum_{g\in G} \chi(g)\overline{\chi(g)} = \sum_{g\in G}\sum_{a=1}^{n}\sum_{b=1}^{n} \pi_{a,a}(g)\overline{\pi_{b,b}(g)} \\ &= \sum_{a=1}^{n}\sum_{b=1}^{n} \langle \pi_{a,a}, \pi_{b,b}\rangle_2 = |G|.\end{aligned}$$

Suppose that $\pi \ncong \phi$. Then, by Lemma 6.57 we have that

$$\begin{aligned}\langle \chi, \psi \rangle_2 &= \sum_{g\in G} \chi(g)\overline{\psi(g)} = \sum_{g\in G}\sum_{a=1}^{n}\sum_{b=1}^{m} \pi_{a,a}(g)\overline{\phi_{b,b}(g)} \\ &= \sum_{a=1}^{n}\sum_{b=1}^{m} \langle \pi_{a,a}, \phi_{b,b}\rangle_2 = 0.\end{aligned}$$

Remark 6.59
Recall from Definition 1.31 that the standard inner product on $L^2(G)$ is given by $\langle f, g\rangle_2 = \sum_{x\in G} f(x)\overline{g(x)}$. It is common in other texts on representation theory to normalize this inner product by dividing by $|G|$. Because of this, some of our results differ from other texts by a factor of $|G|$. For example, in other texts Theorem 6.58 becomes

$$\langle \chi, \psi \rangle_2 = \begin{cases} 1 & \text{if } \pi \cong \phi \\ 0 & \text{if } \pi \ncong \phi \end{cases}.$$

However, we want keep the same inner product throughout the entire book.

Corollary 6.60
Let π be a representation (or matrix representation) of G with character χ. Suppose that

$$\pi \cong a_1\pi_1 \oplus \cdots \oplus a_n\pi_n,$$

where the π_i are pairwise inequivalent irreducible representations (or matrix representations) of G. Let χ_i be the character of π_i for $i = 1, 2, \ldots, n$. Then

1. $\chi = a_1\chi_1 + \cdots + a_n\chi_n$.
2. Suppose that ϕ is an irreducible representation (or matrix representation) of G with character ψ. Then

$$\langle \chi, \psi \rangle_2 = \begin{cases} a_i|G| & \text{if } \phi \cong \pi_i \text{ for some } i. \\ 0 & \text{otherwise.} \end{cases}$$

   In particular, $\langle \chi, \chi_i \rangle_2 = a_i\,|G|$.
3. π is irreducible if and only if $\langle \chi, \chi \rangle_2 = |G|$.

Proof

1. Suppose that π is a matrix representation. If $g \in G$, then

$$\chi(g) = \operatorname{tr}(\pi(g)) = \operatorname{tr}\left(\bigoplus_{i=1}^{n} a_i\pi_i(g)\right)$$
$$= \sum_{i=1}^{n} a_i \operatorname{tr}(\pi_i(g)) = \sum_{i=1}^{n} a_i\chi_i(g).$$

   Now suppose that $\pi : G \to GL(V)$ is a representation. As in Remark 6.40, there exist bases β for V and β_i for the subrepresentations of V corresponding to π_i such that

$$[\pi(g)]_\beta = a_1[\pi_1(g)]_{\beta_1} \oplus \cdots \oplus a_n[\pi_n(g)]_{\beta_n} \tag{22}$$

   for all $g \in G$. Taking the trace of Equation 22 gives $\chi(g) = a_1\chi_1(g) + \cdots + a_n\chi_n(g)$.
2. Note that

$$\langle \chi, \psi \rangle_2 = \langle a_1\chi_1 + \cdots + a_n\chi_n, \psi \rangle_2 = \sum_{i=1}^{n} a_i \langle \chi_i, \psi \rangle_2 .$$

   The result follows from Theorem 6.58.
3. From Theorem 6.58, we have that

$$\langle \chi, \chi \rangle_2 = \left\langle \sum_{i=1}^{n} a_i\chi_i, \sum_{j=1}^{n} a_j\chi_j \right\rangle_2 = \sum_{i=1}^{n} \langle a_i\chi_i, a_i\chi_i \rangle_2 = |G| \sum_{i=1}^{n} a_i^2.$$

   Note that $\sum_{i=1}^{n} a_i^2 = 1$ if and only if π is irreducible. ⓐ

Directly using the definition of "irreducible" to check whether a representation is irreducible can be quite difficult. The point of Corollary 6.60(3) is that, remarkably, it reduces this question the computation of a single number: the inner product of the representation's character with itself. This is often the most efficient way to determine whether a representation is irreducible. The next few examples illustrate this technique.

Example 6.61
In this example, we show that the representation π_2 given in Example 6.53 is irreducible. Let ξ_2 be the character of π_2. The values of χ_2 are given in Example 6.53. Note that

$$\langle \chi_2, \chi_2 \rangle_2 = 2^2 + 0^2 + 0^2 + 0^2 + (-1)^2 + (-1)^2 = 6 = |S_3| .$$

By Corollary 6.60(3), we have that π_2 is irreducible.

We now compute the character of the permutation representation of S_n. We use this information in the subsequent example to show that the permutation representation is irreducible.

Example 6.62
Let ρ, ρ_{per}, V_n, and V_{per} be as in Example 6.34. Let ρ_{const} and V_{const} be as in Example 6.36. Let χ be the character of ρ, χ_{per} the character of ρ_{per}, and χ_{const} the character of ρ_{const}.

Consider the standard basis $\beta = [\mathbf{e}_1, \ldots, \mathbf{e}_n]$ for V_n where

$$\begin{array}{ccccccccc} \mathbf{e}_1 & = & 1\mathbf{x}_1 & + & 0\mathbf{x}_2 & + & \ldots & + & 0\mathbf{x}_n, \\ \mathbf{e}_2 & = & 0\mathbf{x}_1 & + & 1\mathbf{x}_2 & + & \ldots & + & 0\mathbf{x}_n, \\ \vdots & & & & \vdots & & & & \vdots \\ \mathbf{e}_n & = & 0\mathbf{x}_1 & + & 0\mathbf{x}_2 & + & \ldots & + & 1\mathbf{x}_n. \end{array}$$

Given $\sigma \in S_n$, we have that $\rho(\sigma)\mathbf{e}_i = \mathbf{e}_{\sigma(i)}$. Thus

$$[\rho(\sigma)]_\beta = ([\mathbf{e}_{\sigma(1)}]_\beta | \cdots | [\mathbf{e}_{\sigma(n)}]_\beta).$$

The diagonal element in the ith row of $[\rho(\sigma)]_\beta$ is equal to 1 or 0 depending on whether $\sigma(i) = i$ or $\sigma(i) \neq i$, respectively. Recall that a fixed point of σ is a number i such that $\sigma(i) = i$. Let $f(\sigma)$ denote the number of fixed points of σ. Then $\chi(\sigma) = \mathrm{tr}[\rho(\sigma)]_\beta = f(\sigma)$. Recall from Example 6.36 that $V_n \cong V_{const} \oplus V_{per}$. So by Corollary 6.60(1), we have $\chi = \chi_{const} + \chi_{per} = 1 + \chi_{per}$. Hence,

$$\chi_{per}(\sigma) = f(\sigma) - 1.$$

As an example, consider the case $n = 3$. Then

$$\mathbf{e}_1 = 1\mathbf{x}_1 + 0\mathbf{x}_2 + 0\mathbf{x}_3,$$

$$\mathbf{e}_2 = 0\mathbf{x}_1 + 1\mathbf{x}_2 + 0\mathbf{x}_3,$$

$$\mathbf{e}_3 = 0\mathbf{x}_1 + 0\mathbf{x}_2 + 1\mathbf{x}_3.$$

Note that $\rho((1,2))(\mathbf{e}_1) = \mathbf{e}_2$, $\rho((1,2))(\mathbf{e}_2) = \mathbf{e}_1$, and $\rho((1,2))(\mathbf{e}_3) = \mathbf{e}_3$. Thus,

$$[\rho((1,2))]_\beta = \begin{pmatrix} 0 & 1 & 0 \\ 1 & 0 & 0 \\ 0 & 0 & 1 \end{pmatrix}.$$

So $\chi((1,2)) = 1 = f((1,2))$. Thus $\chi_{per}((1,2)) = 1 - 1 = 0$. Similarly, $\chi_{per}((2,3)) = \chi_{per}((1,3)) = 0$, $\chi_{per}((1,2,3)) = \chi_{per}((1,3,2)) = -1$, and $\chi_{per}(()) = 2$. The reader should compare this result to Example 6.53.

Example 6.63
Continuing Example 6.62, we now use Corollary 6.60(3) to show that the permutation representation is irreducible. We compute that

$$\begin{aligned}\langle \chi_{per}, \chi_{per}\rangle &= \sum_{\sigma\in S_n} (f(\sigma) - 1)^2 \\ &= \sum_{\sigma\in S_n} (f(\sigma)^2 - 2f(\sigma) + 1) \\ &= \sum_{\sigma\in S_n} f(\sigma)^2 - 2\sum_{\sigma\in S_n} f(\sigma) + |S_n|.\end{aligned} \tag{23}$$

We now compute $\sum_{\sigma\in S_n} f(\sigma)$ and $\sum_{\sigma\in S_n} f(\sigma)^2$. First, we show by induction that

$$\sum_{\sigma\in S_n} f(\sigma) = n! \tag{24}$$

for all $n \geq 1$. For the base case $(n = 1)$, we have $f(()) = 1!$. Now we assume that our claim is true for all $k < n$ and show that it holds for n, where $n \geq 2$. Let $B_j = \{\sigma \in S_n \mid \text{the orbit of 1 under } \sigma \text{ has size } j\}$. (For example, if $\sigma = (1,5,3)(2,4)$, then the orbit of 1 under σ is $\{1,5,3\}$, so $\sigma \in B_3$.) Note that

$$\sum_{\sigma\in S_n} f(\sigma) = \sum_{\sigma\in B_1} f(\sigma) + \cdots + \sum_{\sigma\in B_n} f(\sigma).$$

B_1 consists of all elements of S_n that fix 1, so regarding elements of B_1 as permutations of $\{2,3,\ldots,n\}$, we can identify B_1 with S_{n-1}. If $\sigma \in B_1$, then $f(\sigma)$ equals 1 plus the number of fixed points on the set $\{2,3,\ldots,n\}$. By the inductive hypothesis, we have that $\sum_{\sigma\in B_1} f(\sigma) = \sum_{\sigma\in S_{n-1}}(1 + f(\sigma)) = (n-1)! + (n-1)!$. Now suppose $2 \leq j \leq n-1$. For any $\sigma \in B_j$, we know that 1 "lives" in a j-cycle $(1, a_1, \ldots, a_{j-1})$. So σ then defines a permutation of $\{1,2,\ldots,n\} \setminus \{1, a_1, \ldots, a_{j-1}\}$. That is, we can regard σ as an element of S_{n-j}. By our inductive hypothesis, then, the sum of $f(\sigma)$, over all σ containing the j-cycle $(1, a_1, \ldots, a_{j-1})$, is $(n-j)!$. But there are $(n-1)(n-2)\cdots(n-j+1)$ possible j-cycles $(1, a_1, \ldots, a_{j-1})$. So

$$\sum_{\sigma\in B_j} f(\sigma) = (n-1)(n-2)\cdots(n-j+1)\cdot(n-j)! = (n-1)!.$$

Finally, elements of B_n are n-cycles and consequently have no fixed points, so $\sum_{\sigma \in B_n} f(\sigma) = 0$. Putting the pieces together, we find that

$$\sum_{\sigma \in S_n} f(\sigma) = [(n-1)! + (n-1)!] + (n-2)[(n-1)!] + 0 = n!.$$

A similar inductive argument shows that

$$\sum_{\sigma \in S_n} f(\sigma)^2 = 2[n!] \tag{25}$$

for all $n \geq 2$. For the base case ($n = 2$), we have $f(())^2 + f((1,2))^2 = 4 = 2[2!]$. Now we assume that our claim is true for all $k < n$ and show that it holds for n, where $n \geq 3$. Taking B_j as before and using the same sort of reasoning, we find that

$$\begin{aligned}\sum_{\sigma \in B_1} f(\sigma)^2 &= \sum_{\sigma \in S_{n-1}} (1 + f(\sigma))^2 \\ &= \sum_{\sigma \in S_{n-1}} (1 + 2f(\sigma) + f(\sigma)^2) \\ &= 5[(n-1)!], \text{ and}\end{aligned}$$

$$\sum_{\sigma \in B_j} f(\sigma)^2 = 2[(n-1)!] \text{ when } 2 \leq j \leq n-2, \text{ and}$$

$$\sum_{\sigma \in B_{n-1}} f(\sigma)^2 = (n-1)!, \text{ and}$$

$$\sum_{\sigma \in B_n} f(\sigma) = 0.$$

Putting the pieces together, we find that $\sum_{\sigma \in S_n} f(\sigma)^2 = 5[(n-1)!] + (n-3)(2)[(n-1)!] + (n-1)! = 2[n!]$.

Combining Equations 23, 24, and 25 gives us $\langle \chi_{per}, \chi_{per} \rangle = |S_n|$. (We assume that $n \geq 2$, else the permutation representation is not defined.) So by Corollary 6.60(3), we see that the permutation representation is irreducible.

Remark 6.64
See Ledermann [84, pp. 89–93] for a proof of the irreducibility of V_{per} using induced representations (which we introduce in Section 6).

Corollary 6.65
Two representations (or matrix representations) of a finite group are equivalent if and only if they have the same characters.

Proof
In Lemma 6.51 we showed that equivalent representations have the same character. We now show the other direction. We prove the result for matrix representations, but the same proof works for representations.

Let ϕ and π be matrix representations of G with characters χ and ψ, respectively. Suppose that $\chi(g) = \psi(g)$ for all $g \in G$. By Maschke's theorem (Theorem 6.43), we have that

$$\phi \cong a_1\phi_1 \oplus \cdots \oplus a_n\phi_n$$

for some pairwise inequivalent irreducible matrix representations ϕ_i of G. Let χ_i be the character of ϕ_i. Without loss of generality, we may assume that

$$\pi \cong b_1\phi_1 \oplus \cdots \oplus b_n\phi_n,$$

using 0s as coefficients if an irrep occurs in ϕ or π and not in both. By Corollary 6.60 we have that

$$a_i = \frac{1}{|G|}\langle \chi, \chi_i \rangle_2 = \frac{1}{|G|}\langle \psi, \chi_i \rangle_2 = b_i$$

for all i. Therefore, $\phi \cong \pi$. ⊛

Remark 6.66
Maschke's theorem shows that any representation can be decomposed as a direct sum of irreducible representations. The proof of Corollary 6.65 shows that this decomposition is essentially unique, up to reordering of terms. In other words, it shows that if G is a finite group and

$$V_1 \oplus \cdots \oplus V_n \cong W_1 \oplus \cdots \oplus W_m,$$

where $V_1, \ldots, V_n$ and $W_1, \ldots, W_m$ are irreps of G, then $n = m$, and, perhaps after reordering the V_j's, we have that $V_1 \cong W_1, \ldots, V_n \cong W_n$.

Remark 6.67
Suppose that $\rho : G \to GL(V)$ is a representation of a group G and that χ_ρ is the character of ρ. By Remark 6.45 there exists a basis β of V so that $X(k) = [\rho(k)]_\beta$ is unitary for all k. In particular, $X(k)^{-1} = \overline{X(k)}^t$ for all k. Hence

$$\overline{\chi_\rho(k)} = \text{tr}(\overline{X(k)}) = \text{tr}(X(k^{-1})^t) = \text{tr}(X(k^{-1})) = \chi_\rho(k^{-1}).$$

Hence, given two characters ψ and χ of G we have that

$$\langle \psi, \chi \rangle_2 = \sum_{k \in K} \psi(k)\chi(k^{-1}). \tag{26}$$

Equation 26 is sometimes useful when calculating inner products of characters.

4. DECOMPOSITION OF THE RIGHT REGULAR REPRESENTATION

In this section we show that if G is a finite group, then the decomposition of the right regular representation of G as a direct sum of irreps contains *every* irrep of G. We begin by calculating the character of the regular representation.

Lemma 6.68
Let G be a finite group with identity element e_G. Let R be the right regular representation of G given in Example 6.13, and let χ_R be the character for R. Then

$$\chi_R(\gamma) = \begin{cases} |G| & \text{if } \gamma = e_G. \\ 0 & \text{otherwise.} \end{cases}$$

Proof
Recall from Remark 1.33 that if $G = \{g_1, g_2, \ldots, g_n\}$, then the standard basis for $L^2(G)$ is given by $\beta = [\delta_{g_1}, \ldots, \delta_{g_n}]$, where $\delta_{g_i}(g) = 1$ if $g = g_i$, and $\delta_{g_i}(g) = 0$ if $g \neq g_i$.

Recall Example 6.27. Let $\gamma \in G$. To calculate the value of $\chi_R(\gamma)$, we must compute the trace of the matrix

$$[R(\gamma)]_\beta = \left([R(\gamma)\delta_{g_1}]_\beta \;\middle|\; [R(\gamma)\delta_{g_2}]_\beta \;\middle|\; \cdots \;\middle|\; [R(\gamma)\delta_{g_n}]_\beta \right). \tag{27}$$

Let $x, g \in G$. Then

$$\begin{aligned}(R(\gamma)\delta_g)(x) = \delta_g(x\gamma) &= \begin{cases} 1 & \text{if } x\gamma = g \\ 0 & \text{otherwise} \end{cases} \\ &= \delta_{g\gamma^{-1}}(x).\end{aligned}$$

So $R(\gamma)\delta_g = \delta_{g\gamma^{-1}}$. Hence, a diagonal entry in the gth column of the matrix $[R(\gamma)]_\beta$ in Equation 27 is nonzero if and only if $R(\gamma)\delta_g = \delta_{g\gamma^{-1}} = \delta_g$. Note that $R(\gamma)\delta_g = \delta_g$ if and only if $\gamma = e_G$.

Hence,

$$\chi_R(\gamma) = \mathrm{tr}[R(\gamma)]_\beta = \sum_{g\in G} \delta_{g\gamma^{-1}}(g) = \begin{cases} |G| & \text{if } \gamma = e_G \\ 0 & \text{otherwise} \end{cases}.$$ ◈

The following theorem tells us that the regular representation R of a finite group G "contains" all of the irreducible representations of G.

Theorem 6.69
Let G be a finite group. Then there are only finitely many irreps of G, up to equivalence. Suppose that $V_1, V_2, \ldots, V_n$ form a complete list of inequivalent

irreducible representations of G. Let $d_i = \dim(V_i)$. Then $L^2(G)$ is orthogonally equivalent to

$$d_1V_1 \oplus \cdots \oplus d_nV_n.$$

Moreover, $|G| = d_1^2 + \cdots + d_n^2$.

Proof

Let χ_R be as in Lemma 6.68, and let e_G be the identity element of G. Suppose that W is an irreducible representation of G with character χ. Then

$$\langle \chi_R, \chi \rangle_2 = \sum_{g \in G} \chi_R(g)\overline{\chi(g)} = \chi_R(e_G)\overline{\chi(e_G)} = |G| \dim(W).$$

Thus by Corollary 6.60, W occurs in the decomposition of $L^2(G)$ exactly $\dim(W)$ times. This implies (see Remark 6.66) that there must be finitely many irreps of G, up to equivalence. Hence $L^2(G)$ is orthogonally equivalent to $d_1V_1 \oplus \cdots \oplus d_nV_n$. Taking the dimensions of both sides yields $|G| = d_1^2 + \cdots + d_n^2$. ◈

Remark 6.70

It can be shown that the number of irreducible representations of G, up to equivalence, is equal to the number of conjugacy classes of G. See any of the representation theory textbooks that are mentioned in the Notes section for a proof.

We now apply Theorem 6.69 to find all of the irreducible representations of $\mathbb{Z}_n$. First we need a lemma.

Lemma 6.71

Suppose a and b are integers with $0 \leq b \leq a \leq n-1$. Let $\chi_c(m) = e^{2\pi icm/n}$. Then

$$\langle \chi_a, \chi_b \rangle_2 = \begin{cases} n & \text{if } a = b. \\ 0 & \text{if } a \neq b. \end{cases}$$

Proof

Let $\xi = e^{2\pi i/n}$. By the definition of the standard inner product $\langle \cdot, \cdot \rangle_2$ on $L^2(\mathbb{Z}_n)$ we have that

$$\langle \chi_a, \chi_b \rangle_2 = \sum_{k=0}^{n-1} \chi_a(k)\overline{\chi_b(k)} = \sum_{k=0}^{n-1} (\xi^{a-b})^k.$$

If $a = b$, then $\xi^{a-b} = 1$ and so $\langle \chi_a, \chi_b \rangle_2 = n$. Suppose that $a \neq b$. Because n does not divide $a - b$, we have that $\xi^{a-b} \neq 1$. Hence,

$$\langle \chi_a, \chi_b \rangle_2 = \sum_{k=0}^{n-1} (\xi^{a-b})^k = \frac{(\xi^{a-b})^n - 1}{\xi^{a-b} - 1} = 0.$$ ◈

In Prop. 6.72 we show how $L^2(\mathbb{Z}_n)$ decomposes into irreducible subrepresentations. We give another proof of this result in Prop. 7.11.

Proposition 6.72
Recall Examples 6.23 and 6.30. Let $\rho_a : \mathbb{Z}_n \to \mathrm{GL}(\mathbb{C})$ be given by $\rho_a(m)z = e^{2\pi iam/n}z$. Then, $\rho_0, \rho_1, \ldots, \rho_{n-1}$ form a complete set of inequivalent, irreducible representations of $\mathbb{Z}_n$.

Proof
Consider the regular representation of $\mathbb{Z}_n$. Given an integer a with $0 \leq a \leq n-1$, let W_a and χ_a be as in Examples 6.23 and 6.30. Recall that $\chi_a(m) = e^{2\pi iam/n}$ and $W_a = \{\alpha\chi_a \mid \alpha \in \mathbb{C}\}$. Note that χ_a is the character of ρ_a.

Suppose that $0 \leq x < y \leq n-1$. By Lemma 6.71, $\left\langle \chi_x, \chi_y \right\rangle_2 = 0$. By Lemma A.16, the vectors in the set $\beta = \{\chi_0, \chi_1, \ldots, \chi_{n-1}\}$ are linearly independent in $L^2(\mathbb{Z}_n)$. Since the dimension of $L^2(\mathbb{Z}_n)$ as a vector space is $|\mathbb{Z}_n| = n$, we see by Proposition A.12 that β is a basis for $L^2(\mathbb{Z}_n)$. Hence $L^2(\mathbb{Z}_n)$ is orthogonally equivalent to $W_0 \oplus W_1 \oplus \cdots \oplus W_n$. Recall from Example 6.23 that $R|_{W_a} \cong \rho_a$. By Theorem 6.69 we have that $\rho_0, \rho_1, \ldots, \rho_{n-1}$ form a complete set of inequivalent, irreducible representations of $\mathbb{Z}_n$. Ⓐ

Remark 6.73
See Proposition 7.13 for a generalization of Proposition 6.72 to any abelian group.

Recall from Chapter 1 that $L_0^2(G)$ is used to calculate the second-largest eigenvalue of a Cayley graph. In the following corollary, we show how $L_0^2(G)$ decomposes into irreducible representations of G.

Corollary 6.74
Let G be a finite group. Suppose $V_1, V_2, \ldots, V_n$ are a complete set of inequivalent irreps of G. Furthermore, suppose that V_1 is the trivial representation of G. Let $d_i = \dim(V_i)$. Then $L_0^2(G)$ is orthogonally equivalent to $d_2V_2 \oplus \ldots \oplus d_nV_n$.

Proof
One-line proof: $L_0^2(G)$ is the orthogonal complement of V_1. We spell this out in more detail. Let f_0 be the function that is equal to 1 on all of G. Recall that

$$\mathbb{C}f_0 = \{\alpha f_0 \mid \alpha \in \mathbb{C}\} = \{f \in L^2(G) \mid f \text{ is a constant function}\}$$

and

$$L_0^2(G) = \{f \in L^2(G) \mid \langle f, f_0\rangle_2 = 0\}.$$

This tells us that $L_0^2(V)^\perp = \mathbb{C}f_0$. Therefore, by Proposition A.27, $L^2(G) = \mathbb{C}f_0 \oplus L_0^2(V)$. If $f \in \mathbb{C}f_0$ and $\gamma, g \in G$, then $(R(\gamma)f)(g) = f(g\gamma) = f(g)$. Thus, $\mathbb{C}f_0$ corresponds to the trivial representation and is invariant under the action of the regular representation R. Therefore, by Theorem 6.69, $L_0^2(V)$ must decompose into the remaining representations that are contained in $L^2(G)$. Ⓐ

5. UNIQUENESS OF INVARIANT INNER PRODUCTS

In this section, we show that there is essentially only one invariant inner product (up to a scalar constant) on an irrep. We will need this fact later in Chapter 8. We will not need this section for Chapter 7.

Definition 6.75 Let V be a vector space. Let $V^* = \{\phi : V \to \mathbb{C} \mid \phi \text{ is linear}\}$. We call V^* the *dual space* of V.

Define addition and scalar multiplication on V^* pointwise, that is,

$$(\phi_1 + \phi_2)(v) = \phi_1(v) + \phi_2(v) \qquad \text{and} \qquad (c\phi)(v) = c\phi(v). \tag{28}$$

With addition and scalar multiplication defined as in Equation 28, we have that V^* is a vector space over $\mathbb{C}$.

Given a representation V, there are several ways to create a new representation from it. One natural space to consider is the dual space V^*. At first, one might think that G acts on V^* by $(g \cdot \phi)(\mathbf{w}) = \phi(g \cdot \mathbf{w})$. However, this does not define an action, because we would then have $[g \cdot (h \cdot \phi)](\mathbf{w}) = (h \cdot \phi)(g \cdot \mathbf{w}) = \phi(hg \cdot \mathbf{w})$, which is not necessarily equal to $(gh \cdot \phi)(\mathbf{w}) = \phi(gh \cdot \mathbf{w})$. We can fix this problem, though, by using g^{-1} instead of g in the following definition.

Definition 6.76 Let V be a representation for some finite group G. Let V^* be the dual space of V, as in Def. 6.75. Define an action of G on V^* as follows. For $\phi \in V^*$, define $g \cdot \phi \in V^*$ by $(g \cdot \phi)(\mathbf{v}) = \phi(g^{-1} \cdot \mathbf{v})$. With this action, we call V^* the *dual representation* of V.

In Exercise 8, we ask the reader to verify that the dual representation is indeed a representation of G.

We can briefly sketch the proof of the main theorem of this section. An inner product on V defines, in a natural way, a map from V to V^*. Using Schur's lemma, we show that if V is irreducible, then any two such maps must differ by a constant. Hence any two inner products on an irrep must differ by a (necessarily positive real) constant. We now proceed to fill in the details in this argument.

Given a G-invariant inner product $\langle \cdot, \cdot \rangle$ on a representation V and a vector $\mathbf{v} \in V$, define $H_{\mathbf{v}} : V \to \mathbb{C}$ by $H_{\mathbf{v}}(\mathbf{w}) = \langle \mathbf{w}, \mathbf{v} \rangle$. Note that $H_{\mathbf{v}}$ is linear, so $H_{\mathbf{v}} \in V^*$. Define $H : V \to V^*$ by

$$H(\mathbf{v}) = H_{\mathbf{v}}. \tag{29}$$

Definition 6.77 Let A, B be vector spaces over $\mathbb{C}$, and let $f : A \to B$. We say that f is *conjugate-linear* if $f(c_1\mathbf{v}_1 + c_2\mathbf{v}_2) = \overline{c_1}f(\mathbf{v}_1) + \overline{c_2}f(\mathbf{v}_2)$ for all $c_1, c_2 \in \mathbb{C}$ and $\mathbf{v}_1, \mathbf{v}_2 \in A$.

Remark 6.78
Conjugate-linearity is sometimes also called *antilinearity* or *semilinearity*.

Lemma 6.79
The map H in (29) has the following properties:

1. H is bijective, and
2. H is conjugate-linear, and
3. H is G-invariant.

Remark 6.80
So H is almost a G-homomorphism; it is close but no cigar because it is conjugate-linear instead of linear.

Proof
(1) Let $\mathbf{v}_1, \ldots, \mathbf{v}_n$ be an orthonormal basis for V. Suppose that $H(\mathbf{v}) = H(\mathbf{u})$. Then $H_{\mathbf{v}}(\mathbf{w}) = H_{\mathbf{u}}(\mathbf{w})$ for all $\mathbf{w} \in V$. Write $\mathbf{v} = a_1\mathbf{v}_1 + \cdots a_n\mathbf{v}_n$ and $\mathbf{u} = b_1\mathbf{v}_1 + \ldots b_n\mathbf{v}_n$. So $\overline{a_j} = \langle \mathbf{v}_j, \mathbf{v}\rangle = H_{\mathbf{v}}(\mathbf{v}_j) = H_{\mathbf{u}}(\mathbf{v}_j) = \langle \mathbf{v}_j, \mathbf{u}\rangle = \overline{b_j}$ for all j. Hence $\mathbf{v} = \mathbf{u}$. Therefore H is injective.

We now show that H is surjective. Let $\phi \in V^*$. Let $a_j = \phi(\mathbf{v}_j)$ for all j. Let $\mathbf{v} = \overline{a_1}\mathbf{v}_1 + \cdots + \overline{a_n}\mathbf{v}_n$. Then $H_{\mathbf{v}}(\mathbf{v}_j) = \phi(\mathbf{v}_j)$ for all j. By linearity, it follows that $H(\mathbf{v}) = H_{\mathbf{v}} = \phi$.

(2) This follows immediately from the definition of inner product (Def. A.13).

(3) By G-invariance of $\langle \cdot, \cdot \rangle$, for all $g \in G$, $\mathbf{v}, \mathbf{w} \in V$ we have that

$$\langle g^{-1} \cdot \mathbf{w}, \mathbf{v}\rangle = \langle \mathbf{w}, g \cdot \mathbf{v}\rangle, \text{ so}$$

$$H_{\mathbf{v}}(g^{-1} \cdot \mathbf{w}) = H_{g\cdot\mathbf{v}}(\mathbf{w}), \text{ so}$$

$$(g \cdot H_{\mathbf{v}})(\mathbf{w}) = H_{g\cdot\mathbf{v}}(\mathbf{w}), \text{ so}$$

$$g \cdot H_{\mathbf{v}} = H_{g\cdot\mathbf{v}}, \text{ so}$$

$$g \cdot (H(\mathbf{v})) = H(g \cdot \mathbf{v}).$$

Remark 6.81
A slicker, less "hands dirty" approach to part (1) of Lemma 6.79 would be to redefine the vector space structure on V^* by taking the scalar product $a\phi$ to be $(a\phi)(\mathbf{v}) = \overline{a}\phi(\mathbf{v})$. With this structure, the map H is not conjugate-linear but genuinely linear. We can then make use of tools from linear algebra: show that V and V^* have the same dimension, and that H has trivial kernel.

Now suppose that $\langle \cdot, \cdot \rangle$ and $\langle \cdot, \cdot \rangle'$ are two G-invariant inner products on V. Let H, H' be the two maps defined as in Equation 29 corresponding to $\langle \cdot, \cdot \rangle$ and $\langle \cdot, \cdot \rangle'$, respectively. Let $\psi = H^{-1} \circ H'$.

Lemma 6.82
ψ is a G-isomorphism.

Proof
By Lemma 6.79, we know that both H and H' are bijective, conjugate-linear, and G-invariant. Therefore H^{-1} is bijective. We now show that H^{-1} is conjugate-linear.

Let $f_1, f_2 \in V^*$. Let $\mathbf{v}_1 = H^{-1}(f_1)$ and $\mathbf{v}_2 = H^{-1}(f_1)$. Thus

$$\begin{aligned} H^{-1}(c_1 f_1 + c_2 f_2) &= H^{-1}(c_1 H(\mathbf{v}_1) + c_2 H(\mathbf{v}_2)) \\ &= H^{-1}(H(\overline{c_1}\mathbf{v}_1 + \overline{c_2}\mathbf{v}_2)) \\ &= \overline{c_1}\mathbf{v}_1 + \overline{c_2}\mathbf{v}_2 \\ &= \overline{c_1}H^{-1}(f_1) + \overline{c_2}H^{-1}(f_2). \end{aligned}$$

A similar argument shows that H^{-1} is G-invariant. It follows that $\psi = H^{-1} \circ H'$ is bijective, linear, and G-invariant. (Note: We are using the fact that the composition of a conjugate-linear map with a conjugate-linear map is *linear*.) ◉

Proposition 6.83
Suppose that V is an irrep for a finite group G and that $\langle \cdot, \cdot \rangle$ and $\langle \cdot, \cdot \rangle'$ are two G-invariant inner products on V. Then $\langle \cdot, \cdot \rangle' = C\langle \cdot, \cdot \rangle$ for some positive real number C. That is, $\langle \mathbf{v}, \mathbf{w} \rangle' = C\langle \mathbf{v}, \mathbf{w} \rangle$ for all $\mathbf{v}, \mathbf{w} \in V$.

Proof
By Lemma 6.82, we know that $\psi = H^{-1} \circ H'$ is a G-isomorphism. Note that $\psi : V \to V$. By Lemma 6.56, we have that $\psi = C \cdot \mathrm{id}_V$ for some nonzero complex number C, where id_V is the identity function on V. Applying H on the left to both sides, we see that $H' = C \cdot H$. Unwinding the definitions, it follows that $\langle \mathbf{v}, \mathbf{w} \rangle' = C\langle \mathbf{v}, \mathbf{w} \rangle$ for all $\mathbf{v}, \mathbf{w} \in V$. Let $\mathbf{v}$ be a nonzero vector in V. Because $\langle \mathbf{v}, \mathbf{v} \rangle'$ and $\langle \mathbf{v}, \mathbf{v} \rangle$ are both positive and real, C is positive and real. ◉

6. INDUCED REPRESENTATIONS

In this section, we show how a representation of a subgroup H of a group G induces a representation of G. We then prove a theorem called *Frobenius reciprocity*, which gives us information about inner products involving the character of an induced representation. This information can help determine which irreps appear in the direct sum decomposition of the induced representation. Later, in Section 4 of Chapter 8, that is precisely how we will make use of Frobenius reciprocity. We will not need this section for Chapter 7.

Definition 6.84 Let G be a finite group and $H < G$. Let $\sigma : H \to GL(W)$ be a representation of H. Define the vector space

$$V = \{f : G \to W \mid f(hg) = \sigma(h)f(g) \text{ for all } g \in G, h \in H\}.$$

Let $\rho : G \to GL(V)$ be defined by the formula $(\rho(x)f)(g) = f(gx)$. The map ρ is called the *induced representation from H up to G* and is denoted by $\mathrm{Ind}_H^G(\sigma)$.

Lemma 6.85
Let G, H, σ, V, and ρ be as in Definition 6.84. Then ρ is a representation of G.

Proof
We leave this proof to the reader because it is similar to the proof that R is a representation. See Example 6.13. ◉

In the following example we construct the regular representation of a group by inducing it from the trivial subgroup.

Example 6.86
Let G be a group and let e_G be the identity element of G. Consider the subgroup $H = \{e_G\}$. Let $\sigma : H \to GL(\mathbb{C})$ be the trivial representation of H. That is, $\sigma(e_G) = I$ where I is the identity function on $\mathbb{C}$. Then $V = L^2(G)$ and $\rho = \mathrm{Ind}_H^G(\sigma)$ satisfies $(\rho(g)f)(x) = f(xg) = (R(g)f)(x)$. Hence, ρ is the regular representation of G.

Let G, H, σ, V, and ρ be as in Definition 6.84. We now define several objects that we use for the remainder of this section. By Lemma 6.42, there exists an inner product $\langle \cdot, \cdot \rangle_W$ with respect to which σ is unitary. Let $g_1, \ldots, g_m$ form a set of representatives for $H\backslash G$. By Proposition A.18 we can find an orthonormal basis $e_1, \ldots, e_s$ of W with respect to $\langle \cdot, \cdot \rangle_W$. Let

$$\tilde{\sigma}(y) = \begin{cases} \sigma(y) & \text{if } y \in H \\ 0 & \text{if } y \notin H \end{cases}.$$

For each $i = 1, \ldots, m$ and $j = 1, \ldots, s$, define the function

$$f_{ij}(x) = \tilde{\sigma}(xg_i^{-1})e_j.$$

The reader should check that $f_{ij} \in V$. We show that the functions f_{ij} form a basis for V. First we need an inner product on V.

Lemma 6.87
Let $G, H, \sigma, V, \rho, g_i$, and $\langle \cdot, \cdot \rangle_W$ be as before. Then

$$\langle y_1, y_2 \rangle_V = \sum_{t=1}^{m} \langle y_1(g_t), y_2(g_t) \rangle_W$$

is a well-defined G-invariant inner product on V.

Proof
We begin by showing that $\langle \cdot, \cdot \rangle_V$ is well defined. Suppose that $g_1', g_2', \ldots, g_m'$ is another set of representatives for $H\backslash G$. Then $h_1 g_1' = g_1$, $h_2 g_2' = g_2, \ldots,$ $h_m g_m' = g_m$ for some $h_i \in H$. Let $y_1, y_2 \in V$. Then

$$\begin{aligned} \sum_{t=1}^{m} \langle y_1(g_t), y_2(g_t) \rangle_W &= \sum_{t=1}^{m} \langle y_1(h_t g_t'), y_2(h_t g_t') \rangle_W \\ &= \sum_{t=1}^{m} \langle \sigma(h_t) y_1(g_t'), \sigma(h_t) y_2(g_t') \rangle_W \\ &= \sum_{t=1}^{m} \langle y_1(g_t'), y_2(g_t') \rangle_W . \end{aligned}$$

Hence $\langle \cdot, \cdot \rangle_V$ is well defined. We leave it to the reader to show that $\langle \cdot, \cdot \rangle_V$ is an inner product.

We now show that $\langle \cdot, \cdot \rangle_V$ is G-invariant. Let $f \in V$. Let $x = hg \in G$. For each t, $g_t x = g_t hg = h_t \hat{g}_t$ for some $h_t \in H$ and $\hat{g}_t \in \{g_1, \ldots, g_m\}$. Note that if $t_1 \neq t_2$, then $\hat{g}_{t_1} \neq \hat{g}_{t_2}$ since $(g_{t_1}x)(g_{t_2}x)^{-1} = g_{t_1}g_{t_2}^{-1} \in H$ iff $t_1 = t_2$. Thus $\{\hat{g}_1, \ldots, \hat{g}_m\} = \{g_1, \ldots, g_m\}$. So

$$\langle \rho(x)f, \rho(x)f \rangle_V = \sum_{t=1}^{m} \langle f(g_t x), f(g_t x) \rangle_W = \sum_{t=1}^{m} \langle \sigma(h_t) f(\hat{g}_t), \sigma(h_t) f(\hat{g}_t) \rangle_W$$
$$= \sum_{t=1}^{m} \langle f(g_t), f(g_t) \rangle_W = \langle f, f \rangle_V .$$

Ⓐ

Lemma 6.88
The functions f_{ij} where $i = 1, \ldots, m$ and $j = 1, \ldots, s$ form an orthonormal basis for V.

Proof
Note that $\tilde{\sigma}(g_a g_b^{-1}) \neq 0$ if and only if $a = b$. Suppose that $1 \leq i, k \leq m$ and $1 \leq j, r \leq s$. Then

$$\left\langle f_{ij}, f_{kr} \right\rangle_V = \sum_{t=1}^{m} \left\langle f_{ij}(g_t), f_{kr}(g_t) \right\rangle_W$$
$$= \sum_{t=1}^{m} \left\langle \tilde{\sigma}(g_t g_i^{-1}) e_j, \tilde{\sigma}(g_t g_k^{-1}) e_r \right\rangle_W$$
$$= \begin{cases} 1 & \text{if } i = k \text{ and } j = r \\ 0 & \text{otherwise} \end{cases}.$$

Hence, the vectors f_{ij} are mutually orthogonal and are each of norm 1. By Proposition A.16 the vectors f_{ij} form a linearly independent set in V.

Let $f \in V$ and $x = hg_r \in G$. Suppose that $f(g_r) = c_1 e_1 + \cdots + c_s e_s$. Note that $f_{ij}(x) = \tilde{\sigma}(hg_r g_i^{-1}) e_j$, which simplifies to $\sigma(h)e_j$ if $i = r$, or to 0 if $i \neq r$. Thus

$$f(x) = \sigma(h) f(g_r) = c_1 \sigma(h) e_1 + \cdots + c_s \sigma(h) e_s$$
$$= \langle f(g_r), e_1 \rangle_W \sigma(h) e_1 + \cdots + \langle f(g_r), e_s \rangle_W \sigma(h) e_s$$
$$= \langle f(g_r), e_1 \rangle_W f_{r1}(x) + \cdots + \langle f(g_r), e_s \rangle_W f_{rs}(x).$$

Hence, $f = \sum_{a=1}^{m} \sum_{b=1}^{s} \langle f(g_a), e_b \rangle_W f_{ab}$. Therefore the vectors f_{ij} span V. Ⓐ

Remark 6.89
By Lemma 6.88, $\dim(V) = \dim(W)[G : H]$.

Example 6.90
In this example we construct the permutation representation of S_3 using the induced representation construction. Let

$$G = S_3 = \{(), (1,2), (2,3), (1,3), (1,2,3), (1,3,2)\}$$

and $H = \{(), (1,2,3), (1,3,2)\}$ be the subgroup generated by $(1,2,3)$. Define the representation $\sigma : H \to GL(1,\mathbb{C})$ by the formula $\sigma((1,2,3)^r) = \xi^r$ where $\xi = e^{2\pi i/3}$. Here we identify the matrix (a) with the complex number a. We do so for the remainder of this example when appropriate. Note that $H\backslash S_3 = \{H, H(2,3)\}$. Let V and $\rho = \mathrm{Ind}_H^G(\sigma)$ be as in Def. 6.84. Set $e_1 = 1$, $g_1 = ()$, and $g_2 = (2,3)$. Let $\beta_V = [f_{11}, f_{21}]$ where $f_{11}(g) = \tilde{\sigma}(gg_1^{-1})e_1 = \tilde{\sigma}(g)\cdot 1$ and $f_{21}(g) = \tilde{\sigma}(gg_2^{-1})e_1 = \tilde{\sigma}(g(2,3))\cdot 1$ are defined as before. Note that

$$(\rho(x)f_{11})(g) = \tilde{\sigma}(gx)\cdot 1$$
$$(\rho(x)f_{21})(g) = \tilde{\sigma}(gx(2,3))\cdot 1.$$

As an example computation, note that

$$(\rho(1,2,3)f_{11})(g) = \begin{cases} \xi & \text{if } g = () \\ \xi^2 & \text{if } g = (1,2,3) \\ 1 & \text{if } g = (1,3,2) \\ 0 & \text{if } g = (1,2) \\ 0 & \text{if } g = (2,3) \\ 0 & \text{if } g = (1,3) \end{cases}$$

and

$$(\rho(1,2,3)f_{21})(g) = \begin{cases} 0 & \text{if } g = () \\ 0 & \text{if } g = (1,2,3) \\ 0 & \text{if } g = (1,3,2) \\ 1 & \text{if } g = (1,2) \\ \xi^2 & \text{if } g = (2,3) \\ \xi & \text{if } g = (1,3). \end{cases}$$

Hence, $\rho(1,2,3)f_{11} = \xi f_{11}$ and $\rho(1,2,3)f_{21} = \xi^2 f_{21}$. The computation and similar computations give that

$$[\rho(())]_{\beta_V} = \begin{pmatrix} 1 & 0 \\ 0 & 1 \end{pmatrix}, \quad [\rho((1,2))]_{\beta_V} = \begin{pmatrix} 0 & \xi \\ \xi^2 & 0 \end{pmatrix}, \quad [\rho((1,3))]_{\beta_V} = \begin{pmatrix} 0 & \xi^2 \\ \xi & 0 \end{pmatrix},$$

$$[\rho((2,3))]_{\beta_V} = \begin{pmatrix} 0 & 1 \\ 1 & 0 \end{pmatrix}, \quad [\rho((1,2,3))]_{\beta_V} = \begin{pmatrix} \xi & 0 \\ 0 & \xi^2 \end{pmatrix}, \quad [\rho((1,3,2))]_{\beta_V} = \begin{pmatrix} \xi^2 & 0 \\ 0 & \xi \end{pmatrix}.$$

Let π_2 be as in Example 6.53. Let $X = \begin{pmatrix} 0 & 1 \\ 1 & 0 \end{pmatrix}$. Note that $[\rho(g)]_{\beta_V} = X\pi_2(g)X^{-1}$. Therefore ρ is equivalent to the permutation representation of S_3. (Alternatively, to establish equivalence we could have showed that ρ and π_2 have the same character.)

In the following proposition, we compute the character of the induced representation.

Proposition 6.91
Let G, H, σ, V, and $\rho = \mathrm{Ind}_H^G(\sigma)$ be as in Definition 6.84. Let

$$\tilde{\chi}_\sigma(x) = \begin{cases} \chi_\sigma(x) & \text{if } x \in H \\ 0 & \text{if } x \notin H \end{cases},$$

where χ_σ is the character of σ. Let χ_ρ be the character of ρ. Then

$$\chi_\rho(g) = \frac{1}{|H|} \sum_{x \in G} \tilde{\chi}_\sigma(xgx^{-1}).$$

Proof
Let $x \in G$. Then $x = hg_k$ for some $h \in H$ and $1 \le k \le m$. Then

$$(\rho(g)f_{ij})(x) = f_{ij}(hg_k g) = \tilde{\sigma}(h)\tilde{\sigma}(g_k g g_i^{-1})e_j.$$

Let β_W be the ordered basis $[e_1, \ldots, e_s]$, and let

$$[\tilde{\sigma}(x)]_{\beta_W} = \begin{pmatrix} \tilde{\sigma}_{11}(x) & \ldots & \tilde{\sigma}_{1s}(x) \\ \tilde{\sigma}_{21}(x) & \ldots & \tilde{\sigma}_{2s}(x) \\ \vdots & \ddots & \vdots \\ \tilde{\sigma}_{s1}(x) & \ldots & \tilde{\sigma}_{ss}(x) \end{pmatrix}.$$

That is, the functions $\tilde{\sigma}_{rj}$ are defined so that $\tilde{\sigma}(x)e_j = \sum_{r=1}^{s} \tilde{\sigma}_{rj}(x)e_r$. Because $h = xg_k^{-1}$ we have that

$$(\rho(g)f_{ij})(x) = \tilde{\sigma}(h)\sum_{r=1}^{s} \tilde{\sigma}_{rj}(g_k g g_i^{-1})e_r = \sum_{r=1}^{s} \tilde{\sigma}_{rj}(g_k g g_i^{-1})\tilde{\sigma}(xg_k^{-1})e_r$$

$$= \sum_{r=1}^{s} \tilde{\sigma}_{rj}(g_k g g_i^{-1}) f_{kr}(x).$$

Note that $f_{ar}(x) = \tilde{\sigma}(hg_k g_a^{-1})e_r \neq 0$ if and only if $a = k$. Hence,

$$(\rho(g)f_{ij})(x) = \sum_{a=1}^{m}\sum_{r=1}^{s} \tilde{\sigma}_{rj}(g_a g g_i^{-1}) f_{ar}(x).$$

Let β_V be the ordered basis $[f_{11}, \ldots, f_{1s}, f_{2s}, \ldots, f_{2s}, \ldots, f_{m1}, \ldots, f_{ms}]$. Then

$$[\rho(g)]_{\beta_V} = \begin{pmatrix} [\tilde{\sigma}(g_1 g g_1^{-1})]_{\beta_W} & [\tilde{\sigma}(g_1 g g_2^{-1})]_{\beta_W} & \cdots & [\tilde{\sigma}(g_1 g g_m^{-1})]_{\beta_W} \\ [\tilde{\sigma}(g_2 g g_1^{-1})]_{\beta_W} & [\tilde{\sigma}(g_2 g g_2^{-1})]_{\beta_W} & \cdots & [\tilde{\sigma}(g_2 g g_m^{-1})]_{\beta_W} \\ \vdots & \vdots & \ddots & \vdots \\ [\tilde{\sigma}(g_m g g_1^{-1})]_{\beta_W} & [\tilde{\sigma}(g_m g g_2^{-1})]_{\beta_W} & \cdots & [\tilde{\sigma}(g_m g g_m^{-1})]_{\beta_W} \end{pmatrix}.$$

This gives us that

$$\operatorname{tr}[\rho(g)]_{\beta_V} = \sum_{r=1}^{m} \operatorname{tr}(\tilde{\sigma}(g_r g g_r^{-1})) = \sum_{r=1}^{m} \tilde{\chi}_\sigma(g_r g g_r^{-1}).$$

If $x = hg_k$ then $\tilde{\chi}_\sigma(xgx^{-1}) = \tilde{\chi}_\sigma(hg_k g g_k^{-1} h^{-1}) = \tilde{\chi}_\sigma(g_k g g_k^{-1})$. Therefore,

$$\chi_\rho(g) = \operatorname{tr}[\rho(g)]_{\beta_V} = \frac{1}{|H|} \sum_{x \in G} \tilde{\chi}_\sigma(xgx^{-1}).$$ ⓐ

Definition 6.92 Let G be a finite group and $H < G$. Let $\phi : G \to GL(V)$ be a representation of G. Define the *restriction* of ϕ to H to be the representation $\operatorname{Res}_H^G(\phi) : H \to GL(V)$, where $\operatorname{Res}_H^G(\phi)(h) = \phi(h)$ for all $h \in H$.

Theorem 6.93 (Frobenius reciprocity)
Let G be a finite group and H a subgroup of G. Suppose that $\sigma : H \to GL(V_1)$ is a representation of H. Let $\phi : G \to GL(V_2)$ be any representation of G. Let $\rho = \operatorname{Ind}_H^G(\sigma)$ and $\tau = \operatorname{Res}_H^G(\phi)$. Then

$$\frac{1}{|G|} \langle \chi_\rho, \chi_\phi \rangle_2 = \frac{1}{|H|} \langle \chi_\sigma, \chi_\tau \rangle_2,$$

where the inner product on the left-hand side is the standard inner product in $L^2(G)$ and the inner product on the right-hand side is the standard inner product in $L^2(H)$.

Proof
By Proposition 6.91 we have that

$$\begin{aligned} \frac{1}{|G|} \langle \chi_\rho, \chi_\phi \rangle_2 &= \frac{1}{|G|\,|H|} \sum_{g \in G} \sum_{x \in G} \tilde{\chi}_\sigma(xgx^{-1}) \overline{\chi_\phi(g)} \\ &= \frac{1}{|G|\,|H|} \sum_{x \in G} \sum_{g \in G} \tilde{\chi}_\sigma(xgx^{-1}) \overline{\chi_\phi(xgx^{-1})} \\ &= \frac{1}{|G|\,|H|} \sum_{x \in G} \sum_{y \in G} \tilde{\chi}_\sigma(y) \overline{\chi_\phi(y)} \end{aligned}$$

$$= \frac{1}{|H|} \sum_{y \in G} \tilde{\chi}_\sigma(y) \overline{\chi_\phi(y)}$$

$$= \frac{1}{|H|} \sum_{h \in H} \chi_\sigma(h) \overline{\chi_\tau(h)} = \frac{1}{|H|} \left\langle \chi_\sigma, \chi_\tau \right\rangle_2 .$$

Remark 6.94

Frobenius reciprocity shows us how to compute the inner product of the character of an induced representation with another character. Instead of performing the computation in G, we can instead "push everything down" to H and take the inner product there, taking advantage of whatever information we have about the original representation on H. The point of computing these inner products is that it tells us about which irreps appear in the direct sum decomposition of the induced representation (see Exercise 10).

Example 6.95

Let G be a finite group, and let $H < G$. Let σ be a nontrivial irrep of H. Let $\rho = \mathrm{Ind}_H^G(\sigma)$ be the induced representation. Let 1 denote the trivial representation of G. Note that 1 restricted to H is the trivial representation of H, which we also denote by 1. Then, by Frobenius reciprocity,

$$\frac{1}{|G|}\langle 1, \chi_\rho \rangle = \frac{1}{|H|}\langle 1, \chi_\sigma \rangle = 0.$$

Hence, by Corollary 6.60, the trivial representation does not appear in the direct sum decomposition of ρ.

NOTE

1. The reader who is interested in learning more representation theory may wish to consult one of the following textbooks. The books by Ledermann [84] and Sagan [121] give readable introductions to the subject of representation theory. In addition, Sagan's book focuses on the representations of the symmetric group. The book by Serre [126] develops the representation theory of finite groups in both characteristic 0 and p. The book by Fulton and Harris [62] begins by presenting the representation theory of finite groups and continues by developing the theory of Lie groups and Lie algebras.

   Sections 1–4 of this chapter were modeled on the presentation in [121]. Section 6 was modeled on the presentation in [129].

EXERCISES

1. Prove Corollary 6.44. (Hint: Let $V = \mathbb{C}^n$ and consider the representation $g \cdot \mathbf{v} = \rho(g)\mathbf{v}$.)
2. Let G be a group and $a, b \in G$. Write $a \sim b$ if a is conjugate to b. Prove that $\sim$ is an equivalence relation on G and that therefore G is partitioned into its conjugacy classes.
3. Compute the conjugacy classes of S_3 and S_4.

4. Prove that the representations in Example 6.53 are all of the irreducible representations of S_3 up to equivalence. (Hint: Use Theorem 6.69.)
5. Prove Corollary 6.55.
6. Let $\beta = [\chi_0, \chi_1, \ldots, \chi_n]$ where χ_a is defined as in Examples 6.23, 6.30, and 6.72. Show that $[R(g)]_\beta = (\chi_0(g)) \oplus (\chi_1(g)) \oplus \cdots \oplus (\chi_{n-1}(g))$.
7. Suppose that $\phi : G \to GL(V)$ and $\rho : G \to \mathrm{GL}(W)$ are equivalent representations of G. Prove that ϕ is irreducible if and only if ρ is irreducible.
8. Show that Def. 6.76 indeed defines a representation of G. Be sure to attend to all the nitpicky details—for example, show that $g \cdot \phi \in V^*$ and that $\phi \mapsto g \cdot \phi$ is linear.
9. Let G be a finite group, and let $H < G$. Let σ be a representation of H that does not contain the trivial representation in its direct sum decomposition. Let $\rho = \mathrm{Ind}_H^G(\sigma)$ be the induced representation. Prove that ρ does not contain the trivial representation in its direct sum decomposition.
10. Let $G = S_3$. Let $H = \{(), (1,2)\}$. Define a representation $\sigma : H \to GL(\mathbb{C})$ by $\sigma(()) \cdot z = z$ and $\sigma((1,2)) \cdot z = -z$. Let $\rho = \mathrm{Ind}_H^G(\sigma)$. Use Frobenius reciprocity and Exercise 4 to compute the decomposition of ρ as a direct sum of irreps.
11. Let $G = \{g_1, g_2, \ldots, g_n\}$ be a finite group. The *group algebra of* G is defined to be

$$\mathbb{C}[G] = \{a_1 g_1 + \cdots + a_n g_n \mid a_1, \ldots, a_n \in \mathbb{C}\}.$$

$\mathbb{C}[G]$ is an n-dimensional vector space over $\mathbb{C}$ with the operations

$$\left(\sum_{i=1}^{n} a_i g_i\right) + \left(\sum_{i=1}^{n} b_i g_i\right) = \sum_{i=1}^{n} (a_i + b_i) g_i$$

and

$$\alpha \left(\sum_{i=1}^{n} a_i g_i\right) = \sum_{i=1}^{n} (\alpha a_i) g_i.$$

(a) The *left regular representation of* G is given by the following action of G on $\mathbb{C}[G]$.

$$g\left(\sum_{i=1}^{n} a_i g_i\right) = \sum_{i=1}^{n} a_i (g g_i).$$

Prove that this is a representation of G.

(b) The right regular representation of G is given by the following action of G on $\mathbb{C}[G]$.

$$g\left(\sum_{i=1}^{n} a_i g_i\right) = \sum_{i=1}^{n} a_i (g_i g^{-1}).$$

Prove that this is a representation of G. Prove that this representation is equivalent to the right regular representation R from Example 6.13.

(c) Prove that the left regular representation is equivalent to the right regular representation.

12. This exercise decomposes $\mathbb{C}[\mathbb{Z}_4]$ (see Exercise 11) into irreps. Consider the vectors

$$\mathbf{w_1} = -\mathbf{0} + \mathbf{1} - \mathbf{2} + \mathbf{3},$$
$$\mathbf{w_2} = -i \cdot \mathbf{0} - \mathbf{1} + i \cdot \mathbf{2} + \mathbf{3},$$
$$\mathbf{w_3} = i \cdot \mathbf{0} - \mathbf{1} - i \cdot \mathbf{2} + \mathbf{3}, \text{ and}$$
$$\mathbf{w_4} = \mathbf{0} + \mathbf{1} + \mathbf{2} + \mathbf{3}$$

of $\mathbb{C}[\mathbb{Z}_4]$. (Note that the elements of $\mathbb{Z}_4$, when regarded as vectors in $\mathbb{C}[\mathbb{Z}_4]$, are in bold.)

Let $W_i = \{\alpha w_i \mid \alpha \in \mathbb{C}\}$ for $i = 1, 2, 3, 4$. So, each W_i is a one-dimensional subspace of V.

(a) Show that W_1, W_2, W_3, and W_4 are $\mathbb{Z}_4$-invariant.
(b) Consider the standard basis $\beta = [\mathbf{0}, \mathbf{1}, \mathbf{2}, \mathbf{3}]$ of $\mathbb{C}[\mathbb{Z}_4]$. Compute the matrices $\phi(g)$ for the left regular representation of $\mathbb{Z}_4$ (see Exercise 11) in terms of the basis β.
(c) Let $\beta' = [\mathbf{w}_1, \mathbf{w}_2, \mathbf{w}_3, \mathbf{w}_4]$. Show that β' is a basis for $\mathbb{C}[\mathbb{Z}_4]$. Compute the matrices $\pi(g)$ for the left regular representation of $\mathbb{Z}_4$ in terms of the basis β'.
(d) Show that $\pi(g)$ from (c) can be decomposed into a direct sum of $\chi_0, \chi_1, \chi_2, \chi_3$ (see Example 6.17).
(e) Compute the change-of-basis matrix X from β' to the standard basis β (see Appendix A).
(f) Compute the matrices $X^{-1}\phi(g)X$ for $g \in \mathbb{Z}_4$. You should get the same matrices you got in (c).
(g) Can you see how to generalize this example to $\mathbb{C}[\mathbb{Z}_n]$?

7

Representation Theory and Eigenvalues of Cayley Graphs

In the previous chapter, we learned some of the basics from the representation theory of finite groups. In this chapter, we apply this theory to obtain bounds on eigenvalues of Cayley graphs.

1. DECOMPOSING THE ADJACENCY OPERATOR INTO IRREPS

Consider a Cayley graph $\mathrm{Cay}(G, \Gamma)$ with adjacency operator A. Let $f \in L^2(G)$, and let R be the right regular representation of G. Then

$$(Af)(x) = \sum_{\gamma \in \Gamma} f(x\gamma) = \sum_{\gamma \in \Gamma} (R(\gamma)f)(x). \tag{30}$$

Equation 30 establishes the link between representation theory and Cayley graphs that will enable us to get at the eigenvalues of $\mathrm{Cay}(G, \Gamma)$. The key fact is in the following proposition.

Proposition 7.1
Let G be a finite group, $\Gamma \overset{s}{\subset} G$, and A the adjacency operator of $\mathrm{Cay}(G, \Gamma)$. If $\pi_1, \ldots, \pi_k$ is a complete set of inequivalent matrix irreps of G, then

$$A \cong d_1 M_{\pi_1} \oplus d_2 M_{\pi_2} \oplus \cdots \oplus d_k M_{\pi_k},$$

where d_i is the dimension of π_i and $M_\pi = \sum_{\gamma \in \Gamma} \pi(\gamma)$.

Proof
Let R be the right regular representation of G. Then, by Theorem 6.69 and Equation 30, we have

$$A = \sum_{\gamma \in \Gamma} R(\gamma) \cong \sum_{\gamma \in \Gamma} (d_1 \pi_1(\gamma) \oplus \ldots \oplus d_k \pi_k(\gamma)).$$ ⊛

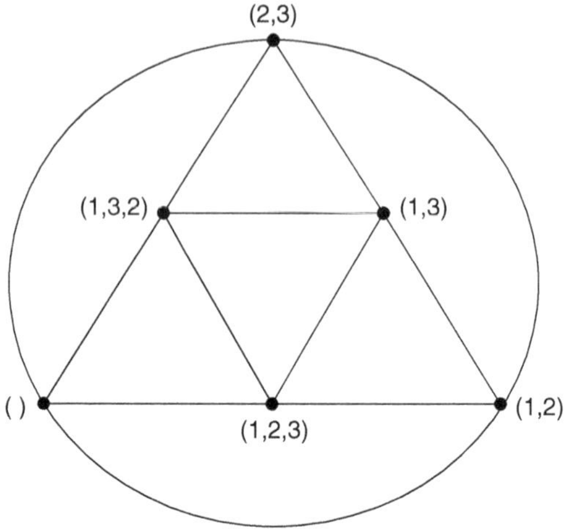

Figure 7.1 $\mathrm{Cay}(S_3, \Gamma)$

Example 7.2
Let $X = \mathrm{Cay}(S_3, \Gamma)$ where

$$\Gamma = \{(1,2), (2,3), (1,2,3), (1,3,2)\},$$

and let A be the adjacency operator of X. This graph is shown in Figure 7.1.

To compute the eigenvalues of X, we use the same notation as in Example 6.53 for the three irreducible representations of S_3. Because

$$\sum_{\gamma\in\Gamma} \pi_0(\gamma) = (4), \sum_{\gamma\in\Gamma} \pi_1(\gamma) = (0), \text{ and}$$

$$\sum_{\gamma\in\Gamma} \pi_2(\gamma) = \begin{pmatrix} 0 & \xi^2 \\ \xi & 0 \end{pmatrix} + \begin{pmatrix} 0 & 1 \\ 1 & 0 \end{pmatrix} + \begin{pmatrix} \xi^2 & 0 \\ 0 & \xi \end{pmatrix} + \begin{pmatrix} \xi & 0 \\ 0 & \xi^2 \end{pmatrix}$$

$$= \begin{pmatrix} \xi+\xi^2 & 1+\xi^2 \\ 1+\xi & \xi+\xi^2 \end{pmatrix} = \begin{pmatrix} -1 & -\xi \\ -\xi^2 & -1 \end{pmatrix},$$

by Proposition 7.1 we have that

$$A \cong \left(\begin{array}{c|c|cc|cc} 4 & 0 & 0 & 0 & 0 & 0 \\ \hline 0 & 0 & 0 & 0 & 0 & 0 \\ \hline 0 & 0 & -1 & -\xi & 0 & 0 \\ 0 & 0 & -\xi^2 & -1 & 0 & 0 \\ \hline 0 & 0 & 0 & 0 & -1 & -\xi \\ 0 & 0 & 0 & 0 & -\xi^2 & -1 \end{array}\right).$$

The characteristic polynomial of the matrix $\begin{pmatrix} -1 & -\xi \\ -\xi^2 & -1 \end{pmatrix}$ is $x(x+2)$.

Thus, the eigenvalues coming from the one-dimensional representations of S_3 are 4 and 0, whereas the eigenvalues coming from the two-dimensional representations of S_3 are -2 and 0 (each counted with multiplicity 2). Therefore, the spectrum of X is given by the multi-set $\{-2, -2, 0, 0, 0, 4\}$.

The following corollary shows that adding generators to a Cayley graph never decreases the spectral gap.

Corollary 7.3
Let G be a finite group and let $\Gamma, \Gamma' \circledS G$ such that $\Gamma \subset \Gamma'$. Let $X = \text{Cay}(G, \Gamma)$ and $X' = \text{Cay}(G, \Gamma')$. Then $|\Gamma| - \lambda_1(X) \leq |\Gamma'| - \lambda_1(X')$.

Proof
Let A and A' be the adjacency operators for X and X', respectively. Recall from Example 6.13 that R is unitary with respect to the standard inner product on $L^2(G)$. If $f \in L^2(G)$ satisfies $\|f\|_2 = 1$, then by the Cauchy-Schwarz inequality (Proposition A.20) we have

$$\left|\langle R(\gamma)f, f\rangle_2\right| \leq \|R(\gamma)f\|_2 \|f\|_2 = \|f\|_2^2 = 1 \tag{31}$$

for all $\gamma \in \Gamma'$. Also if $\gamma = \gamma^{-1}$, then $\langle R(\gamma)f, f\rangle_2 = \langle f, R(\gamma)f\rangle_2 = \overline{\langle R(\gamma)f, f\rangle_2}$, so

$$\langle R(\gamma)f, f\rangle_2 \text{ is real.} \tag{32}$$

For any γ, we have

$$\begin{aligned}\langle R(\gamma)f, f\rangle_2 + \langle R(\gamma^{-1})f, f\rangle_2 &= \langle f, R(\gamma^{-1})f\rangle_2 + \langle f, R(\gamma)f\rangle_2 \\ &= \overline{\langle R(\gamma^{-1})f, f\rangle_2} + \overline{\langle R(\gamma)f, f\rangle_2} \\ &= \overline{\langle R(\gamma)f, f\rangle_2 + \langle R(\gamma^{-1})f, f\rangle_2},\end{aligned}$$

so

$$\langle R(\gamma)f, f\rangle_2 + \langle R(\gamma^{-1})f, f\rangle_2 \text{ is real.} \tag{33}$$

Let $f \in L^2(G, \mathbb{R})$. Then, by Equation 30 and Equation 31, we have

$$\begin{aligned}|\Gamma| - \langle Af, f\rangle_2 &= |\Gamma| - \sum_{\gamma\in\Gamma} \langle R(\gamma)f, f\rangle_2 = \sum_{\gamma\in\Gamma} \left(1 - \langle R(\gamma)f, f\rangle_2\right) \\ &\leq \sum_{\gamma'\in\Gamma'} \left(1 - \langle R(\gamma')f, f\rangle_2\right) = |\Gamma'| - \sum_{\gamma'\in\Gamma'} \langle R(\gamma')f, f\rangle_2 \\ &= |\Gamma'| - \langle A'f, f\rangle_2 .\end{aligned}$$

Note: To establish the inequality, we need to know that $\sum_{\gamma\in\Gamma} \langle R(\gamma)f, f\rangle_2$ and $\sum_{\gamma'\in\Gamma'} \langle R(\gamma')f, f\rangle_2$ are real. Here we use Equations 32 and 33 and the fact that Γ and Γ' are symmetric.

Therefore, by Proposition 1.82, we have

$$|\Gamma| - \lambda_1(X) \leq |\Gamma'| - \lambda_1(X').$$

2. UNIONS OF CONJUGACY CLASSES

In this section, we describe how to find the eigenvalues of Cayley graphs Cay(G, Γ) where Γ is a union of conjugacy classes. For these types of Cayley graphs, the eigenvalues can be expressed as a sum over the irreducible characters of the group G. This allows us to bound the eigenvalues by bounding the characters of the group.

Proposition 7.4
Let G be a finite group and $\Gamma \Subset G$ such that $g\Gamma g^{-1} = \{g\gamma g^{-1} \mid \gamma \in \Gamma\} = \Gamma$ for all $g \in G$. Let $X = \text{Cay}(G, \Gamma)$ and let A be the adjacency operator of X.

Let $\rho_1, \dots, \rho_r$ be a complete set of inequivalent irreps of G; let χ_i be the character of ρ_i; and let d_i be the degree of ρ_i. Then the eigenvalues of A are given by

$$\mu_i = \frac{1}{d_i} \sum_{\gamma \in \Gamma} \chi_i(\gamma), \quad i = 1, \dots, r,$$

where each eigenvalue μ_i occurs with multiplicity d_i^2.

Remark 7.5
We are somewhat sloppy when we say that μ_i occurs with multiplicity d_i^2. For it is quite possible that $\mu_i = \mu_j$ for some $i \neq j$, in which case the multiplicities "combine." For maximum precision, we ought to say that μ_i occurs with multiplicity $\sum d_j^2$, where the sum is over all j such that $\mu_i = \mu_j$. We hope the reader forgives our choice to err on the side of expository clarity.

Proof of Proposition 7.4
As in Theorem 6.69, decompose $L^2(G) = \oplus_{i=1}^r d_i V_i$, where $V_1, \dots, V_r$ form a complete list of inequivalent irreps of G, and $R(\gamma)\mathbf{v}_i = \rho_i(\gamma)\mathbf{v}_i$ for all $\mathbf{v}_i \in V_i$. That is, the restriction of $R(\gamma)$ to V_i is ρ_i.

Because Γ is closed under conjugation, we have that

$$\begin{aligned} AR(g) &= \sum_{\gamma \in \Gamma} R(\gamma)R(g) = \sum_{\gamma \in \Gamma} R(\gamma g) \\ &= \sum_{\gamma \in \Gamma} R((g\gamma g^{-1})g) = \sum_{\gamma \in \Gamma} R(g\gamma) = R(g)A \end{aligned}$$

for all $g \in G$.

Fix some i with $1 \leq i \leq r$. Since $A = \sum_{\gamma \in \Gamma} R(\gamma)$ and $R : V_i \to V_i$, we have that $A : V_i \to V_i$. If we pick a basis β for V_i, we have that $[R(g)]_\beta [A]_\beta = [A]_\beta [R(g)]_\beta$ for all $g \in G$. Since R restricted to V_i is the irreducible representation ρ_i, by Corollary 6.56, we must have that $[A]_\beta = \mu_i I_{d_i}$, where I_{d_i} is the $d_i \times d_i$ identity matrix and $\mu_i \in \mathbb{C}$. Considering the trace of the restriction of A to V_i, denoted by $A|_{V_i}$, we have that

$$d_i\mu_i = \operatorname{tr}(A|_{V_i}) = \operatorname{tr}\left(\sum_{\gamma\in\Gamma}\rho_i(\gamma)\right) = \sum_{\gamma\in\Gamma}\chi_i(\gamma).$$

Thus, $\mu_i = \frac{1}{d_i}\sum_{\gamma\in\Gamma}\chi_i(\gamma)$.

Doing this for each i, we see that

$$A \cong d_1(\mu_1 I_{d_1}) \oplus \cdots \oplus d_r(\mu_r I_{d_r}).$$

Therefore, the eigenvalues of A are $\mu_1, \mu_2, \ldots, \mu_r$, where μ_i has multiplicity d_i^2 for $i = 1, \ldots, r$.

Remark 7.6

Suppose that G is a finite group and $\Gamma \circledS G$ such that $\Gamma = \cup_{i=1}^{n} K_{g_i}$ for some $g_1, \ldots, g_n \in G$, where K_{g_i} is the conjugacy class of g_i. Because characters are constant on conjugacy classes, the eigenvalue μ_i is given by

$$\mu_i = \frac{1}{d_i}\sum_{\gamma\in\Gamma}\chi_i(\gamma) = \frac{1}{d_i}\sum_{j=1}^{n}\sum_{h\in K_{g_j}}\chi_i(h) = \frac{1}{d_j}\sum_{j=1}^{n}\left|K_{g_j}\right|\chi_i(g_j).$$

Example 7.7

Consider the Cayley graph $X = \mathrm{Cay}(S_3, K_{(1,2)})$ in Figure 7.2, where $K_{(1,2)} = \{(1,2), (1,3), (2,3)\}$ is one of the conjugacy classes of S_3. Recall the character table of S_3 given in Example 6.53. The eigenvalues

$$\mu_0 = \frac{1}{1}\left|K_{(1,2)}\right|\chi_0((1,2)) = 3$$

and

$$\mu_1 = \frac{1}{1}\left|K_{(1,2)}\right|\chi_1((1,2)) = -3$$

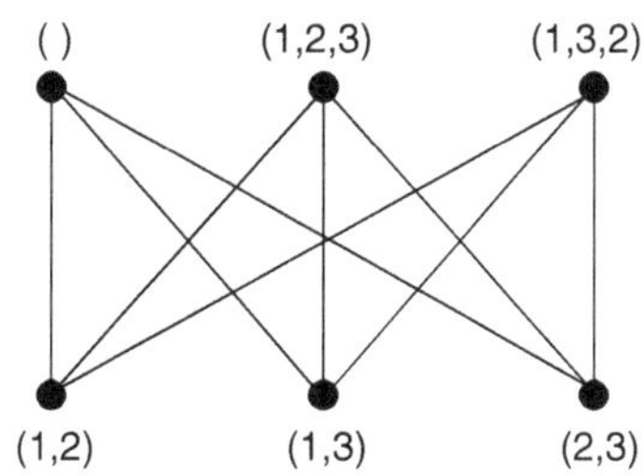

Figure 7.2 $\mathrm{Cay}(S_3, K_{(1,2)})$

occur with multiplicity 1. The eigenvalue

$$\mu_2 = \frac{1}{2}\left|K_{(1,2)}\right| \chi_2((1,2)) = 0$$

occurs with multiplicity 4. Hence, the spectrum of X is given by the multiset $\{-3, 0, 0, 0, 0, 3\}$.

3. AN UPPER BOUND ON $\lambda(X)$

As we saw in the second proof of the Alon-Boppana theorem (see Chapter 3), taking traces of large powers of the adjacency operator gives us information about a graph's eigenvalues. As we saw in Prop. 7.1, the adjacency operator of a Cayley graph decomposes according to irreps. Putting these pieces together, we would expect that irreducible characters should give us information about a Cayley graph's eigenvalues. We present one result along these lines in Prop. 7.8. This result is due to Loh and Schulman [86].

Proposition 7.8
Let G be a finite group, $\Gamma \Subset G$, and $X = \mathrm{Cay}(G, \Gamma)$. Furthermore, assume that X is nonbipartite.

Let $\chi_1, \chi_2, \ldots, \chi_r$ be a complete list of irreducible, inequivalent, nontrivial characters for G. Let d_i be the degree of χ_i for $i = 1, \ldots, r$. For any positive integer m, we have

$$\lambda(X) \leq \left(\sum_{k=1}^{r} d_k \sum_{g\in G} N_g\, \chi_k(g)\right)^{1/2m} \tag{34}$$

where N_g is the number of ways to express g as a product of $2m$ (not necessarily) distinct elements of Γ.

Remark 7.9
Note that the expression within the $2m$th root in the right-hand side of Equation 34 must be a positive real number for the equation to make sense. We point out why this fact is true in the proof.

Proof of Proposition 7.8
Let $d = |\Gamma|$ and $n = |X|$. For each $i = 1, \ldots, r$ let π_i be an irreducible matrix representation corresponding to χ_i. From Proposition 7.1,

$$A \cong (d) \oplus d_1 M_{\pi_1} \oplus \cdots \oplus d_r M_{\pi_r}$$

where $M_{\pi_i} = \sum_{\gamma\in\Gamma} \pi_i(\gamma)$. Note the first component, (d), in the decomposition of A. This is where the eigenvalue d is hiding. The remaining eigenvalues of A lie in the matrices M_{π_i}. Since X is nonbipartite, $-d$ does not occur as an eigenvalue of the matrices $M_{\pi_1}, M_{\pi_2}, \ldots, M_{\pi_r}$.

Let

$$\hat{A} = d_1 M_{\pi_1} \oplus \cdots \oplus d_r M_{\pi_r}$$

be the nontrivial portion of A. The eigenvalues of $\hat{A}$ are $\lambda_{n-1}(X), \ldots, \lambda_1(X)$. Taking the $2m$th power of $\hat{A}$ gives

$$\hat{A}^{2m} = d_1 M_{\pi_1}^{2m} \oplus \cdots \oplus d_r M_{\pi_r}^{2m},$$

where

$$M_{\pi_i}^{2m} = \left(\sum_{\gamma \in \Gamma} \pi_i(\gamma) \right)^{2m} = \sum_{\gamma_1, \ldots, \gamma_{2m} \in \Gamma} \pi_i(\gamma_1 \cdots \gamma_{2m}).$$

Because $\lambda(X)$ occurs as one of the eigenvalues $\lambda_1(X), \ldots, \lambda_{n-1}(X)$, we have that

$$\lambda(X)^{2m} \leq \lambda_1(X)^{2m} + \cdots + \lambda_{n-1}(X)^{2m}.$$

By Lemma A.62, the sum of the $2m$th powers of the eigenvalues of $\hat{A}$ is equal to the trace of $\hat{A}^{2m}$. Hence,

$$\begin{aligned}
\lambda(X)^{2m} &\leq \operatorname{tr}(\hat{A}^{2m}) \\
&= d_1 \operatorname{tr}(M_{\pi_1}^{2m}) + \cdots + d_r \operatorname{tr}(M_{\pi_r}^{2m}) \\
&= \sum_{k=1}^{r} d_k \left(\sum_{\gamma_1, \ldots, \gamma_{2m} \in \Gamma} \chi_k(\gamma_1 \cdots \gamma_{2m}) \right) \\
&= \sum_{k=1}^{r} d_k \sum_{g \in G} N_g \; \chi_k(g).
\end{aligned}$$

Take the $2m$th root of both sides of the equation to get the result. Note that $\sum_{k=1}^{r} d_k \sum_{g \in G} N_g \; \chi_k(g)$ is a positive real number since it is equal to the sum of the $2m$th powers of the eigenvalues of $\hat{A}$. ◈

Remark 7.10
For a bipartite, connected graph define $g \in G$ to be "even" if its word norm in Γ is even and "odd" otherwise. Let

$$\pi(g) = \begin{cases} (1) & \text{if } g \text{ is even.} \\ (-1) & \text{if } g \text{ is odd.} \end{cases}$$

Because G is bipartite, π is a representation of G, and all of the elements of Γ are odd. The representation π gives the eigenvalue $-d$ (see Exercise 1). Modifying the proof of Prop. 7.8 to eliminate M_π from $\hat{A}$ gives a similar statement for bipartite Cayley graphs. If $\mathrm{Cay}(G, \Gamma)$ is not connected, then $\lambda(X) = d$, so there's no need to estimate $\lambda(X)$ in this case.

4. EIGENVALUES OF CAYLEY GRAPHS ON ABELIAN GROUPS

In this section, we describe the representations of any finite abelian group and give a formula for the eigenvalues of a Cayley graph constructed from $\mathbb{Z}_n$.

Recall from Example 6.17 that $\phi_a(k) = (e^{2\pi iak/n})$ is a one-dimensional matrix representation of $\mathbb{Z}_n$ with character $\chi_a(k) = e^{2\pi iak/n}$. We give another proof that these are all the irreducible matrix representations of $\mathbb{Z}_n$. (The first proof was given in Proposition 6.72.)

Proposition 7.11
The irreducible matrix representations of $\mathbb{Z}_n$ are given by

$$\phi_a(k) = \left(e^{\frac{2\pi iak}{n}}\right)$$

for $a = 0, 1, 2, \ldots, n-1$.

Proof
Each ϕ_a is one-dimensional and therefore irreducible. By Lemma 6.71 and Theorem 6.58, the ϕ_a are inequivalent.

Theorem 6.69 applied to $L^2(\mathbb{Z}_n)$ and the fact that $|\mathbb{Z}_n| = \sum_{i=0}^{n-1} 1^2 = \sum_{i=0}^{n-1} \text{degree}(\phi_a)^2$ tell us that we have a complete set of inequivalent, irreducible representations of $\mathbb{Z}_n$. ⍟

Remark 7.12
Note that two one-dimensional matrix representations are inequivalent if and only if they are unequal. This is an easy way to show that the representations given above are inequivalent. For higher dimensional representations, however, one frequently uses the inner product of the two corresponding characters to show that the representations are inequivalent.

For a general abelian group, we have the following. (Recall that every finite abelian group is a direct product of cyclic subgroups.)

Proposition 7.13
Suppose $G = \mathbb{Z}_{n_1} \times \mathbb{Z}_{n_2} \times \cdots \times \mathbb{Z}_{n_r}$. Then the set of irreducible representations of G is

$$\{\phi_a \mid a = (a_1, \ldots, a_r) \in \mathbb{Z}_{n_1} \times \mathbb{Z}_{n_2} \times \cdots \times \mathbb{Z}_{n_r}\},$$

where

$$\phi_a(k_1, k_2, \ldots, k_r) = \left(e^{\frac{2\pi ia_1k_1}{n_1}} e^{\frac{2\pi ia_2k_2}{n_2}} \cdots e^{\frac{2\pi ia_rk_r}{n_r}}\right).$$

Proof
We leave this to the reader as an exercise. ⍟

In the following corollary we give a formula for the eigenvalues of a Cayley graph on $\mathbb{Z}_n$. We leave it as an exercise for the reader to give the corresponding corollary for general finite abelian groups.

Corollary 7.14
Let $\Gamma \Subset \mathbb{Z}_n$ and $\xi = e^{2\pi i/n}$. The eigenvalues of $\text{Cay}(\mathbb{Z}_n, \Gamma)$ are given by

$$\lambda_a = \sum_{\gamma \in \Gamma} \xi^{a\gamma}$$

where $a = 0, 1, \ldots, n-1$.

Proof
This follows from Proposition 7.1 and Proposition 7.11.

Example 7.15
Let $X_n = \text{Cay}(\mathbb{Z}_{2n}, \Gamma_n)$ where $\Gamma_n = \{1, -1, n\}$. For example, the graph $\text{Cay}(\mathbb{Z}_8, \{1, -1, 4\})$ is shown in Figure 7.3. By Corollary 7.14, the eigenvalues of X_n are given by

$$\lambda_a = \sum_{\gamma \in \Gamma_n} e^{\frac{2\pi a}{2n}\gamma} = e^{\frac{\pi ia}{n}} + e^{\frac{-\pi ia}{n}} + e^{\pi ia} = 2\cos\left(\frac{\pi a}{n}\right) + (-1)^a,$$

where $a = 0, 1, \ldots, 2n-1$. Therefore, $\lambda_1(X_n) = 2\cos(\frac{2\pi}{n}) + 1 \to 3$ as $n \to \infty$. Because each X_n has degree 3, we see by Corollary 1.87 that (X_n) is not a family of expanders.

The previous example illustrates the general fact that no sequence of abelian groups yields an expander family. One can construct infinite families of Ramanujan graphs with abelian groups, but one must let the degrees go to infinity as the size of the graphs go to infinity. We give such a family in Section 6 of this chapter.

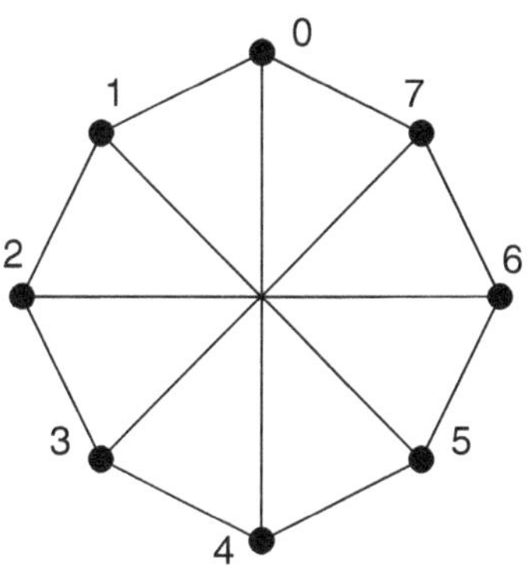

Figure 7.3 $\text{Cay}(\mathbb{Z}_8, \{1, -1, 4\})$

5. EIGENVALUES OF CAYLEY GRAPHS ON DIHEDRAL GROUPS

Here we show how to find the eigenvalues of a Cayley graph generated by a dihedral group D_n where n is even. The case when n is odd is similar and is left to the exercises.

Recall that the dihedral group is given by

$$D_n = \{1, r, r^2, \ldots, r^{n-1}, s, sr, \ldots, sr^{n-1}\},$$

where $r^n = 1$, $s^2 = 1$, and $rs = sr^{-1}$. From these equations, one can show that $r^i s = sr^{-i}$ for any integer i.

Proposition 7.16
If n is even, then a complete list of inequivalent, irreducible matrix representations of the dihedral group D_n is given as follows.

There are 4 matrix representations of degree 1:

	r^k	sr^k
ψ_1	1	1
ψ_2	1	-1
ψ_3	$(-1)^k$	$(-1)^k$
ψ_4	$(-1)^k$	$(-1)^{k+1}$

There are $\frac{n}{2} - 1$ matrix representations of degree 2:

$$\pi_h(r^k) = \begin{pmatrix} \xi^{hk} & 0 \\ 0 & \xi^{-hk} \end{pmatrix}, \quad \pi_h(sr^k) = \begin{pmatrix} 0 & \xi^{-hk} \\ \xi^{hk} & 0 \end{pmatrix},$$

where h is an integer with $1 \leq h \leq n/2 - 1$ and $\xi = e^{2\pi i/n}$.

Proof
We leave it to the reader to show that the ψ_i and π_h are representations of D_n, that is, that they are group homomorphisms. See Exercise 5.

It is clear that the one-dimensional representations ψ_i are irreducible and inequivalent. Let χ_h denote the character of π_h. Suppose that $1 \leq h_1 \leq h_2 \leq n/2 - 1$. Then,

$$\begin{aligned} \langle \chi_{h_1}, \chi_{h_2} \rangle_2 &= \sum_{k=0}^{n-1} \chi_{h_1}(r^k)\overline{\chi_{h_2}(r^k)} + \sum_{k=0}^{n-1} \chi_{h_1}(sr^k)\overline{\chi_{h_2}(sr^k)} \\ &= \sum_{k=0}^{n-1} \left(\xi^{h_1 k} + \xi^{-h_1 k}\right)\left(\xi^{-h_2 k} + \xi^{h_2 k}\right) \\ &= \sum_{k=0}^{n-1} \left(\xi^{(h_1 - h_2)k} + \xi^{(h_1 + h_2)k} + \xi^{-(h_1 + h_2)k} + \xi^{(h_2 - h_1)k}\right). \end{aligned}$$

Since $0 < h_1 \leq h_2 < n/2$, we have $0 < h_1 + h_2 < n$. Thus, $\xi^{h_1+h_2} \neq 1$. Hence,

$$\sum_{k=0}^{n-1} \xi^{(h_1+h_2)k} = \frac{\xi^{(h_1+h_2)n} - 1}{\xi^{h_1+h_2} - 1} = 0.$$

Similarly $\sum_{k=0}^{n-1} \xi^{-(h_1+h_2)k} = 0$. Therefore,

$$\langle \chi_{h_1}, \chi_{h_2} \rangle_2 = \sum_{k=0}^{n-1} \left(\xi^{(h_1-h_2)k} + \xi^{(h_2-h_1)k} \right). \tag{35}$$

If $h_1 \neq h_2$, then $0 < h_2 - h_1 < n$, so $\sum_{k=0}^{n-1} \xi^{(h_2-h_1)k} = \sum_{k=0}^{n-1} \xi^{(h_1-h_2)k} = 0$. So, applying Lemma 6.71 to Equation 35 gives that

$$\langle \chi_{h_1}, \chi_{h_2} \rangle_2 = \begin{cases} |D_n| & \text{if } h_1 = h_2. \\ 0 & \text{if } h_1 \neq h_2. \end{cases}$$

Hence, by Corollary 6.60 and Theorem 6.58, the two-dimensional representations are irreducible and inequivalent.

Theorem 6.69 applied to $L^2(D_n)$, together with the fact that $2n = 1^2 + 1^2 + 1^2 + 1^2 + \sum_{i=1}^{n/2-1} 2^2$, tells us that we have a complete list of inequivalent, irreducible representations of D_n. ⓐ

Corollary 7.17

Let ψ_i and π_h be as in Proposition 7.16. Let n be even, and let $\Gamma \circledS D_n$. For each $i = 0, 1, 2, 3$, let

$$\lambda_{\psi_i} = \sum_{\gamma \in \Gamma} \psi_i(\gamma).$$

For each $1 \leq h \leq n/2 - 1$, define the matrix

$$M_h = \sum_{\gamma \in \Gamma} \pi_h(\gamma).$$

Let $\lambda_h^{(1)}$ and $\lambda_h^{(2)}$ denote the eigenvalues of M_h. Then the spectrum of $\mathrm{Cay}(D_n, \Gamma)$ is

$$\{\lambda_{\psi_1}, \lambda_{\psi_2}, \lambda_{\psi_3}, \lambda_{\psi_4},$$
$$\lambda_1^{(1)}, \lambda_1^{(1)}, \lambda_1^{(2)}, \lambda_1^{(2)}, \lambda_2^{(1)}, \lambda_2^{(1)}, \lambda_2^{(2)}, \lambda_2^{(2)}, \ldots,$$
$$\lambda_{n/2-1}^{(1)}, \lambda_{n/2-1}^{(1)}, \lambda_{n/2-1}^{(2)}, \lambda_{n/2-1}^{(2)}\}.$$

Proof

This follows from Props. 7.1 and 7.16. ⓐ

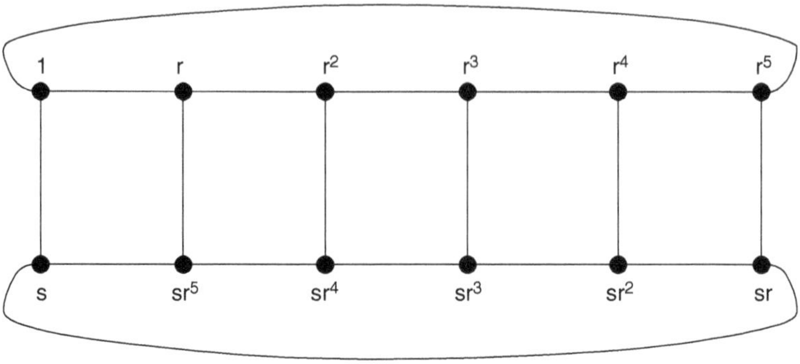

Figure 7.4 $\mathrm{Cay}(D_6, \{r, r^5, s\})$

Example 7.18

Let n be an even positive integer. Let $\Gamma_n = \{r, r^{-1}, s\} \subset D_n$. Let $X_n = \mathrm{Cay}(D_n, \Gamma)$. Note that $r^{-1} = r^{n-1}$ and that X_n is 3-regular. Let ψ_i, λ_{ψ_i}, M_h, π_h, and $\lambda_i^{(j)}$ be as in Prop. 7.17.

For example, if $n = 6$, then

$$D_6 = \{1, r, r^2, r^3, r^4, r^5, s, sr, sr^2, sr^3, sr^4, sr^5\}$$

and $\Gamma = \{r, r^5, s\}$. Figure 7.4 shows a picture of $X = \mathrm{Cay}(D_6, \Gamma)$. The formula $r^i s = sr^{-i} = sr^{6-i}$ shows that s induces the edge between the vertices r^i and sr^{6-i}.

To compute the eigenvalues of X_n we employ Corollary 7.17. We begin by computing the eigenvalues λ_{ψ_i}. Note that $\lambda_{\psi_i} = \psi_i(r) + \psi_i(r^{n-1}) + \psi_i(s)$ for $i = 1, 2, 3, 4$. Hence

$$\begin{aligned} \lambda_{\psi_1} &= 1 + 1 + 1 = 3, \\ \lambda_{\psi_2} &= 1 + 1 - 1 = 1, \\ \lambda_{\psi_3} &= (-1)^1 + (-1)^{n-1} + (-1)^0 = -1, \\ \text{and} \quad \lambda_{\psi_4} &= (-1)^1 + (-1)^{n-1} + (-1)^{0+1} = -3. \end{aligned}$$

We now compute the eigenvalues $\lambda_i^{(j)}$. Note that $M_h = \pi_h(r) + \pi_h(r^{n-1}) + \pi_h(s)$ for $1 \leq h \leq n/2 - 1$. Let $\xi = e^{2\pi i/n}$. Note that $\xi^{-1} = \xi^{n-1}$. Thus

$$\begin{aligned} M_h &= \begin{pmatrix} \xi^h & 0 \\ 0 & \xi^{-h} \end{pmatrix} + \begin{pmatrix} \xi^{(n-1)h} & 0 \\ 0 & \xi^{-h(n-1)} \end{pmatrix} + \begin{pmatrix} 0 & 1 \\ 1 & 0 \end{pmatrix} \\ &= \begin{pmatrix} \xi^h + \xi^{-h} & 1 \\ 1 & \xi^{-h} + \xi^h \end{pmatrix}. \end{aligned}$$

Let $\alpha_h = \xi^h + \xi^{-h}$.

The characteristic polynomial of M_h is

$$\begin{aligned} p_{M_h}(x) = \det(M_h - xI) &= \det \begin{pmatrix} \alpha_h - x & 1 \\ 1 & \alpha_h - x \end{pmatrix} \\ &= x^2 - 2\alpha_h x + (\alpha_h^2 - 1) \\ &= (x - (1 + \alpha_h))(x - (-1 + \alpha_h)). \end{aligned}$$

Hence the eigenvalues of M_h are $\lambda_h^{(1)} = 1 + \alpha_h$ and $\lambda_h^{(2)} = -1 + \alpha_h$. Note that

$$\begin{aligned} \alpha_h &= \xi^h + \xi^{-h} \\ &= \cos\left(\frac{2\pi h}{n}\right) + i \sin\left(\frac{2\pi h}{n}\right) + \cos\left(-\frac{2\pi h}{n}\right) + i \sin\left(-\frac{2\pi h}{n}\right) \\ &= 2\cos\left(\frac{2\pi h}{n}\right). \end{aligned}$$

The eigenvalues of M_h are

$$\lambda_h^{(1)} = 1 + 2\cos(2\pi h/n) \quad \text{and} \quad \lambda_h^{(2)} = -1 + 2\cos(2\pi h/n).$$

We show that as h takes on its values between 1 and $n/2 - 1$ there are relationships between the eigenvalues of the different M_h. Specifically,

$$\begin{aligned} -\lambda_h^{(1)} &= -1 - 2\cos(2\pi h/n) \\ &= -1 + 2\cos(\pi - 2\pi h/n) \\ &= -1 + 2\cos\left(\frac{2\pi}{n}(n/2 - h)\right) \\ &= \lambda_{n/2-h}^{(2)}. \end{aligned}$$

As the value of h increases from 1 to $n/2 - 1$, the value of $n/2 - h$ decreases from $n/2 - 1$ to 1. Therefore, the spectrum of X_n is

$$\begin{gathered} \{-3, -1, 1, 3, \\ 1+2\cos(2\pi/n), 1+2\cos(2\pi/n), -1-2\cos(2\pi/n), -1-2\cos(2\pi/n) \\ 1+2\cos(4\pi/n), 1+2\cos(4\pi/n), -1-2\cos(4\pi/n), -1-2\cos(4\pi/n), \ldots \\ 1+2\cos(2\pi(n/2-1)/n), 1+2\cos(2\pi(n/2-1)/n), -1-2\cos(2\pi(n/2-1)/n), -1-2\cos(2\pi(n/2-1)/n)\}. \end{gathered}$$

Note in particular that $\lambda_1^{(1)} = 1 + 2\cos(2\pi/n) \to 3$ as $n \to \infty$. Since each X_n has degree 3, by Corollary 1.87, we see that (X_n) is not a family of expanders, where n ranges over the positive even integers.

Remark 7.19
In fact, it can be shown that no matter what generating sets you choose, you can never construct an expander family using dihedral groups. This fact is proven in Chapter 4.

One can, however, construct unbounded sequences of Ramanujan graphs using dihedral groups—as long as you let the degree go to infinity. See Note 4 for an interesting example of such a construction.

Example 7.20
Consider the graph $X_6 = \mathrm{Cay}(D_6, \{r, r^{-1}, s\})$ shown in Figure 7.4. Using the notation of Example 7.18, we have that

$$\begin{aligned}
\lambda_1^{(1)} &= 1 + 2\cos(2\pi/6) = 2,\\
-\lambda_1^{(1)} &= -2,\\
\lambda_2^{(1)} &= 1 + 2\cos(4\pi/6) = 0,\\
\text{and}\qquad -\lambda_2^{(1)} &= 0.
\end{aligned}$$

Thus, the spectrum of X_6 is the multiset

$$\{-3, -1, -2, -2, 0, 0, 0, 0, 2, 2, 1, 3\}.$$

6. PALEY GRAPHS

Recall from Chapter 3 that Ramanujan graphs have asymptotically optimal spectral gap. Lubotzky, Philips, Sarnak [89], Morgenstern [100], and Chiu [37] constructed families of d-regular Ramanujan graphs for any fixed d of the form one plus a prime power. The proofs that these graphs are Ramanujan go well beyond the scope of this book.

In this section, we focus on a family of Ramanujan graphs where we relax the condition that the degree be constant and instead let the degrees go to infinity. One trivial family of this sort is the sequence (K_n) of complete graphs (see Example 1.52). The example we present here is less trivial. We produce a sequence of $(p-1)/2$-regular Cayley graphs (X_p), where p indexes over the primes congruent to one modulo four. The eigenvalues of X_p are related to Gauss sums. By deriving an upper bound on the absolute value of any Gauss sum, we prove that each X_p is Ramanujan. Note that (X_p) is not an expander family, because the degree of X_p goes to infinity as p goes to infinity.

Throughout this section, let

$$\mathbb{Z}_p^\times = \mathbb{Z}_p \setminus \{0\} \text{ and } \Gamma_p = \{x^2 \mid x \in \mathbb{Z}_p^\times\}.$$

Note that each element of Γ_p occurs with multiplicity 1; that is, we are thinking of Γ_p as a set, not a multiset.

The proof of the following lemma requires familiarity with elementary field theory. If you like, skip the proof and simply accept the statement of the lemma for now.

Lemma 7.21
If p is a prime and $p \equiv 1 (\bmod 4)$, then there exists an $x \in \mathbb{Z}_p^\times$ such that $x^2 = -1$.

Proof
Since $\mathbb{Z}_p$ is a finite field, we have that $\mathbb{Z}_p^\times$ is a cyclic group under multiplication. Therefore, $\mathbb{Z}_p^\times = \langle g \rangle$ for some $g \in \mathbb{Z}_p^\times$. Consider the equation $(x-1)(x+1) = x^2 - 1 = 0$ in $\mathbb{Z}_p$. If $x \in \mathbb{Z}_p$ solves this equation, then (because p is a prime) either $x - 1 = 0$ or $x + 1 = 0$. Thus, the only solutions to $x^2 = 1$ are $x = \pm 1$. Since $g^0 = 1$ and $g^{(p-1)/2}$ both solve $x^2 = 1$, we must have that $g^{(p-1)/2} = -1$. Because $p \equiv 1 (\bmod 4)$, we may set $x = g^{(p-1)/4}$. Then $x^2 = -1$. ◈

Lemma 7.22
$|\Gamma_p| = \frac{p-1}{2}$.

Proof
Consider the group homomorphism $\phi : \mathbb{Z}_p^\times \to \mathbb{Z}_p^\times$ given by $\phi(x) = x^2$. As in Lemma 7.21, we get that the only solutions to $x^2 = 1$ are $x = \pm 1$. Therefore, $|\Gamma_p| = |\text{Im}(\phi)| = |\mathbb{Z}_p^\times|/|\text{Ker}(\phi)| = (p-1)/2$. ◈

Proposition 7.23
Let p be a prime with $p \equiv 1 (\bmod 4)$.

1. Γ_p is a symmetric subset of $\mathbb{Z}_p$.
2. The graph $\text{Cay}(\mathbb{Z}_p, \Gamma_p)$ is connected and $\frac{p-1}{2}$-regular.

Proof
(1) Since $p \equiv 1 (\bmod 4)$, there exists an $x \in \mathbb{Z}_p^\times$ with $x^2 = -1$. Therefore, if $\gamma = y^2 \in \Gamma_p$, then $-\gamma = (xy)^2 \in \Gamma_p$.

(2) X_p is connected because $1 \in \Gamma_p$. Lemma 7.22 implies that X_p is $(p-1)/(2)$-regular. ◈

Definition 7.24 Let p be a prime with $p \equiv 1 (\bmod 4)$. The graph $\text{Cay}(\mathbb{Z}_p, \Gamma_p)$ is called a *Paley graph*.

Remark 7.25
Some books define $\text{Paley}(p) = \text{Cay}(\mathbb{Z}_p, \Gamma_p)$.

Example 7.26
Consider the graph $\text{Cay}(\mathbb{Z}_{13}, \Gamma_{13})$ shown in Figure 7.5. Note that $\Gamma_{13} = \{1, 3, 4, 9, 10, 12\}$.

Remark 7.27
There are infinitely many primes of the form $4k + 1$ by Dirichlet's theorem [104, p. 401]. So Paley graphs indeed form an infinite family.

To estimate the eigenvalues of X_p, we first relate them to Gauss sums. Then we estimate the absolute value of the Gauss sums.

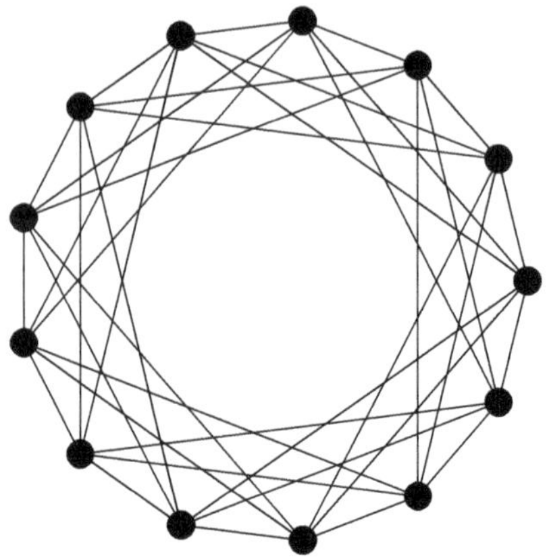

Figure 7.5

Definition 7.28 Let $\xi = e^{2\pi i/p}$. Then

$$G_p(a) = \sum_{k=0}^{p-1} \xi^{ak^2}$$

is called a *Gauss sum.*

By Corollary 7.14, the eigenvalues of the Cayley graph $\mathrm{Cay}(\mathbb{Z}_p, \Gamma_p)$ are given by

$$\lambda_a = \sum_{k \in \Gamma_p} \xi^{ak} = \frac{1}{2}\sum_{k=1}^{p-1} \xi^{ak^2} = \frac{1}{2}\, G_p(a) - \frac{1}{2} \tag{36}$$

where $a = 0, 1, \ldots, p-1$ and $\xi = e^{2\pi i/p}$.

Henceforth, we focus our attention on bounding $G_p(a)$ from above. First, we derive a more useful expression for $G_p(a)$. This formula involves the quadratic reciprocity symbol, or Legendre symbol, which we now introduce.

Definition 7.29 Let $a \in \mathbb{Z}$ and p an odd prime. The *Legendre symbol* is defined to be

$$\left(\frac{a}{p}\right) = \begin{cases} 0 & \text{if } p \text{ divides } a. \\ 1 & \text{if } x^2 \equiv a (\mathrm{mod}\, p) \text{ has a solution.} \\ -1 & \text{if } x^2 \equiv a (\mathrm{mod}\, p) \text{ does not have a solution.} \end{cases}$$

Remark 7.30
The Legendre symbol $\left(\frac{a}{p}\right)$ looks like a fraction, so you may experience some confusion at first if you interpret it as "a divided by p."

Example 7.31
$\left(\frac{-1}{5}\right) = 1$ since $2^2 \equiv -1 (\text{mod } 5)$. But $\left(\frac{2}{5}\right) = -1$ because $x^2 \equiv 2 (\text{mod } 5)$ has no solutions.

The following lemma enumerates some of the properties of the Legendre symbol.

Lemma 7.32
Let $a, b \in \mathbb{Z}$ and p be an odd prime.

1. $\left(\frac{a^2}{p}\right) = 1.$
2. If $a \equiv b (\text{mod } p)$, then $\left(\frac{a}{p}\right) = \left(\frac{b}{p}\right)$.
3. $\left(\frac{a}{p}\right)\left(\frac{b}{p}\right) = \left(\frac{ab}{p}\right)$.

Proof
The reader can work these out as exercises or consult the textbook by Niven, Zuckerman, and Montgomery [104, p. 132]. ◬

Proposition 7.33
If $a \not\equiv 0 (\text{mod } p)$, then

$$G_p(a) = \sum_{k=1}^{p-1} \left(\frac{k}{p}\right) \xi^{ak},$$

where $\xi = e^{2\pi i/p}$.

Proof
Let $T_p = \mathbb{Z}_p^\times \setminus \Gamma_p$. Note that Γ_p is the set of squares in $\mathbb{Z}_p^\times$ and T_p is the set of nonsquares in $\mathbb{Z}_p^\times$. Thus, $\left(\frac{k}{p}\right) = 1$ for all $k \in \Gamma_p$, and $\left(\frac{k}{p}\right) = -1$ for all $k \in T_p$. Therefore,

$$\begin{aligned}\sum_{k=1}^{p-1} \left(\frac{k}{p}\right) \xi^{ak} &= \sum_{k \in \Gamma_p} \xi^{ak} - \sum_{k \in T_p} \xi^{ak} \\ &= 2\sum_{k \in \Gamma_p} \xi^{ak} + 1 - \sum_{k=0}^{p-1} \xi^{ak}.\end{aligned}$$

By the geometric sum formula, we have that $\sum_{k=0}^{p-1} \xi^{ak} = \frac{(\xi^{ak})^p - 1}{\xi^{ak} - 1} = 0$. Thus

$$\begin{aligned}\sum_{k=1}^{p-1} \left(\frac{k}{p}\right) \xi^{ak} &= 2\sum_{k \in \Gamma_p} \xi^{ak} + 1 \\ &= \sum_{k=0}^{p-1} \xi^{ak^2} = G_p(a).\end{aligned}$$

◬

Proposition 7.34
If $a \not\equiv 0 (\mathrm{mod}\, p)$, then

1. $G_p(a) = \left(\frac{a}{p}\right) G_p(1)$.
2. $|G_p(a)| = \sqrt{p}$.

Proof
Let $\xi = e^{2\pi i/p}$.

(1) Because $\mathbb{Z}_p^\times$ is a group under multiplication, for all $a \in \mathbb{Z}_p^\times$ the map $a \to ak$ is a bijection. Hence, by Lemma 7.32,

$$\left(\frac{a}{p}\right) G_p(a) = \left(\frac{a}{p}\right) \sum_{k=1}^{p-1} \left(\frac{k}{p}\right) \xi^{ak} = \sum_{k=1}^{p-1} \left(\frac{ak}{p}\right) \xi^{ak} = \sum_{k=1}^{p-1} \left(\frac{k}{p}\right) \xi^{k} = G_p(1).$$

(2) By part (1) it is sufficient to show that $|G_p(1)| = \sqrt{p}$. We have that

$$\begin{aligned} G_p(1)^2 &= \left(\sum_{m=1}^{p-1} \left(\frac{m}{p}\right) \xi^m\right) \left(\sum_{n=1}^{p-1} \left(\frac{n}{p}\right) \xi^n\right) \\ &= \sum_{m=1}^{p-1} \sum_{n=1}^{p-1} \left(\frac{m}{p}\right) \xi^m \left(\frac{mn}{p}\right) \xi^{mn} \\ &= \sum_{n=1}^{p-1} \sum_{m=1}^{p-1} \left(\frac{m^2 n}{p}\right) \xi^{m+mn} \\ &= \sum_{n=1}^{p-1} \sum_{m=1}^{p-1} \left(\frac{n}{p}\right) \xi^{m+mn}, \end{aligned}$$

where the last equality is by Lemma 7.32. By the geometric sum formula, we have that

$$\sum_{m=1}^{p-1} \left(\xi^{1+n}\right)^m = \begin{cases} p-1 & \text{if } n = p-1 \\ -1 & \text{if } n = 1, 2, \ldots, p-2 \end{cases}.$$

Let T_p be defined as in Proposition 7.33. By Lemma 7.22, we have that $\left|\Gamma_p\right| = \left|T_p\right| = \frac{p-1}{2}$. Therefore,

$$\begin{aligned} G_p(1)^2 &= -\sum_{n=1}^{p-2} \left(\frac{n}{p}\right) + (p-1)\left(\frac{p-1}{p}\right) \\ &= -\sum_{n=1}^{p-1} \left(\frac{n}{p}\right) + p\left(\frac{p-1}{p}\right) \end{aligned}$$

$$= -\left[\sum_{n\in\Gamma_p} 1 + \sum_{n\in T_p}(-1)\right] + p\left(\frac{p-1}{p}\right)$$

$$= p\left(\frac{p-1}{p}\right) = \left(\frac{-1}{p}\right)p = p,$$

where the last equality is by Lemma 7.21. Hence, $|G_p(1)| = \sqrt{p}$.

Remark 7.35
In Prop. 7.34 we calculated the absolute value of $G_p(1)$. One can calculate the exact value of this sum. See Note 7.

Proposition 7.36
Recall Def. 3.7. If $p \equiv 1 \pmod 4$, then $\mathrm{Cay}(\mathbb{Z}_p, \Gamma_p)$ is a Ramanujan graph.

Proof
If $a = 0$, then $\lambda_0 = \frac{p-1}{2}$ is the trivial eigenvalue. If $1 \leq a \leq p-1$, then by Exercise 2,

$$|\lambda_a| = \left|\frac{1}{2}G_p(a) - \frac{1}{2}\right| \leq \frac{1}{2}\left(\sqrt{p}+1\right) \leq 2\sqrt{\frac{p-1}{2} - 1}.$$

Remark 7.37
Paley graphs have a lot of edges—the degree is about half the order (number of vertices). Moreover, the generating set Γ_p is all "mixed up," insofar as the group law on $\mathbb{Z}_p$ is addition, but Γ_p is defined in terms of the multiplicative structure. The fact that they are Ramanujan reflects these properties.

NOTES

1. Prop. 7.8 is quite difficult to use to get a good eigenvalue bound for any particular family of Cayley graphs, since the numbers N_g are nontrivial to compute or estimate. Loh and Schulman [86] use Prop. 7.8 to obtain results about *random* Cayley graphs. See Notes 11 and 12 of Chapter 1.
2. Friedman, Murty, and Tillich [60] give an asymptotic lower bound for the second-largest eigenvalue of a Cayley graph on an abelian group: For a fixed d, a d-regular Cayley graph on an abelian group of order n has the second-largest eigenvalue bounded below by $d - O(dn^{-4/d})$, where the implied constant is absolute. They give two proofs of this fact. One proof uses a sphere packing bound and the second proof uses graph covers and an eigenvalue "pushing" argument.
3. In [118], Roichman gives the following results on expansion properties of symmetric groups and alternating groups. The proofs combine spectral and representation-theoretic techniques. Let X be a graph with vertex set V and edge set E. We make the following conventions in this note. We say that X is an ϵ-expander if every set of vertices A with $|A| \leq |V|/2$ satisfies $\left|N(A)\right| \geq (1+\epsilon)|A|$, where

$$N(A) = \{y \in V \mid y \in A \text{ or there exists an } x \in A \text{ such that } \mathrm{dist}(x,y) = 1\}.$$

Let C be a conjugacy class of the symmetric group S_n. Recall that C consists of all of the permutations of a given cycle type. Hence we may define the support of C, denoted by $\text{supp}(C)$, to be the number of nonfixed digits under the action of a permutation in C. If C has "large" support, then each element of C "moves" a lot of letters. If the permutations in C are all even, then we say that C is an even conjugacy class. If the permutations in C are all odd, we say that C is an odd conjugacy class. For $n \geq 5$, the subgroup generated by C, denoted by $\langle C \rangle$, is S_n if C is odd, and A_n if C is even.

Recall Definition B.4. For each $n \geq 5$, let C_n be a conjugacy class of S_n. Roichman proves the following. If $\text{supp}(C_n) = o(\sqrt{n})$, then there does not exist $\epsilon > 0$ such that $(\text{Cay}(\langle C_n \rangle, C_n))$ is a family of ϵ-expanders. If $\text{supp}(C_n) = \Omega(n)$, then there is an $\epsilon > 0$ such that $(\text{Cay}(\langle C_n \rangle, C_n))$ is a family of ϵ-expanders.

4. We present Schellwat's construction [124] of a sequence of Ramanujan Cayley graphs on dihedral groups. Let p be an odd prime and $\mathbb{E}$ be a finite extension of $\mathbb{Z}_p$ of degree 2. Thus, $|\mathbb{E}| = p^2$, and $\mathbb{E} = \mathbb{Z}_p(\theta)$ where θ satisfies some monic, irreducible quadratic polynomial over $\mathbb{Z}_p$. Because $\mathbb{E}^\times = \mathbb{E} \setminus \{0\}$ is cyclic, there is an element $g \in \mathbb{E}$ such that $\mathbb{E}^\times = \langle g \rangle$. Let

$$\hat{\Gamma}_p = \{z + \theta \mid z \in \mathbb{Z}_p\}.$$

For each $z \in \mathbb{Z}_p$ we can write $z + \theta = g^{a_z}$, where a_z is uniquely determined and $1 \leq a_z \leq p^2 - 1$. Set

$$\Gamma_p = \{sr^{a_z} \mid z \in \mathbb{Z}_p\}.$$

Let $X_p = \text{Cay}(D_{p^2-1}, \Gamma_p)$. Then X_p is a connected, bipartite, p-regular graph on $2p^2 - 2$ vertices. Moreover, Schellwat shows that $\left|\lambda(X_p)\right| \leq \sqrt{p} \leq 2\sqrt{p-1}$, so X_p is Ramanujan.

For example, let $p = 3$. The polynomial $p(x) = x^2 + x + 2$ is irreducible over $\mathbb{Z}_3$. Let $\mathbb{E} = \{a + b\theta \mid a, b \in \mathbb{Z}_3\}$ where θ is a root of $p(x)$. Therefore, $\theta^2 = 2\theta + 1$. Then, $\mathbb{E}^\times = \langle 1 + \theta \rangle$ and the powers of $1 + \theta$ are as follows:

n	1	2	3	4	5	6	7	8
$(1+\theta)^n$	$1+\theta$	$2+\theta$	2θ	2	$2+2\theta$	$1+2\theta$	θ	1

Therefore,

$$\hat{\Gamma}_3 = \{\theta, 1+\theta, 2+\theta\} = \{(1+\theta)^7, (1+\theta)^1, (1+\theta)^2\}$$

and $X_3 = \text{Cay}(D_8, \{sr, sr^2, sr^7\})$.

5. Terras and her students (see [129]) have studied the spectrum and properties of finite upper half-plane graphs. Let $q = p^r$, where p is an odd prime, and δ a nonsquare element of the finite field with q elements, $\mathbb{F}_q$. The finite upper half-plane H_q is given by $H_q = \{x + y\sqrt{\delta} \mid x, y \in \mathbb{F}_q, y \neq 0\}$. Given $z = x + y\sqrt{\delta}$

and $w = u + v\sqrt{\delta}$ define the "distance" between z and w to be

$$\text{dist}(z, w) = \frac{\text{N}(z - w)}{\text{Im}(z)\text{Im}(w)} = \frac{(x-u)^2 - \delta(y-v)^2}{yv}.$$

For each $a \in \mathbb{F}_q$, construct a graph $X_q(\delta, a)$ as follows: the vertices of the graph are the elements of H_q. Two vertices z and w are adjacent if and only if $\text{dist}(z, w) = a$. If $a \neq 0, 4\delta$, then $X_q(\delta, a)$ is a $(q+1)$-regular, connected, Ramanujan graph. Finite upper half plane graphs are examples of finite symmetric spaces. See [81] for another example.

6. Let $\mathbb{F}_q$ be the finite field of size $q = p^r$ where p is an odd prime. The qth Platonic graph, $G^*(q)$, has vertex set

$$V = \{(\alpha, \beta) \mid \alpha, \beta \in \mathbb{F}_q, (\alpha, \beta) \neq (0, 0)\}/(\pm 1).$$

Two vertices (α, β) and (γ, δ) are adjacent if and only if $\det \begin{pmatrix} \alpha & \beta \\ \gamma & \delta \end{pmatrix} = \pm 1$. The spectra of the Platonic graphs are determined by Dedeo, Lanphier, and Minei [47]. They show that the spectrum of $G^*(q)$ contains the eigenvalue q with multiplicity 1, -1 with multiplicity q, and $\pm\sqrt{q}$ each with multiplicity $(q+1)(q-3)/4$. This implies that $G^*(q)$ is a q-regular Ramanujan graph. In addition, they show that the isoperimetric constant of $G^*(q)$ satisfies

$$\frac{q}{2} - \frac{\sqrt{q}}{2} \leq h(G^*(q)) \leq \begin{cases} \frac{q(q-1)}{2(q+1)} & \text{if } q \equiv 1 (\text{mod } 4) \\ \frac{q}{2} & \text{if } q \equiv 3 (\text{mod } 4). \end{cases}$$

7. We followed the presentation given in [102] for the proofs of Propositions 7.33 and 7.34.

The proof that Paley graphs are Ramanujan used character sums that are called Gauss sums. Gauss sums are ubiquitous in modern number theory. For an introduction to Gauss sums and number theory in general, see [72]. One can show that the following is true for any positive integer n:

$$G_n = \sum_{k=0}^{n-1} e^{\frac{2\pi i k^2}{n}} = \begin{cases} (1+i)\sqrt{n} & , \text{ if } n \equiv 0 (\text{mod } 4) \\ \sqrt{n} & , \text{ if } n \equiv 1 (\text{mod } 4) \\ 0 & , \text{ if } n \equiv 2 (\text{mod } 4) \\ i\sqrt{n} & , \text{ if } n \equiv 3 (\text{mod } 4) \end{cases}.$$

For a proof of this fact using residue theory see [18, p. 258]. For a proof when n is a prime, see [129, p. 144]. One can use Gauss sums to prove the quadratic reciprocity law and count points on curves over finite fields [72].

8. Paley graphs are examples of *strongly regular graphs*—that is, regular graphs for which there exist integers x, y such that any two adjacent vertices have x common neighbors and any two nonadjacent vertices have y common neighbors. The strong regularity condition implies that the adjacency operator A satisfies

a quadratic equation with coefficients involving x and y. From this equation we can find the eigenvalues of A. (See [66] for a discussion of this approach. Paley graphs there are defined more generally for any finite field, not just $\mathbb{Z}_p$.) We saw in Section 6, though, that the nontrivial eigenvalues of A are related to Gauss sums. It may well be possible, then, to evaluate Gauss sums through strictly graph-theoretic means, that is, by using strong regularity to compute eigenvalues of Paley graphs.

EXERCISES

1. Suppose that ρ is a representation of a finite group G of degree 1. Let $\Gamma \Subset G$. Let A be the adjacency matrix of $\mathrm{Cay}(G, \Gamma)$. Prove that the eigenvalue associated with ρ is $\sum_{\gamma \in \Gamma} \mathrm{tr}(\rho(\gamma))$.

2. Prove that $\frac{1}{2}\left(\sqrt{p}+1\right) \leq 2\sqrt{\frac{p-1}{2}} - 1$ for all $p \geq 5$. This completes the proof of Proposition 7.36.

3. Prove Proposition 7.13.

4. State and prove an analogue to Corollary 7.14 for an arbitrary finite abelian group.

5. Let ψ_i and π_h be as in Proposition 7.16. Show that these functions are representations of D_n. That is, show that they are group homomorphisms.

6. Calculate the eigenvalues of C_n and K_n by thinking of them as Cayley graphs on $\mathbb{Z}_n$.

7. Let $X_n = \mathrm{Cay}(\mathbb{Z}_n, B(r))$, where

$$B(r) = \{-r, -r+1, \ldots, -2, -1, 0, 1, 2, \ldots, r-1, r\}.$$

   Calculate the eigenvalues of X_n. Use your calculation to show that if r is fixed, then $(X_n)_{n=2r+1}^{\infty}$ does not form an expander family.

8. Compute the eigenvalues of

$$\mathrm{Cay}(\mathbb{Z}_{2n}, \{1, 3, 5, \ldots, 2n-1\}).$$

   When is the graph Ramanujan?

9. Let $c \geq 2$ be a positive integer and $n \equiv 0 (\mathrm{mod}\, c)$. Compute the eigenvalues of $\mathrm{Cay}(\mathbb{Z}_n, \Gamma)$ where

$$\Gamma = \{1, -1\} \cup \{kc \mid 0 < k < n/c\}.$$

   When is the graph Ramanujan?

10. Prove Proposition A.64 using Corollary 7.14. (Hint: Every circulant matrix is an adjacency matrix for a certain Cayley graph of $\mathbb{Z}_n$.)

11. In Proposition 7.16 prove π_h is an irrep by showing that the two-dimensional representations have no invariant subspaces.

12. Let G be a finite group. Let $\gamma \in G$ such that $\gamma^2 = e$. Let Γ be the conjugacy class of γ. Define $\theta : G \to \text{Aut}(G)$ by $[\theta(a)](x) = axa^{-1}$. Let $\Upsilon = \{a\gamma a^{-1}\gamma b\gamma b^{-1} \mid (a, b) \in G \times G\}$. Use Props. 7.4 and 5.39 to give an upper bound for $\lambda(\text{Cay}(G \rtimes_\theta G, \Upsilon))$ in terms of the irreducible characters of G.

13. Prove the following. If n is odd, then a complete list of inequivalent irreps of the dihedral group D_n is given as follows.
There are two representations of degree 1 given by the following table.

	r^k	sr^k
ψ_1	1	1
ψ_2	1	-1

There are $\frac{1}{2}(n-1)$ representations of degree 2 given by

$$\pi_h(r^k) = \begin{pmatrix} \xi^{hk} & 0 \\ 0 & \xi^{-hk} \end{pmatrix}, \quad \pi_h(sr^k) = \begin{pmatrix} 0 & \xi^{-hk} \\ \xi^{hk} & 0 \end{pmatrix},$$

where $h \in \mathbb{Z}$ and $1 \le h \le (n-1)/2$ and $\xi = e^{2\pi i/n}$.

14. Use Exercise 13 to find the eigenvalues of $X_n = \text{Cay}(D_n, \{r, r^{-1}, s\})$ when n is odd. Show that $\lambda_1(X_n) \to 3$ as $n \to \infty$.

The following definition will be needed in Exercises 15–17.

Definition 7.38 Let G be a finite group and $f, g \in L^2(G)$. Define the *convolution* of f and g, denoted by $f * g$, as follows. For $x \in G$,

$$(f * g)(x) = \sum_{t \in G} f(xt^{-1})g(t) = \sum_{t \in G} f(t)g(t^{-1}x).$$

Note that $f * g \in L^2(G)$.

15. Let

$$Z[L^2(G)] = \{f \in L^2(G) \mid f * g = g * f \text{ for all } g \in L^2(G)\}$$

and let

$$S = \{f \in L^2(G) \mid f(g) = f(xgx^{-1}) \text{ for all } x, g \in G\}.$$

This exercise shows that $Z[L^2(G)] = S$. That is, $Z[L^2(G)]$ consists of the functions that are constant on conjugacy classes. Do this by showing the following.

(a) Let $f \in S$ and $h \in L^2(G)$. Show that $(h * f)(x) = (f * h)(x)$ for all $x \in G$.
(b) Now suppose $f \in Z[L^2(G)]$ and $x, a \in G$. Define $\delta_a \in L^2(G)$ where $\delta_a(x) = 1$ if $x = a$, and $\delta_a(x) = 0$ if $x \neq a$. Show that $(f * \delta_a)(x) = f(xa^{-1})$ and $(\delta_a * f)(x) = f(a^{-1}x)$. This implies that $f(aga^{-1}) = f(g)$ for all $g, a \in G$.

16. See Exercise 15 for notation. Let G be a group and $g \in G$.

(a) Prove that $R(g)f = f * \delta_{g^{-1}}$.
(b) $\delta_a * \delta_b = \delta_{ab}$ for all $a, b \in G$.
(c) Conclude that $R(gh)f = R(g)(R(h)f)$ using the convolution definition of the action of R.

17. Let X and Γ be as in Proposition 7.4. See Exercise 15 for notation. Prove that the adjacency operator A of X satisfies $Af = \sum_{\gamma \in \Gamma} f * \delta_{\gamma^{-1}}$ for all $f \in L^2(G)$.

8

Kazhdan Constants

Let $X = \text{Cay}(G, \Gamma)$ be a finite Cayley graph. We have seen in previous chapters that the expansion constant of X is closely related to its spectral gap, which in turn is determined by the irreducible representations of G. Nevertheless, the spectral gap can still be quite difficult to compute or even estimate. In this chapter, we introduce a representation-theoretic invariant of the pair (G, Γ), namely, the Kazhdan constant. A sequence $(\text{Cay}(G_n, \Gamma_n))$ of finite Cayley graphs with fixed degree forms a family of expanders iff the corresponding Kazhdan constants are uniformly bounded away from zero. Although it can still be quite difficult to explicitly compute Kazhdan constants, the inequalities in this chapter provide some techniques for finding lower bounds. This has been one of the chief methods for demonstrating that certain sequences of Cayley graphs are expander families—see the Notes section for references to examples in the literature of this approach in action.

1. KAZHDAN CONSTANT BASICS

In this section, we define the Kazhdan constant and give some basic facts about it.

Let G be a finite group, and let $\Gamma \subset G$. Let ρ be a unitary representation of G on some representation space V with G-invariant inner product $\langle \cdot, \cdot \rangle$. Recall that the norm $\|\cdot\|$ associated with $\langle \cdot, \cdot \rangle$ is defined by $\|\mathbf{v}\| = \langle \mathbf{v}, \mathbf{v} \rangle$. When $\Gamma \neq \emptyset$, we define

$$\kappa(G, \Gamma, \rho, \langle \cdot, \cdot \rangle) = \min_{\|\mathbf{v}\|=1} \max_{\gamma \in \Gamma} \left\| \rho(\gamma)\mathbf{v} - \mathbf{v} \right\|. \tag{37}$$

As a special case, define $\kappa(G, \emptyset, \rho, \langle \cdot, \cdot \rangle) = 0$. Note that whether Γ is a set or a multiset is immaterial; either way, we maximize over the elements of Γ. Also note that we do not require Γ to be symmetric.

Remark 8.1

One may well ask whether the minimum in the left-hand side of Equation 37 necessarily exists. It does, and here's why. First, identify V with $\mathbb{C}^n$ by choosing a basis for V. Then note that $\left\| \rho(\gamma)\mathbf{v} - \mathbf{v} \right\|$ defines a continuous function, because when you write everything down in terms of the basis, after the dust settles, the only operations involved are addition, subtraction, multiplication, and a square root of a non-negative number. Since Γ has finitely many elements, we are therefore taking the max of a finite number of continuous functions; hence $\max_{\gamma \in \Gamma} \|\rho(\gamma)\mathbf{v} - \mathbf{v}\|$ is continuous. The unit sphere in V is closed and bounded hence compact. A continuous function on a compact set attains a minimum, and so Equation 37 is legitimate. If we were to work with infinite-dimensional representations, we would need to take an infimum instead of a minimum.

Proposition 8.2
Let G, Γ, ρ be as above. Let $\langle \cdot, \cdot \rangle$ and $\langle \cdot, \cdot \rangle'$ be two G-invariant inner products on V with norms $\|\cdot\|$ and $\|\cdot\|'$, respectively. Then

$$\kappa(G, \Gamma, \rho, \langle \cdot, \cdot \rangle) = \kappa(G, \Gamma, \rho, \langle \cdot, \cdot \rangle').$$

Somewhat sketchy proof
We present the main ideas here; the reader should check details and fill in the gaps. Using $\langle \cdot, \cdot \rangle$, decompose V into an orthogonal direct sum of irreps

$$V = V_1 \oplus V_2 \oplus \cdots \oplus V_n,$$

as in the proof of Maschke's theorem (Theorem 6.43).

Also do the same with $\langle \cdot, \cdot \rangle'$ to get

$$V = V_1' \oplus V_2' \oplus \cdots \oplus V_m'.$$

Then, by Remark 6.66, we have that $n = m$ and (perhaps after re-ordering terms) $V_1 \cong V_1', \ldots, V_n \cong V_n'$. Using the foregoing isomorphisms, let $\phi : V \to V$ be a G-isomorphism with $\phi(V_1) = V_1', \ldots, \phi(V_n) = V_n'$.

Define an inner product $\langle \cdot, \cdot \rangle''$ on V by

$$\langle \mathbf{v}, \mathbf{w} \rangle'' = \langle \phi(\mathbf{v}), \phi(\mathbf{w}) \rangle'.$$

(Check that this really is an inner product.) Note that $V_1 \oplus V_2 \oplus \cdots \oplus V_n$ is an orthogonal direct sum with respect to $\langle \cdot, \cdot \rangle''$. By Prop. 6.83, we know that there exist positive real numbers $C_1, \ldots, C_n$ such that

$$\langle \mathbf{v}, \mathbf{w} \rangle'' = C_i \langle \mathbf{v}, \mathbf{w} \rangle$$

whenever $\mathbf{v}, \mathbf{w} \in V_i$. So the map

$$\mathbf{v}_1 + \cdots + \mathbf{v}_n \mapsto \frac{\phi(\mathbf{v}_1)}{\sqrt{C_1}} + \cdots + \frac{\phi(\mathbf{v}_n)}{\sqrt{C_n}}$$

defines a bijection from $\{\|\mathbf{v}\| = 1\}$ to $\{\|\mathbf{v}\|' = 1\}$. (Check this.) Observe that by Prop. A.21

$$\begin{aligned}\left\| \rho(\gamma) \cdot \sum_{i=1}^{n} \mathbf{v}_i - \sum_{i=1}^{n} \mathbf{v}_i \right\|^2 &= \sum_{i=1}^{n} \left\| \rho(\gamma) \cdot \mathbf{v}_i - \mathbf{v}_i \right\|^2 \\ &= \sum_{i=1}^{n} C_i^{-1} \left(\left\| \rho(\gamma) \cdot \mathbf{v}_i - \mathbf{v}_i \right\|'' \right)^2 \\ &= \sum_{i=1}^{n} C_i^{-1} \left(\left\| \rho(\gamma) \cdot \phi(\mathbf{v}_i) - \phi(\mathbf{v}_i) \right\|' \right)^2 \\ &= \left(\left\| \rho(\gamma) \cdot \sum_{i=1}^{n} \frac{\phi(\mathbf{v}_i)}{\sqrt{C_i}} - \sum_{i=1}^{n} \frac{\phi(\mathbf{v}_i)}{\sqrt{C_i}} \right\|' \right)^2.\end{aligned}$$

The result follows. ⊛

Prop. 8.2 shows that the quantity $\kappa(G, \Gamma, \rho) := \kappa(G, \Gamma, \rho, \langle \cdot, \cdot \rangle)$ is independent of the choice of G-invariant inner product. (Recall from Lemma 6.42 that such an inner product necessarily exists.)

Remark 8.3

Recall Remark 8.1. The fact that a minimum exists implies that given any unitary irrep $\rho : G \to GL(V)$, there exists a unit vector $\mathbf{v} \in V$ and $\gamma \in \Gamma$ such that $\left\| \rho(\gamma) \cdot \mathbf{v} - \mathbf{v} \right\| = \kappa(G, \Gamma, \rho)$.

Lemma 8.4

If ϕ and ρ are equivalent representations, then $\kappa(G, \Gamma, \phi) = \kappa(G, \Gamma, \rho)$.

Proof

The proof is utterly standard: use the isomorphism between ϕ and ρ as a "dictionary" to "translate" back and forth between the two representations. For the sake of readers unfamiliar with this sort of argument, we fill in the details.

Let V, W be the representation spaces for ϕ, ρ, respectively. Let ψ be a G-isomorphism from V to W. Let $\langle \cdot, \cdot \rangle$ be a G-invariant inner product on W. Define

$$\langle \mathbf{v}_1, \mathbf{v}_2 \rangle' = \langle \psi(\mathbf{v}_1), \psi(\mathbf{v}_2) \rangle.$$

It is straightforward to verify that $\langle \cdot, \cdot \rangle'$ defines a G-invariant inner product on V. Note that $\|\mathbf{v}\|' = 1$ iff $\left\| \psi(\mathbf{v}) \right\| = 1$. So

$$\begin{aligned}\kappa(G, \Gamma, \phi) &= \min_{\|\mathbf{v}\|'=1} \max_{\gamma \in \Gamma} \left\| \phi(\gamma)\mathbf{v} - \mathbf{v} \right\|' \\ &= \min_{\|\mathbf{v}\|'=1} \max_{\gamma \in \Gamma} \left\| \psi(\phi(\gamma)\mathbf{v} - \mathbf{v}) \right\| \\ &= \min_{\|\mathbf{v}\|'=1} \max_{\gamma \in \Gamma} \left\| \rho(\gamma)\psi(\mathbf{v}) - \psi(\mathbf{v}) \right\| \\ &= \min_{\|\mathbf{w}\|=1} \max_{\gamma \in \Gamma} \left\| \rho(\gamma)\mathbf{w} - \mathbf{w} \right\| = \kappa(G, \Gamma, \rho).\end{aligned}$$

⊛

Definition 8.5 Let G be a finite nontrivial group, and let $\Gamma \subset G$. Define

$$\kappa(G,\Gamma) = \min_{\rho} \left\{\kappa(G,\Gamma,\rho)\right\},$$

where the minimum is over all irreducible, nontrivial, unitary representations of G. The quantity $\kappa(G,\Gamma)$ is called the *Kazhdan constant* of the pair (G,Γ).

Remark 8.6
Recall from Theorem 6.69 that G has only finitely many irreps, up to equivalence. Moreover, Corollary 6.74 guarantees that there are nontrivial irreps of G. Therefore, by Lemma 8.4, the Kazhdan constant minimizes over a finite nonempty set of values; hence, a minimum exists.

What does Def. 8.5 say in English? If κ is the Kazhdan constant of the pair (G,Γ), then it says the following. Given any nontrivial unitary irrep ρ and any unit vector $\mathbf{v}$ in its representation space, there exists an element in Γ that moves $\mathbf{v}$ a distance of at least κ. Moreover, κ is the largest number for which the preceding statement is true.

The following computation occurs frequently in examples in the sequel.

Lemma 8.7
Let $\xi = e^{2\pi i\theta}$. Then $|\xi - 1| = 2\sin\frac{\theta}{2}$.

Proof
Interpret $|\xi - 1|$ as the distance from ξ to 1 in the complex plane, as in Figure 8.1. Rotate the unit circle clockwise by an angle of $\theta/2$, to obtain a picture like that in Figure 8.2. The result follows from elementary trigonometry. ∎

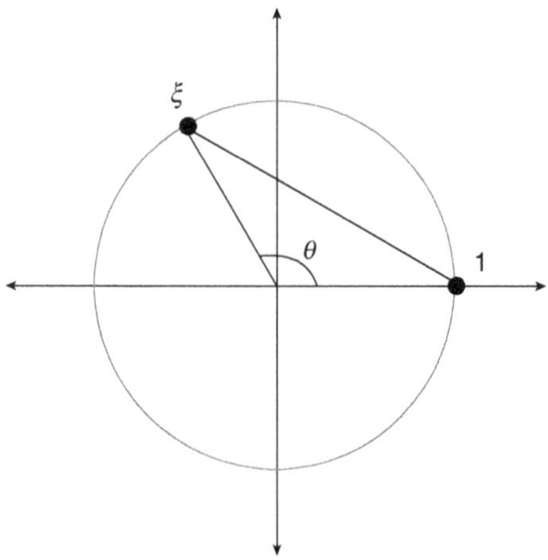

Figure 8.1

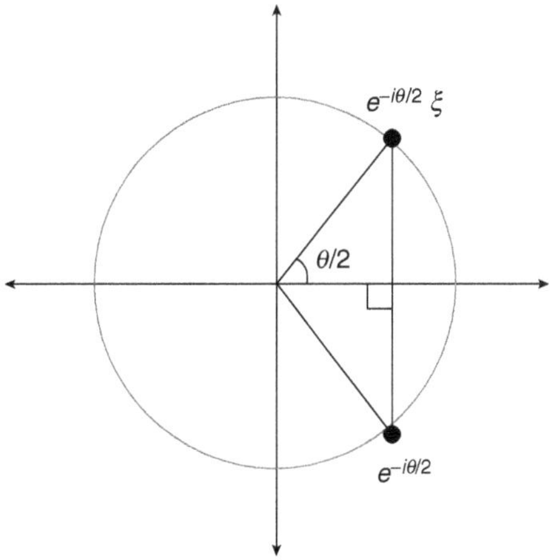

Figure 8.2

Example 8.8
Let $G = \mathbb{Z}_5$, and let $\Gamma = \{1\}$. Let $\xi = e^{2\pi i/5}$. Define $\rho_a : G \to GL(\mathbb{C})$ by $\rho_a(k)z = \xi^{ak}z$. We know from Prop. 6.72 that up to equivalence, ρ_1, ρ_2, ρ_3, and ρ_4 are the only nontrivial unitary irreps of G. Note that $\rho_1(1)$ is counterclockwise rotation by an angle of 72°; $\rho_2(1)$ is counterclockwise rotation by an angle of 144°; $\rho_3(1)$ is clockwise rotation by an angle of 144°; and $\rho_4(1)$ is clockwise rotation by an angle of 72°. Let $\mathbf{v}$ be any unit vector in $\mathbb{C}$. (That is, $\mathbf{v}$ lies on the unit circle in the complex plane.) Then

$$\left\|\rho_1(1)\mathbf{v} - \mathbf{v}\right\| = \left\|\rho_4(1)\mathbf{v} - \mathbf{v}\right\| = 2\sin 36^\circ, \text{ and}$$
$$\left\|\rho_2(1)\mathbf{v} - \mathbf{v}\right\| = \left\|\rho_3(1)\mathbf{v} - \mathbf{v}\right\| = 2\sin 72^\circ,$$

by Lemma 8.7. Therefore $\kappa(G, \Gamma) = \min\{2\sin 36^\circ, 2\sin 72^\circ\} = 2\sin 36^\circ$.

Example 8.9
Generalizing Example 8.8, we now show that $\kappa(\mathbb{Z}_n, \{1\}) = 2\sin(\frac{\pi}{n})$. Prop. 6.72 tells us that up to equivalence, the irreducible representations of $\mathbb{Z}_n$ are the maps $\rho_a : \mathbb{Z}_n \to GL(\mathbb{C})$ defined by $\rho_a(k)z = \xi^{ak}z$, where $\xi = \exp(\frac{2\pi i}{n})$ and $0 \le a \le n-1$. If $1 \le a \le n-1$, then

$$\kappa(\mathbb{Z}_n, \{1\}, \rho_a) = \|\xi^a - 1\| = 2\sin\left(\frac{\pi a}{n}\right)$$

by Lemma 8.7. Therefore,

$$\kappa(\mathbb{Z}_n, \{1\}) = \min_{a=1,2,\ldots,n-1} \kappa(\mathbb{Z}_n, \{1\}, \rho_a) = 2\sin\left(\frac{\pi}{n}\right).$$

Example 8.10
Let $G = S_n$, and let $\Gamma \subset S_n$. Recall Example 6.19. Let π be the alternating representation acting on $\mathbb{C}$—using Remark 6.20, we regard π as a representation, not as a matrix representation. Let $\mathbf{v}$ be a complex number of norm 1 in $\mathbb{C}$.

For any $\gamma \in \Gamma$, we have that

$$\left\|\pi(\gamma)\mathbf{v} - \mathbf{v}\right\| = \begin{cases} \|-\mathbf{v} - \mathbf{v}\| = 2 & \text{if } \gamma \text{ is odd} \\ \|\mathbf{v} - \mathbf{v}\| = 0 & \text{if } \gamma \text{ is even} \end{cases}.$$

Hence

$$\kappa(G, \Gamma, \pi) = \begin{cases} 2 & \text{if } \Gamma \text{ contains an odd permutation} \\ 0 & \text{if } \Gamma \subset A_n \end{cases}.$$

Example 8.11
Let $G = S_3$, and let $\Gamma = \{(1, 2), (1, 2, 3)\}$. We now compute $\kappa(G, \Gamma)$. Let π_2 be as in Example 6.53. We know from Exercise 4 of Chapter 6 that π_2 and the alternating representation π_1 are, up to equivalence, the only two nontrivial irreps of S_3. (Using Remark 6.20, regard π_1 and π_2 as representations, not matrix representations.) By Example 8.10, we have that $\kappa(G, \Gamma, \pi_1) = 2$. We now compute $\kappa(G, \Gamma, \pi_2)$. Note that π_2 is unitary with respect to the standard inner product on $\mathbb{C}^2$. Let $(a, b)^t$ be a unit vector in $\mathbb{C}^2$. Then

$$\begin{aligned} \left\|\pi_2((1,2,3)) \cdot (a,b)^t - (a,b)^t\right\|^2 &= \left\|(\xi^2 a - a, \xi b - b)^t\right\|^2 \\ &= |\xi^2 a - a|^2 + |\xi b - b|^2 \\ &= |\xi a - a|^2 + |\xi b - b|^2 && (\text{factor out } -\xi^2) \\ &= |\xi - 1|^2(|a|^2 + |b|^2) \\ &= |\xi - 1|^2 && (\text{since } (a,b)^t \text{ is a unit vector.}) \\ &= 3. && (\text{by Lemma 8.7}) \end{aligned}$$

So $\left\|\pi_2((1, 2, 3)) \cdot (a, b)^t - (a, b)^t\right\| = \sqrt{3}$ for any unit vector $\mathbf{v}$ in $\mathbb{C}^2$. Therefore, $\kappa(G, \Gamma, \pi_2) \geq \sqrt{3}$. Compute that

$$\left\|\pi_2((1,2)) \cdot (1,0)^t - (1,0)^t\right\| = \sqrt{2}.$$

Hence

$$\max_{\gamma \in \Gamma} \left\|\pi_2(\gamma) \cdot (1,0)^t - (1,0)^t\right\| = \sqrt{3},$$

so $\kappa(G, \Gamma, \pi_2) \leq \sqrt{3}$. Therefore $\kappa(G, \Gamma, \pi_2) = \sqrt{3}$. So $\kappa(G, \Gamma) = \min\{2, \sqrt{3}\} = \sqrt{3}$.

Example 8.12
In this example, we show that

$$\kappa(\mathbb{Z}_n, \mathbb{Z}_n) = \begin{cases} 2 & \text{if } n \text{ is a power of } 2 \\ 2\cos\left(\frac{\pi}{2p}\right) & \text{otherwise, where } p \text{ is the smallest odd prime dividing } n \end{cases}.$$

Recall that the irreducible representations of $\mathbb{Z}_n$ are the maps $\rho_a : \mathbb{Z}_n \to GL(\mathbb{C})$ defined by $\rho_a(k)z = \xi^{ak}z$ where $\xi = \exp(\frac{2\pi i}{n})$ and $0 \leq a \leq n-1$.

Given $1 \leq a \leq n-1$, we have that

$$\kappa(\mathbb{Z}_n, \mathbb{Z}_n, \rho_a) = \max_{k=1,2,\ldots,n-1} \left\| \exp\left(\frac{2\pi iak}{n}\right) - 1 \right\|.$$

If $n = 2^m$ is a power of 2, then $\kappa(\mathbb{Z}_n, \mathbb{Z}_n, \rho_a) = 2$ for all $1 \leq a \leq n-1$. (To see this, write $a = 2^b \ell$ where ℓ is an odd integer, and plug in $k = 2^{m-b-1}$.) So $\kappa(\mathbb{Z}_n, \mathbb{Z}_n) = 2$ in this case.

Now suppose that n is not a power of 2 and that the prime factorization of n is $n = 2^e p_1^{e_1} p_2^{e_2} \cdots p_j^{e_j}$, where $p_1 < p_2 < \cdots < p_j$. Given $1 \leq a \leq n-1$, let $n_a = n/g$, where $g = \gcd(n, a)$. Then

$$\begin{aligned}
\kappa(\mathbb{Z}_n, \mathbb{Z}_n, \rho_a) &= \max_{k=1,2,\ldots,n-1} \left\| \exp\left(\frac{2\pi iak}{n}\right) - 1 \right\| \\
&= \max_{k=1,2,\ldots,n-1} \left\| \exp\left(\frac{2\pi i(a/g)k}{n_a}\right) - 1 \right\| \\
&= \max_{k=1,2,\ldots,n_a-1} \left\| \exp\left(\frac{2\pi i(a/g)k}{n_a}\right) - 1 \right\|.
\end{aligned}$$

Because $\gcd(a/g, n_a) = 1$, the map $k \mapsto (a/g)k$ is just a reordering of $\mathbb{Z}_{n_a}$. Therefore,

$$\kappa(\mathbb{Z}_n, \mathbb{Z}_n, \rho_a) = \max_{k=1,2,\ldots,n_a-1} \left\| \exp\left(\frac{2\pi ik}{n_a}\right) - 1 \right\|.$$

The maximum occurs when $\exp(2\pi ik/n_a)$ is as close to -1 as possible. See Figure 8.3 to visualize this in the case $n_a = 5$. When n_a is even, this occurs at $k = n_a/2$, which gives $\kappa(\mathbb{Z}_n, \mathbb{Z}_n, \rho_a) = 2$. When n_a is odd, this occurs at $k = (n_a - 1)/2$ (or at $k = (n_a + 1)/2$; think about where the roots of unity

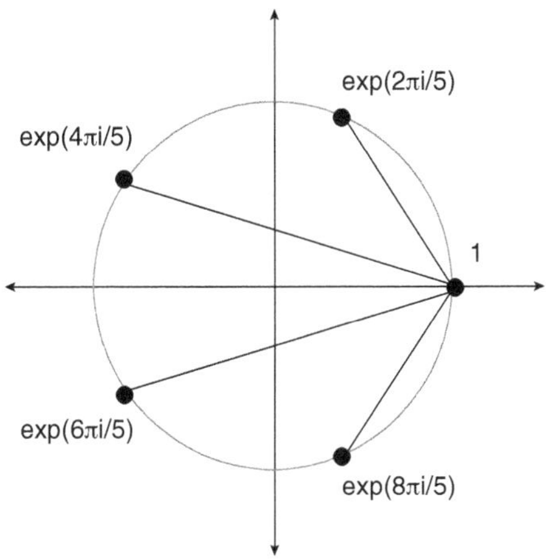

Figure 8.3

lie in the complex plane). Thus, $\kappa(\mathbb{Z}_n, \mathbb{Z}_n, \rho_a) = \left|\exp(\pi i(n_a - 1)/n_a) - 1\right|$ in this case.

Therefore, the minimum $\kappa(\mathbb{Z}_n, \mathbb{Z}_n, \rho_a)$ occurs when n_a is odd and as small as possible. Since n_a is a divisor of n, the smallest possible such value is p_1. We obtain $n_a = p_1$ by letting $a = n/p_1$. So by Lemma 8.7,

$$\kappa(\mathbb{Z}_n, \mathbb{Z}_n) = \left\| \exp\left(\frac{\pi i(p_1 - 1)}{p_1}\right) - 1 \right\| = 2\sin\left(\frac{\pi(p_1 - 1)}{2p_1}\right) = 2\cos\left(\frac{\pi}{2p_1}\right).$$

Remark 8.13
Note that $\kappa(\mathbb{Z}_n, \mathbb{Z}_n \setminus \{0\}) = \kappa(\mathbb{Z}_n, \mathbb{Z}_n)$. Note also that $\mathrm{Cay}(\mathbb{Z}_n, \mathbb{Z}_n \setminus \{0\}) = K_n$, where K_n is the complete graph on n vertices.

Proposition 8.14
Let G be a finite nontrivial group, and let $\Gamma \subset G$. Then $\kappa(G, \Gamma) \leq 2$.

Proof
If π is any unitary irrep on a vector space V, and $\mathbf{v}$ is any unit vector in V, then by the triangle inequality, we have $\|\pi(\gamma)\mathbf{v} - \mathbf{v}\| \leq \|\pi(\gamma)\mathbf{v}\| + \|\mathbf{v}\| = 1 + 1 = 2$ for all $\gamma \in \Gamma$.

The following proposition states one of the most useful properties of the Kazhdan constant: it allows us to bound the Kazhdan constant with respect to one generating set in terms of another generating set.

Proposition 8.15
Let $\rho : G \to GL(V)$ be a unitary representation of a finite nontrivial group G. Let $\Gamma, \Gamma' \subset G$. Suppose that every element of Γ' can be expressed as a word in Γ of length $\leq n$. Then

1. $\kappa(G, \Gamma, \rho) \geq \kappa(G, \Gamma', \rho)/n$, and therefore
2. $\kappa(G, \Gamma) \geq \kappa(G, \Gamma')/n$.

Proof
We first prove (1). For notational convenience, we write $\gamma\mathbf{v}$ instead of $\rho(\gamma)\mathbf{v}$. Let $\mathbf{v}$ be a unit vector in V such that $\kappa(G, \Gamma, \rho) = \|\gamma\mathbf{v} - \mathbf{v}\|$ for some $\gamma \in \Gamma$. Let $\gamma' \in \Gamma'$. By hypothesis, we know that $\gamma' = \gamma_1\gamma_2\cdots\gamma_j$ for some $\gamma_1, \gamma_2, \ldots, \gamma_j \in \Gamma$ and $j \leq n$. Define $\alpha_k = \gamma_1\gamma_2\cdots\gamma_k$ for $k = 1, \ldots, j$. By the triangle inequality and G-invariance of the norm, we have

$$\begin{aligned}
\|\gamma'\mathbf{v} - \mathbf{v}\| &= \left\|\alpha_j\mathbf{v} - \mathbf{v}\right\| \\
&= \left\|(\alpha_j\mathbf{v} - \alpha_{j-1}\mathbf{v}) + (\alpha_{j-1}\mathbf{v} - \alpha_{j-2}\mathbf{v}) + \cdots + (\alpha_2\mathbf{v} - \alpha_1\mathbf{v}) + (\alpha_1\mathbf{v} - \mathbf{v})\right\| \\
&\leq \left\|\alpha_j\mathbf{v} - \alpha_{j-1}\mathbf{v}\right\| + \left\|\alpha_{j-1}\mathbf{v} - \alpha_{j-2}\mathbf{v}\right\| + \cdots + \left\|\alpha_2\mathbf{v} - \alpha_1\mathbf{v}\right\| + \left\|\alpha_1\mathbf{v} - \mathbf{v}\right\| \\
&= \left\|\gamma_j\mathbf{v} - \mathbf{v}\right\| + \left\|\gamma_{j-1}\mathbf{v} - \mathbf{v}\right\| + \cdots + \left\|\gamma_2\mathbf{v} - \mathbf{v}\right\| + \left\|\gamma_1\mathbf{v} - \mathbf{v}\right\|
\end{aligned}$$

$$\leq j\kappa(G, \Gamma, \rho)$$
$$\leq n\kappa(G, \Gamma, \rho).$$

By the definition of $\kappa(G, \Gamma', \rho)$, then, we have that $n\kappa(G, \Gamma, \rho) \geq \kappa(G, \Gamma', \rho)$.

Minimizing over all nontrivial irreps ρ, (2) follows. ∎

2. THE KAZHDAN CONSTANT, THE ISOPERIMETRIC CONSTANT, AND THE SPECTRAL GAP

The goal of this section is to show that a sequence $(\text{Cay}(G_n, \Gamma_n))$ of finite Cayley graphs forms a family of expanders iff $\kappa(G_n, \Gamma_n)$ is bounded away from zero. We do so by relating the Kazhdan constant to the expansion constant and the spectral gap.

Recall from Theorem 6.69 and Corollary 6.74 that if G is a finite group, then the regular representation R is unitary with respect to the standard inner product on $L^2(G)$, and that there is a decomposition $L^2(G) = \mathbb{C}f_0 \oplus L_0^2(G)$ where f_0 is a nonzero constant function on G and $L_0^2(G)$ is the set of functions on G orthogonal to f_0. Let 1 denote the trivial representation of G.

Lemma 8.16

Let G be a finite nontrivial group, and let $\Gamma \subset G$. Let $d = |\Gamma|$. Let $\kappa = \kappa(G, \Gamma)$.

1. Let $\rho : G \to GL(V)$ be a representation of G not containing 1, that is, such that $\langle \chi_1, \chi_\rho \rangle = 0$, where χ_1 and χ_ρ are the characters of 1 and ρ, respectively. Then $\kappa(G, \Gamma, \rho) \geq \frac{\kappa}{\sqrt{d}}$.
2. Let $\hat{R}$ be the restriction of R to $L_0^2(G)$, and let $\hat{\kappa} = \kappa(G, \Gamma, \hat{R})$. Then $\kappa \geq \hat{\kappa} \geq \frac{\kappa}{\sqrt{d}}$.

Proof

(1) Let $\mathbf{v}$ be a unit vector in V. It will suffice to show that there exists $\gamma \in \Gamma$ such that $\left\|\rho(\gamma)\mathbf{v} - \mathbf{v}\right\|^2 \geq \frac{\kappa^2}{d}$. We are given that V decomposes as an orthogonal direct sum of nontrivial irreducible representations $V_1 \oplus \cdots \oplus V_n$, with corresponding irreps $\rho_1, \ldots, \rho_n$. (That is, $\rho|_{V_i} = \rho_i$.) Write $\mathbf{v} = \mathbf{v}_1 + \cdots + \mathbf{v}_n$ where $\mathbf{v}_j \in V_j$ for all j. For each j, choose $\gamma_j \in \Gamma$ such that $\left\|\rho_j(\gamma_j)\mathbf{v}_j - \mathbf{v}_j\right\| \geq \kappa \left\|\mathbf{v}_j\right\|$. (The definition of the Kazhdan constant guarantees the existence of γ_j; see Exercise 1.) Let $\alpha_1, \ldots, \alpha_d$ be the elements of Γ, and let $\mathbf{w}_\ell = \sum_{\gamma_j = \alpha_\ell} \mathbf{v}_j$. Note that $\mathbf{v} = \sum_{\ell=1}^d \mathbf{w}_\ell$, and $\mathbf{w}_\ell$ is orthogonal to $\mathbf{w}_{\ell'}$ if $\ell \neq \ell'$. It follows that $1 = \|\mathbf{v}\|^2 = \sum_{\ell=1}^d \left\|\mathbf{w}_\ell\right\|^2$. Therefore $\left\|\mathbf{w}_{\ell_0}\right\|^2 \geq \frac{1}{d}$ for some ℓ_0. Consequently, by Prop. A.21,

$$\begin{aligned}
\left\|\rho(\alpha_{\ell_0})\mathbf{v} - \mathbf{v}\right\|^2 &\geq \left\|\rho(\alpha_{\ell_0})\mathbf{w}_{\ell_0} - \mathbf{w}_{\ell_0}\right\|^2 \\
&= \sum_{\gamma_j = \alpha_{\ell_0}} \left\|\rho_j(\gamma_j)\mathbf{v}_j - \mathbf{v}_j\right\|^2 \\
&\geq \sum_{\gamma_j = \alpha_{\ell_0}} \kappa^2 \left\|\mathbf{v}_j^2\right\|
\end{aligned}$$

$$= \kappa^2 \left\| \mathbf{w}_{\ell_0}^2 \right\|$$
$$\geq \frac{\kappa^2}{d}.$$

(2) First we show that $\kappa \geq \hat{\kappa}$. Let $\rho : G \to GL(V)$ be a nontrivial irrep. Let $\mathbf{v}$ be a unit vector in V. By Corollary 6.74, we can assume that V is a subrepresentation of $L_0^2(G)$. Since $\mathbf{v}$ is a unit vector in $L_0^2(G)$, by the definition of $\hat{\kappa}$ there exists $\gamma_0 \in \Gamma$ such that $\left\| R(\gamma_0)\mathbf{v} - \mathbf{v} \right\| \geq \hat{\kappa}$. Because $R(\gamma)\mathbf{v} = \rho(\gamma)\mathbf{v}$ for all $\gamma \in \Gamma$, it follows that $\kappa \geq \hat{\kappa}$.

For the other half of the inequality, we have by Corollary 6.74 that $L_0^2(G)$ decomposes as an orthogonal direct sum of nontrivial irreducible representation spaces. Now apply (1). ⊛

The next proposition relates the Kazhdan constant of (G, Γ) to the isoperimetric constant of the Cayley graph $\mathrm{Cay}(G, \Gamma)$.

Proposition 8.17
Let G be a finite nontrivial group, and let $\Gamma \Subset G$. Let $d = |\Gamma|$, let $\kappa = \kappa(G, \Gamma)$, and let $h = h(\mathrm{Cay}(G, \Gamma))$. Then $h \geq \frac{\kappa^2}{4d}$.

Proof
Let $A \subset G$ such that $|A| \leq \frac{1}{2} |G|$, and let $B = G \setminus A$. Let $\hat{\kappa}$ be as in Lemma 8.16. We will show that $\frac{|\partial A|}{|A|} \geq \frac{\hat{\kappa}^2}{4}$. The result will then follow from Lemma 8.16 and the definition of the isoperimetric constant.

The following is a standard trick for producing an element of $L_0^2(G)$, given a subset of G. Define $\tilde{f} \in L^2(G)$ by

$$\tilde{f}(g) = \begin{cases} |B| & \text{if } g \in A. \\ -|A| & \text{if } g \in B. \end{cases}$$

Note that $\tilde{f} \in L_0^2(G)$. Let $f = \frac{\tilde{f}}{\|\tilde{f}\|}$. Then f is a unit vector in $L_0^2(G)$. So there exists $\gamma \in \Gamma$ such that $\left\| R(\gamma)f - f \right\|^2 \geq \hat{\kappa}^2$. Recall that $(R(\gamma)\tilde{f})(g) = \tilde{f}(g\gamma)$. So

$$|\tilde{f}(g\gamma) - \tilde{f}(g)| = \begin{cases} |B| + |A| & \text{if } g\gamma \in A \text{ and } g \in B, \text{ or if } g \in A \text{ and } g\gamma \in B. \\ 0 & \text{otherwise.} \end{cases}$$

Let E_γ be the set (not multiset) of edges in $\mathrm{Cay}(G, \Gamma)$ of the form $\{g, g\gamma\}$ where either $g \in A$ and $g\gamma \in B$ or else $g\gamma \in A$ and $g \in B$. Then

$$\left\| R(\gamma)\tilde{f} - \tilde{f} \right\|^2 = \sum_{g \in G} |R(\gamma)\tilde{f}(g) - \tilde{f}(g)|^2$$
$$= \sum_{g \in G} |\tilde{f}(g\gamma) - \tilde{f}(g)|^2$$

$$= \begin{cases} 0 & \text{if } \gamma \text{ is the identity element} \\ 2\left|E_\gamma\right|(|B|+|A|)^2 & \text{if } \gamma \text{ has order } 2 \\ \left|E_\gamma\right|(|B|+|A|)^2 & \text{if } \gamma \text{ has order } > 2 \end{cases}$$

$$\leq 2\left|E_\gamma\right|(|B|+|A|)^2.$$

Hence

$$\begin{aligned} \hat{\kappa}^2 &\leq \left\|R(\gamma)f - f\right\|^2 \\ &= \frac{\left\|R(\gamma)\tilde{f} - \tilde{f}\right\|^2}{\left\|\tilde{f}\right\|^2} \\ &\leq \frac{2\left|E_\gamma\right|(|B|+|A|)^2}{\|\tilde{f}\|^2} \\ &= \frac{2\left|E_\gamma\right||G|^2}{\|\tilde{f}\|^2}. \end{aligned}$$

Because $\|\tilde{f}\|^2 = |A|\,|B|^2 + |B|\,|A|^2 = |A|\,|B|\,|G|$ and $|B| \geq \frac{|G|}{2}$, we have that

$$\left|E_\gamma\right| \geq \frac{\hat{\kappa}^2\|\tilde{f}\|^2}{2\,|G|^2} = \frac{\hat{\kappa}^2\,|A|\,|B|}{2\,|G|} \geq \frac{\hat{\kappa}^2\,|A|}{4}.$$

Since $|\partial A| \geq \left|E_\gamma\right|$, we have that $\frac{|\partial A|}{|A|} \geq \frac{\hat{\kappa}^2}{4}$, as desired. ◈

We now relate the Kazhdan constant of (G, Γ) to the spectral gap of the Cayley graph $\mathrm{Cay}(G, \Gamma)$.

Proposition 8.18
Let G be a finite nontrivial group, and let $\Gamma \circledS G$. Let $d = |\Gamma|$, let $\kappa = \kappa(G, \Gamma)$, and let λ_1 be the second-largest eigenvalue of $\mathrm{Cay}(G, \Gamma)$. Then $\kappa \geq \sqrt{\frac{2(d-\lambda_1)}{d}}$.

Proof
By Lemma 8.16, it is sufficient to show that $\hat{\kappa} \geq \sqrt{\frac{2(d-\lambda_1)}{d}}$.

Let $\mathbf{v}$ be a unit vector in $L^2_0(G)$, and let R be the right regular representation of G. We write $\gamma\mathbf{v}$ for $R(\gamma)\mathbf{v}$. We must show that there exists $\gamma_0 \in \Gamma$ such that $\left\|\gamma_0\mathbf{v} - \mathbf{v}\right\| \geq \sqrt{\frac{2(d-\lambda_1)}{d}}$. Let A be the adjacency operator of $\mathrm{Cay}(G, \Gamma)$. Using the Rayleigh-Ritz theorem (see Remark 1.90), we have that

$$\lambda_1 \geq \langle A\mathbf{v}, \mathbf{v}\rangle = \sum_{\gamma\in\Gamma}\langle\gamma\mathbf{v}, \mathbf{v}\rangle = \sum_{\gamma\in\Gamma}\mathrm{Re}(\langle\gamma\mathbf{v}, \mathbf{v}\rangle).$$

Consequently,

$$d - \lambda_1 \leq d - \sum_{\gamma \in \Gamma} \mathrm{Re}(\langle \gamma \mathbf{v}, \mathbf{v} \rangle) = \sum_{\gamma \in \Gamma} (1 - \mathrm{Re}(\langle \gamma \mathbf{v}, \mathbf{v} \rangle)).$$

Hence there exists $\gamma_0 \in \Gamma$ such that

$$1 - \mathrm{Re}(\langle \gamma_0 \mathbf{v}, \mathbf{v} \rangle) \geq \frac{d - \lambda_1}{d}.$$

It follows that

$$\begin{aligned} \|\gamma_0 \mathbf{v} - \mathbf{v}\| &= \sqrt{\langle \gamma_0 \mathbf{v} - \mathbf{v}, \gamma_0 \mathbf{v} - \mathbf{v} \rangle} \\ &= \sqrt{2(1 - \mathrm{Re}(\langle \gamma_0 \mathbf{v}, \mathbf{v} \rangle))} \\ &\geq \sqrt{\frac{2(d - \lambda_1)}{d}}. \end{aligned}$$

We can now prove the main result of this section.

Theorem 8.19
Let d be a positive integer. Let (G_n) be a sequence of groups with $|G_n| \to \infty$. For each n, let $\Gamma_n \subseteq G_n$ such that $|\Gamma_n| = d$. Then $(\mathrm{Cay}(G_n, \Gamma_n))$ is an expander family iff there exists $\epsilon > 0$ such that $\kappa(G_n, \Gamma_n) \geq \epsilon$ for all n.

Proof
Combine Props. 8.17 and 8.18 with Corollary 1.87 and the definition of an expander family.

Example 8.20
In Example 8.9, we saw that $\kappa(\mathbb{Z}_n, \{1\}) = 2\sin(\frac{\pi}{n})$. Since $2\sin(\frac{\pi}{n}) \to 0$, by Theorem 8.19 and Exercise 3 it follows that $(\mathrm{Cay}(\mathbb{Z}_n, \{-1, 1\}))$ is not a family of expanders. Hence we obtain yet another proof of the fact that cycle graphs do not form an expander family.

We now revisit bubble-sort graphs one last time (having seen them previously in Examples 1.78, 1.91, and 4.21). This time, we use Kazhdan constants to prove for the fourth and final time that they do not form an expander family.

Example 8.21
Let $\Gamma = \{\sigma, \sigma^{-1}, \tau\}$ where $\sigma = (1, 2, \ldots, n)$ and $\tau = (1, 2)$. Let $X_n = \mathrm{Cay}(S_n, \Gamma)$. We will show that (X_n) is not an expander family. Take notation as in Example 6.34. Let $S^1(V_{per})$ be the set of all unit vectors in V_{per}. We have that

$$\kappa(S_n, \Gamma, \rho) = \min_{\mathbf{v} \in S^1(V_{per})} \max_{\gamma \in \{\tau, \sigma, \sigma^{-1}\}} \|\rho(\gamma)\mathbf{v} - \mathbf{v}\|.$$

Consider the vector $\mathbf{w} = n^{-1/2}(\mathbf{x}_1 + \xi \mathbf{x}_2 + \xi^2 \mathbf{x}_3 + \cdots + \xi^{n-1} \mathbf{x}_n)$, where $\xi = \exp(2\pi i/n)$. Note that $1 + \xi + \cdots + \xi^{n-1} = (\xi^n - 1)/(\xi - 1) = 0$.

So $\mathbf{w} \in V_{per}$. Also note that $\|\mathbf{w}\| = 1$, so $\mathbf{w} \in S^1(V_{per})$. We compute $\|\rho(\gamma)\mathbf{w} - \mathbf{w}\|$ for the different values of γ.

If $\gamma = \tau$, then $\sqrt{n}\,\left[\,\rho(\tau)\mathbf{w}\,\right] = \xi\mathbf{x}_1 + \mathbf{x}_2 + \xi^2\mathbf{x}_3 + \cdots + \xi^{n-1}\mathbf{x}_n$. Thus,

$$\begin{aligned}\|\sqrt{n}\left[\,\rho(\tau)\mathbf{w} - \mathbf{w}\,\right]\|^2 &= \|(\xi - 1)\mathbf{x}_1 + (1-\xi)\mathbf{x}_2\|^2\\ &= |\xi - 1|^2 + |1-\xi|^2 = 2\,|1-\xi|^2.\end{aligned}$$

So, $\|\rho(\tau)\mathbf{w} - \mathbf{w}\| = (\sqrt{2/n})\,|1-\xi| = \sqrt{8/n}\sin(\pi/n)$ by Lemma 8.7.

If $\gamma = \sigma$, then

$$\begin{aligned}&\|\sqrt{n}\left[\,\rho(\sigma)\mathbf{w} - \mathbf{w}\,\right]\|^2\\ &\quad= \|(\xi^{n-1} - 1)\mathbf{x}_1 + (1-\xi)\mathbf{x}_2 + (\xi - \xi^2)\mathbf{x}_3 + \cdots + (\xi^{n-2} - \xi^{n-1})\mathbf{x}_n\|^2\\ &\quad= |\xi^{n-1} - 1|^2 + |1-\xi|^2 + |\xi - \xi^2|^2 + \cdots + |\xi^{n-2} - \xi^{n-1}|^2\\ &\quad= |\xi^{n-1}|^2|1-\xi|^2 + |1-\xi|^2 + |\xi|^2|1-\xi|^2 + \cdots + |\xi^{n-2}|^2|1-\xi|^2\\ &\quad= n\,|1-\xi|^2.\end{aligned}$$

So, $\|\rho(\sigma)\mathbf{w} - \mathbf{w}\| = |1-\xi| = 2\sin(\pi/n)$, by Lemma 8.7. Because ρ is unitary (see Example 6.34), we have that

$$\|\rho(\sigma^{-1})\mathbf{w} - \mathbf{w}\| = \|\rho(\sigma)\left(\rho(\sigma^{-1})\mathbf{w} - \mathbf{w}\right)\| = \|\rho(\sigma)\mathbf{w} - \mathbf{w}\|.$$

From Example 6.63, we know that ρ is irreducible. So by Def. 8.5, we know that $\kappa(S_n, \Gamma) \le \kappa(S_n, \Gamma, \rho)$. Summing everything up, we see that

$$\kappa(S_n, \Gamma) \le \max\left\{2\sin(\pi/n),\ \sqrt{8/n}\sin(\pi/n)\right\} \to 0$$

as $n \to \infty$. So by Theorem 8.19, therefore (X_n) is not an expander family.

Prop. 8.18 implies the following lower bound on the Kazhdan constant of a group with respect to itself.

Corollary 8.22
Let G be a finite nontrivial group. Then $\kappa(G, G) > \sqrt{2}$.

Proof
First note that if ρ is any unitary irrep, then $\left\|\rho(e)\mathbf{v} - \mathbf{v}\right\| = 0$. It follows that $\kappa(G, G) = \kappa(G, \Gamma)$, where $\Gamma = G \setminus \{e\}$. Let $n = |G|$, and observe that $\mathrm{Cay}(G, \Gamma)$ is the complete graph K_n. The degree of K_n is $n - 1$. By Example 1.52 and Prop. 8.18, then, we have that

$$\kappa(G, \Gamma) \ge \sqrt{\frac{2((n-1) - (-1))}{n-1}} > \sqrt{2}.$$

⌂

Remark 8.23
The following is an alternate proof of Prop. 8.22. Assume temporarily that $\kappa(G, G) \leq \sqrt{2}$. Let ρ be a nontrivial unitary irrep on a vector space V, and let $\mathbf{v}$ be a unit vector in V. Then

$$\left\| \rho(g)\mathbf{v} - \mathbf{v} \right\| \leq \sqrt{2} \tag{38}$$

for all $g \in G$. In other words, regarding $\mathbf{v}$ as the "North Pole" of the unit sphere in V, the inequality in Equation 38 tells us that the elements $\rho(g)\mathbf{v}$ all lie in the "northern hemisphere." Let $\mathbf{w} = \sum_{g \in G} \rho(g)\mathbf{v}$. Let P be the plane in V through the "equator." Then $\mathbf{w}$ lies strictly on the same side of P as $\mathbf{v}$. It's "strictly" because $\rho(e)\mathbf{v} = \mathbf{v}$. (To see this algebraically, choose an orthonormal basis with $\mathbf{v}$ as one of the basis vectors, and consider the orthogonal projection of $\mathbf{w}$ onto $\mathbf{v}$.) So $\mathbf{w} \neq \mathbf{0}$. But $\rho(g)\mathbf{w} = \mathbf{w}$ for all $g \in G$. This is a contradiction, as a nontrivial irrep cannot have a nonzero fixed vector. (For if ρ has degree 1, then this would imply that ρ is trivial; and if ρ has degree > 1, then $\mathbb{C}\mathbf{w}$ would be a proper subrepresentation, which would violate irreducibility.)

Our proof of Prop. 8.22, however, yields the slightly stronger inequality:

$$\kappa(G, G) \geq \sqrt{\frac{2\,|G|}{|G| - 1}}.$$

3. ABELIAN GROUPS NEVER YIELD EXPANDER FAMILIES: A REPRESENTATION-THEORETIC PROOF

Corollary 4.26 states that no sequence of abelian groups yields an expander family. In this section, we use Kazhdan constants to provide an alternate proof of this fact.

We begin with a technical lemma.

Lemma 8.24
Let θ be a real number. Then $|e^{i\theta} - 1| \leq |\theta|$.

Proof
Note that $\cos(x) \geq 1 - \frac{x^2}{2}$ for all real x. (Set $f(x) = \cos(x) - 1 + x^2/2$. Then f is even, $f(0) = 0$, and $f'(x) > 0$ when $x > 0$. Hence $f(x) \geq 0$ for all x.) Thus,

$$\begin{aligned} \left|e^{i\theta} - 1\right|^2 &= (\cos(\theta) - 1)^2 + (\sin(\theta))^2 \\ &= 2 - 2\cos(\theta) \\ &\leq 2 - 2\left(1 - \theta^2/2\right) \\ &= \theta^2. \end{aligned}$$

Proposition 8.25
Let G be a nontrivial finite abelian group. Let $\Gamma \subset G$, and let $d = |\Gamma|$. Then

$$\kappa(G, \Gamma) \leq \frac{2\pi}{|G|^{1/d} - 1}.$$

Proof
Let $\kappa = \kappa(G, \Gamma)$. If $\kappa = 0$, we're done, so assume $\kappa > 0$. Let $C = \lceil 2\pi/\kappa \rceil$. (That is, C is the smallest integer greater than or equal to $2\pi/\kappa$.) We first claim that $|G| \leq C^d$; we will prove this by contradiction. Assume temporarily that $|G| > C^d$. By the Fundamental Theorem of Finite Abelian Groups, we know that G is of the form $\mathbb{Z}_{n_1} \times \mathbb{Z}_{n_2} \times \cdots \times \mathbb{Z}_{n_r}$. In Prop. 7.13, we determined the set S of all irreps of G. In particular, we saw that if ρ is any irrep of G and $g \in G$, then $\rho(g)$ is of the form $e^{i\theta}$ for some $\theta \in [0, 2\pi)$. (Here we identify $GL(\mathbb{C})$ with $\mathbb{C}^*$, the set of nonzero elements in $\mathbb{C}$, by the identification $\rho \mapsto \rho(e_G) \cdot 1$.) For any element z on the unit circle in $\mathbb{C}$, let $\ell(z)$ denote the unique element $\theta \in [0, 2\pi)$ such that $e^{i\theta} = z$. Define $\phi : S \to [0, 2\pi)^d$ by

$$\phi(\rho) = (\ell(\rho(\gamma_1)), \ell(\rho(\gamma_2)), \ldots, \ell(\rho(\gamma_d))),$$

where $\gamma_1, \ldots, \gamma_d$ are the elements of Γ.

Suppose $a_1, \ldots, a_d$ are integers with $1 \leq a_j \leq C$ for all j, and that $a = (a_1, \ldots, a_d)$. Define

$$B_a := \left\{ (r_1, \ldots, r_d) \in [0, 2\pi)^d \ : \ \frac{2\pi(a_j - 1)}{C} \leq r_j < \frac{2\pi a_j}{C} \right\}.$$

Then $[0, 2\pi)^d$ is the disjoint union of the C^d sets B_a. Figure 8.4 shows the picture in the case $C = 3$, $d = 2$.

By Prop. 7.13, we know that $|S| = |G|$. By the Pigeonhole Principle, there must be two distinct irreps ρ_1, ρ_2 such that $\rho_1, \rho_2 \in B_a$ for some d-tuple $a = (a_1, \ldots, a_d)$. (That is, there must be two distinct irreps that map to the same "box" B_a.)

Define $\rho : G \to GL(1, \mathbb{C})$ by $\rho(g) = \rho_1(g)/\rho_2(g)$. Note that

$$\rho(gh) = \frac{\rho_1(gh)}{\rho_2(gh)} = \frac{\rho_1(g)\rho_1(h)}{\rho_2(g)\rho_2(h)} = \rho(g)\rho(h).$$

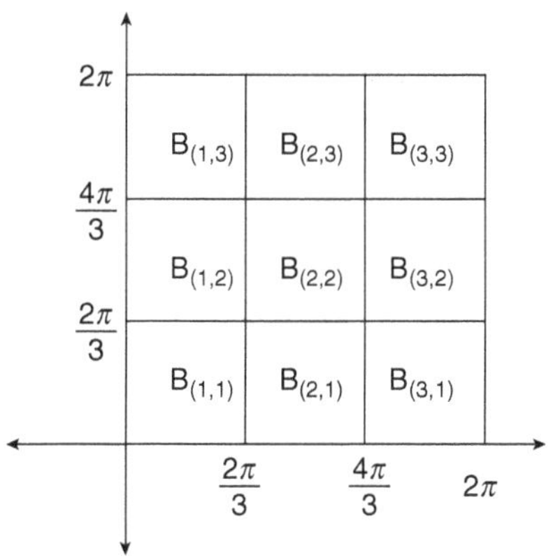

Figure 8.4 We partition $[0, 2\pi)^d$ into C^d boxes B_a.

So ρ is a representation of G. Because ρ has degree 1, ρ is an irrep. Moreover, since ρ_1 and ρ_2 are distinct, it follows that ρ is nontrivial.

Since $\phi(\rho_1), \phi(\rho_2) \in B_a$, for all j we know that

$$|\ell(\rho_1(\gamma_j)) - \ell(\rho_2(\gamma_j))| < 2\pi/C.$$

Note that

$$\rho(\gamma_j) = \exp(i(\ell(\rho_1(\gamma_j)) - \ell(\rho_2(\gamma_j)))).$$

By Lemma 8.24, $|\rho(\gamma_j) - 1| < 2\pi/C$ for all j. Therefore $\kappa(G, \Gamma, \rho) < 2\pi/C$, so by Def. 8.5, we have that $\kappa < 2\pi/C$. Hence $C < 2\pi/\kappa$, which contradicts the definition of C. Therefore $|G| \leq C^d$.

The result now follows by observing that $C \leq 2\pi/\kappa + 1$, so $|G| \leq (2\pi/\kappa + 1)^d$. Then solve for κ. ∎

If (G_n) is any sequence of abelian groups with $|G_n| \to \infty$ and $\Gamma_n \Subset G_n$ for all n with $|\Gamma_n| = d$ for some fixed non-negative integer d, then by Prop 8.25, we see that $\kappa(G_n, \Gamma_n) \to 0$. Therefore, by Theorem 8.19, it follows that (G_n) does not yield an expander family. Thus we recover Corollary 4.26, which states that no sequence of abelian groups yields an expander family.

4. KAZHDAN CONSTANTS, SUBGROUPS, AND QUOTIENTS

Let G be a finite group. In previous chapters, we've seen that the isoperimetric constant, the second-largest eigenvalue, and the diameter of a Cayley graph on G are controlled in part by the corresponding invariants for subgroups and quotients of G. In this section, we obtain analogous results for Kazhdan constants. Using Theorem 8.19, one can use these results to recover the Subgroups and Quotients Nonexpansion Principles. Together with the results of the previous section, we have a purely representation-theoretic proof of Theorem 4.47, which states that no sequence of solvable groups with bounded derived length yields an expander family.

We begin with quotients.

Definition 8.26 Let G, H be groups, $\rho : H \to GL(V)$ a representation, and $\phi : G \to H$ a homomorphism. Let $\rho^* = \rho \circ \phi$. Then $\rho^* : G \to GL(V)$ is a representation of G, called the *pullback of ρ via ϕ*.

Lemma 8.27
Let G be a finite group, and let $N \lhd G$. Suppose $\rho : G/N \to GL(V)$ is an irrep of G/N. Let $\phi : G \to G/N$ be the canonical homomorphism. (That is, $\phi(g) = gN$.) Let ρ^* be the pullback of ρ via ϕ. Then ρ^* is an irrep of G.

Proof
One-line proof: Any subspace invariant under the action of G will also be invariant under the action of G/N. Let's spell this out in more detail. Assume temporarily that there exists a proper nontrivial subspace W of V such that $\rho^*(g) \cdot \mathbf{w} \in W$ for all $g \in G$, $\mathbf{w} \in W$. Then $\rho(\phi(g)) \cdot \mathbf{w} \in W$ for all $g \in G$, $\mathbf{w} \in W$. But ϕ is onto, so $\rho(gN) \cdot \mathbf{w} \in W$ for all $gN \in G/N$, $\mathbf{w} \in W$.

This violates irreducibility of ρ, so we have a contradiction. Therefore no such W exists, that is, ρ^* is irreducible.

Note that if an inner product is G/N-invariant under a representation ρ, then it is also G-invariant under the pullback ρ^*.

Proposition 8.28
Let G be a finite nontrivial group, and let N be a finite nontrivial normal subgroup of G. Let $\Gamma \circledS G$, and let $\overline{\Gamma}$ be the image of Γ under the canonical homomorphism. Then

$$\kappa(G, \Gamma) \leq \kappa(G/N, \overline{\Gamma}).$$

Proof
Let $\rho : G/N \to GL(V)$ be a nontrivial irrep of G/N. Recall Remark 8.3. Let $\mathbf{v}$ be a unit vector in V and $\gamma_0 N \in \overline{\Gamma}$ such that $\|\rho(\gamma_0 N)\mathbf{v} - \mathbf{v}\| = \kappa(G/N, \overline{\Gamma}, \rho)$. Let ρ^* be the pullback of ρ via the canonical homomorphism. Then for all $\gamma \in \Gamma$, we have that $\|\rho^*(\gamma)\mathbf{v} - \mathbf{v}\| = \|\rho(\phi(\gamma))\mathbf{v} - \mathbf{v}\|$. So, by Lemma 8.27 and Def. 8.5, $\kappa(G, \Gamma) \leq \kappa(G, \Gamma, \rho^*) = \kappa(G/N, \overline{\Gamma}, \rho)$. Because ρ was arbitrary, the result follows.

Remark 8.29
With Prop. 8.28 and Theorem 8.19, we recover the Quotients Nonexpansion Principle.

We now turn our attention to subgroups. Our main tools will be induced representations (Section 6.6) and Schreier generators (Def. 2.30).

Proposition 8.30
Let G be a finite nontrivial group, and let H be a nontrivial subgroup of G. Let $\Gamma \subset G$, and let $d = |\Gamma|$. Let $n = [G : H]$, and let $T = \{t_1, \ldots, t_n\}$ be a set of transversals for H in G. Let $\hat{\Gamma}$ be the corresponding set of Schreier generators. Then

$$\kappa(G, \Gamma) \leq d^{1/2}\kappa(H, \hat{\Gamma}).$$

Proof
Let $\sigma : H \to GL(W)$ be a nontrivial irrep of H; $\langle \cdot, \cdot \rangle$ an H-invariant inner product on W; and $\mathbf{w}$ a unit vector in W such that

$$\kappa(H, \hat{\Gamma}) = \max_{\widehat{(t,\gamma)} \in \hat{\Gamma}} \left\| \sigma\left(\widehat{(t,\gamma)}\right) \cdot \mathbf{w} - \mathbf{w} \right\|_W . \tag{39}$$

(Recall Remark 8.3 to see why such σ and $\mathbf{w}$ must exist.) Let $\rho = \operatorname{Ind}_H^G(\sigma)$ be the induced representation. Define $f \in V$ by $f(ht) = n^{-1/2}\sigma(h)\mathbf{w}$ for all $h \in H, t \in T$. (Be sure to verify that $f \in V$, as we claimed.) Using the inner product from Lemma 6.87, note that

$$\|f\|_V^2 = \sum_{j=1}^{n} n^{-1} \|\mathbf{w}\|_W^2 = 1,$$

so f is a unit vector in V. Let $\gamma \in \Gamma$. We have that

$$\begin{aligned}
\|\rho(\gamma)\cdot f - f\|_V^2 &= \sum_{j=1}^{n} \left\| f(t_j\gamma) - f(t_j) \right\|_W^2 \\
&= \sum_{j=1}^{n} \left\| f\left(\widehat{(t_j,\gamma)}\overline{t_j\gamma}\right) - f(t_j) \right\|_W^2 \\
&= \sum_{j=1}^{n} \left\| \sigma\left(\widehat{(t_j,\gamma)}\right)\cdot f\left(\overline{t_j\gamma}\right) - f(t_j) \right\|_W^2 \\
&= \sum_{j=1}^{n} \left\| \sigma\left(\widehat{(t_j,\gamma)}\right)\cdot (n^{-1/2}\mathbf{w}) - n^{-1/2}\mathbf{w} \right\|_W^2 \\
&= \frac{1}{n}\sum_{j=1}^{n} \left\| \sigma\left(\widehat{(t_j,\gamma)}\right)\cdot \mathbf{w} - \mathbf{w} \right\|_W^2 .
\end{aligned}$$

Therefore, for some j we must have

$$\|\rho(\gamma)\cdot f - f\|_V^2 \le \left\| \sigma\left(\widehat{(t_j,\gamma)}\right)\cdot \mathbf{w} - \mathbf{w} \right\|_W^2 ,$$

from which it follows, by Equation 39, that

$$\|\rho(\gamma)\cdot f - f\|_V \le \kappa(H,\hat{\Gamma}).$$

Since γ was arbitrary, we have that $\kappa(G,\Gamma,\rho) \le \kappa(H,\hat{\Gamma})$. (Note that here we use that $\langle\cdot,\cdot\rangle_V$ is G-invariant, as shown in Lemma 6.87.) From Example 6.95, we know that ρ does not contain the trivial representation. Lemma 8.16(1) then implies the desired result. ⓐ

Example 8.31
Suppose G is of the form $\mathbb{Z}_a \rtimes \mathbb{Z}_b$, and that $\Gamma \subset G$ with $|\Gamma| = 4$. By Props. 8.25 and 8.28, we find that $\kappa(G,\Gamma) \le 2\pi/(b^{1/4}-1)$. Using Prop. 8.30 in place of Prop. 8.28, we obtain the estimate $\kappa(G,\Gamma) \le 2\pi/(a^{1/4b}-1)$. So if either a or b is large, then $\kappa(G,\Gamma)$ must be small.

Remark 8.32
With Prop. 8.30 and Theorem 8.19, we recover the Subgroups Nonexpansion Principle. So using Props. 8.25, 8.28, and 8.30 as in the proof of Theorem 4.47, we obtain an entirely independent proof, using Kazhdan constants, of the fact that no sequence of solvable groups with bounded derived length yields an expander family.

NOTES

1. In 2005, Kassabov [76] showed that there exist symmetric generating sets Γ_n and $\tilde{\Gamma}_n$ of the symmetric group S_n and the alternating group A_n, respectively, such that $(\text{Cay}(S_n, \Gamma_n))$ and $(\text{Cay}(A_n, \tilde{\Gamma}_n))$ are families of bounded-degree expander graphs. The proof makes heavy use of Kazhdan constants.
2. For finite groups, the Kazhdan constant has been explicitly computed in only very few cases. In [16], Bacher and de la Harpe determine the Kazhdan constants of several finite groups for specific sets of generating sets. Examples 8.9 and 8.12 are taken from that paper. For dihedral groups, they show that $\kappa(D_n, \{s, t\}) = 2\sin(\pi/2n)$ for $n \geq 3$, and $\kappa(D_2, \{s, t\}) = \sqrt{2}$. For symmetric groups, they show that $\kappa(S_n, \Gamma) = \sqrt{24/(n^3 - n)}$ where Γ is the set of transpositions $\{(1, 2), (2, 3), \ldots, (n-1, n)\}$. In his master's thesis, Derbidge [48] shows that

$$\kappa(\mathbb{Z}_{2n}, \Gamma) = \begin{cases} \sqrt{2} & \text{if n is even} \\ 2\cos\left(\frac{\pi}{2p}\right) & \text{if n is odd, where p is the smallest odd prime dividing n} \end{cases},$$

where $\Gamma = \{1, 3, \ldots, 2n-1\}$ is the set of odd elements of $\mathbb{Z}_{2n}$.
3. Props. 8.25 and 8.30 appear (modulo cosmetic changes) in [90]. Indeed, nearly every theorem in this chapter can be found either in [87] or [90].
4. In [107], Pak and Żuk prove an inequality that relates the Kazhdan constant for a group G to the spectral gap for a coset graph of G.
5. There are several competing definitions in the literature for the Kazhdan constant and competing notations for each. For infinite discrete groups one must consider infinite-dimensional representations on Hilbert spaces, as in [16] and [87]. Sometimes one considers the "average Kazhdan constant," as in [107].
6. In [128], Shalom discusses Kazhdan constants for semisimple groups and their lattices. The paper [127] contains bounds on Kazhdan constants for the groups $\text{SL}_n(R)$ for various rings R, including one for $R = \mathbb{Z}$ and Γ equal to the set of all elementary matrices with 1 or -1 off the diagonal. In [75], Kassabov discusses asymptotics of Kazhdan constants for $\text{SL}_n(\mathbb{Z})$.
7. In [106], Osin shows that the Kazhdan constant of a word-hyperbolic group vanishes.
8. One variant on the Kazhdan constant $\kappa(G, \Gamma)$ is the "relative Kazhdan constant" $\kappa(G, H, \Gamma)$, defined for a subgroup H of G. (See, for example, [76] for a definition.) In [32], Burger finds lower bounds for various relative Kazhdan constant related to $SL_3(\mathbb{Z})$. See also [127] for various results pertaining to relative Kazhdan constants.
9. The original expander families of Margulis [92] made use of the fact that $SL_n(\mathbb{Z})$ possesses Kazhdan's Property (T) for $n \geq 3$. Property (T), and the related Property τ, can be formulated in terms of Kazhdan constants. See [87] for more details, [19] for more on Property (T) in a much more general context, and [78] for the original definition of Property (T), in terms of the Fell topology of the unitary dual of a group. The paper [134] deals with Kazhdan constants of discrete groups; in it, Żuk finds a condition that allows one to prove that many such groups have Property (T). In [65], Gelander and Żuk demonstrate the existence

of finitely generated groups with Property (T) but without a uniform positive Kazhdan constant.

EXERCISES

Unless otherwise stated, let G be a finite nontrivial group.

1. Let $\rho : G \to V$ be a unitary irrep. Let $\mathbf{v} \in V$. Show that there exists $\gamma \in \Gamma$ such that $\left\| \rho(\gamma)\mathbf{v} - \mathbf{v} \right\| \geq \kappa(G, \Gamma) \left\| \mathbf{v} \right\|$.
2. Prove that if $\Gamma \subset \Gamma' \subset G$, then $\kappa(G, \Gamma') \geq \kappa(G, \Gamma)$.
3. Let $\Gamma \subset G$. Let $\Gamma' = \{x \in G \mid x \in \Gamma \text{ or } x^{-1} \in \Gamma\}$. Prove that $\kappa(G, \Gamma') = \kappa(G, \Gamma)$.
4. Let $\Gamma = \{(1, 2), (2, 3)\} \subset S_3$.
    a. Prove that $\kappa(G, \Gamma) \leq 1$. (Hint: Consider the irrep of S_3 from Example 6.53. Find a unit vector that is moved a distance of 1 by both $(1, 2)$ and $(2, 3)$.)
    b. Prove that $\kappa(G, \Gamma) = 1$.
5. Prove that if $\Gamma \Subset G$, then $\text{Cay}(G, \Gamma)$ is connected iff $\kappa(G, \Gamma) \neq 0$. (Hint: Use Props. 8.17 and 8.18. Alternatively, let $\rho = \text{Ind}_{\langle\Gamma\rangle}^{G} 1$ be the induced representation of the trivial representation up from the subgroup generated by Γ, use Frobenius reciprocity to show that if Γ does not generate G then ρ must contain a nontrivial irrep ϕ, and then show that $\kappa(G, \Gamma, \phi) = 0$.)
6. Let $\Gamma \Subset G$. Prove that

$$\kappa(G, \Gamma) > \frac{\sqrt{2}}{\text{diam}(\text{Cay}(G, \Gamma))}.$$

7. Let G, Γ be as in Example 8.11. Let $\hat{\kappa}$ be as in Lemma 8.16. Prove that $\hat{\kappa} \neq \kappa(G, \Gamma)$. (Hint: Use the Intermediate Value Theorem.)

STUDENT RESEARCH PROJECT IDEAS

1. Generalize the results of Derbidge [48]—see Note 2. For example, compute the Kazhdan constant for $\mathbb{Z}_{an}$ with respect to the set of all multiples of a.
2. Is the Kazhdan constant a graph invariant? In other words, if $\text{Cay}(G_1, \Gamma_1)$ is isomorphic to $\text{Cay}(G_2, \Gamma_2)$, does it follow that $\kappa(G_1, \Gamma_1) = \kappa(G_2, \Gamma_2)$? Note that if $\Gamma = G \setminus \{e\}$, then $\text{Cay}(G, \Gamma)$ is a complete graph. One first case to check would be, does $\text{Cay}(G, G)$ depend only on $|G|$?
3. Consider the pairs (G_n, Γ_n) that arise in the construction of Paley graphs. We conjecture that $\kappa(G_n, \Gamma_n) \to 2$ as $n \to \infty$. Prove or disprove this conjecture. (Quadratic reciprocity will probably be of use.)
4. Attempt to simplify the computation, in [16], of $\kappa(D_n, \{r, s\})$. Generalize this result to other semidirect products of the form $\mathbb{Z}_p \rtimes \mathbb{Z}_q$, where p, q are primes with $q | p - 1$.

Appendix A

Linear Algebra

This appendix provides some notation and theorems from linear algebra that we use throughout the text. This is intended to be a refresher and a quick reference, not a way of learning this material for the first time. We used the textbook by Friedberg, Insel, and Spence [56] as the main reference for this appendix. We assume that the reader is familiar with vector spaces.

To save space, we write column vectors as the transpose of a row vector; that is,

$$(x_1, x_2, \ldots, x_n)^t = \begin{pmatrix} x_1 \\ x_2 \\ \vdots \\ x_n \end{pmatrix}.$$

We use ^{t} to denote the transpose of a matrix. That is,

$$\begin{pmatrix} a_{11} & a_{12} & \cdots & a_{1n} \\ a_{21} & a_{22} & \cdots & a_{2n} \\ \vdots & \vdots & \ddots & \vdots \\ a_{n1} & a_{n2} & \cdots & a_{nn} \end{pmatrix}^t = \begin{pmatrix} a_{11} & a_{21} & \cdots & a_{n1} \\ a_{12} & a_{22} & \cdots & a_{n2} \\ \vdots & \vdots & \ddots & \vdots \\ a_{1n} & a_{2n} & \cdots & a_{nn} \end{pmatrix}.$$

Throughout this appendix, all vector spaces are over the field of complex numbers $\mathbb{C}$.

1. DIMENSION OF A VECTOR SPACE

In this section we develop the notion of the dimension of a vector space.

Definition A.1 Let V be a vector space. The vectors $\mathbf{v}_1, \ldots, \mathbf{v_n}$ are said to be *linearly dependent* if there exist complex numbers $c_1, \ldots, c_n$, not all 0, such that

$$c_1\mathbf{v}_1 + c_2\mathbf{v}_2 + \cdots + c_n\mathbf{v}_n = \mathbf{0}.$$

If the vectors are not linearly dependent, then we say that they are *linearly independent.*

Example A.2
In $\mathbb{C}^3$, the vectors $(1, 2, i)^t$, $(2, 3, 2 + 2i)^t$, and $(0, -1, 2)^t$ are linearly dependent because

$$2 \cdot (1, 2, i)^t - (2, 3, 2 + 2i)^t + (0, -1, 2)^t = (0, 0, 0)^t.$$

Recall that the vectors $(1, 0, 0)^t$, $(0, 1, 0)^t$, and $(0, 0, 1)^t$ are linearly independent.

Definition A.3 Let V be a vector space and $\mathbf{v}_1, \ldots, \mathbf{v}_n$ be vectors from V. The *span* of $\mathbf{v}_1, \ldots, \mathbf{v}_n$ is defined to be the subspace

$$\text{span}\{\mathbf{v}_1, \ldots, \mathbf{v}_n\} = \{c_1\mathbf{v}_1 + c_2\mathbf{v}_2 + \cdots + c_n\mathbf{v}_n \mid c_1, c_2, \ldots, c_n \in \mathbb{C}\}$$

of V. We say that $\mathbf{v}_1, \ldots, \mathbf{v}_n$ *span* V if $V = \text{span}\{\mathbf{v}_1, \ldots, \mathbf{v}_n\}$.

Example A.4
The vectors $(1, 0, 0)^t$, $(0, 1, 0)^t$, and $(0, 0, 1)^t$ span $\mathbb{C}^3$ because every vector $(a, b, c)^t \in \mathbb{C}^3$ can be written in the form

$$(a, b, c)^t = a \cdot (1, 0, 0)^t + b \cdot (0, 1, 0)^t + c \cdot (0, 0, 1)^t.$$

We now recall the definition of a basis of a vector space.

Definition A.5 Let V be a vector space and $\mathbf{v}_1, \ldots, \mathbf{v}_n$ be vectors from V. We say that $\mathbf{v}_1, \ldots, \mathbf{v}_n$ is a *basis* of V if

1. the vectors $\mathbf{v}_1, \ldots, \mathbf{v}_n$ span V, and
2. the vectors $\mathbf{v}_1, \ldots, \mathbf{v}_n$ are linearly independent.

Example A.6
$(1, 0, 0)^t$, $(0, 1, 0)^t$, and $(0, 0, 1)^t$ form a basis for $\mathbb{C}^3$.

Remark A.7
The reader should show that the vectors $\mathbf{v}_1, \ldots, \mathbf{v}_n$ form a basis for a vector space V if and only if every vector v in V can be expressed uniquely in the form

$$v = c_1\mathbf{v}_1 + c_2\mathbf{v}_2 + \cdots + c_n\mathbf{v}_n$$

for some $c_1, \ldots, c_n$ in $\mathbb{C}$.

Proposition A.8
Let V be a vector space. If there exists a basis for V with n elements, then every basis for V has n elements.

Proof
See [56, p. 44].

The preceeding proposition allows us to define the dimension of a vector space.

Definition A.9 Let V be a vector space. Suppose that V has a basis with n elements. Then we say that V is *finite-dimensional* and that the *dimension* of V as a vector space over $\mathbb{C}$ is n.

Example A.10
$\mathbb{C}^3$ has dimension 3. In general, $\mathbb{C}^n$ is an n-dimensional vector space.

Remark A.11
The reader should verify the following. Let V be a vector space. Suppose that $\mathbf{v}_1, \mathbf{v}_2, \ldots, \mathbf{v}_m$ are linearly independent vectors in V. Then the dimension of $\text{span}\{\mathbf{v}_1, \mathbf{v}_2, \ldots, \mathbf{v}_m\}$ is m.

The following proposition is useful.

Proposition A.12
Let V be a vector space of dimension n. If $\mathbf{v}_1, \mathbf{v}_2, \ldots, \mathbf{v}_\mathbf{n}$ are linearly independent vectors of V, then they form a basis for V.

Proof
See [56, p. 45].

2. INNER PRODUCT SPACES, DIRECT SUM OF SUBSPACES

Definition A.13 Let V be a vector space. An *inner product* (or *Hermitian inner product*) over V is a function $\langle \cdot, \cdot \rangle : V \times V \to F$ such that for all $\mathbf{x}, \mathbf{y}, \mathbf{z} \in V$ and $\alpha \in \mathbb{C}$ we have

1. $\langle \mathbf{x}, \mathbf{y} \rangle = \overline{\langle \mathbf{y}, \mathbf{x} \rangle}$.
2. $\langle \alpha\mathbf{x}, \mathbf{y} \rangle = \alpha \langle \mathbf{x}, \mathbf{y} \rangle$.
3. $\langle \mathbf{x} + \mathbf{y}, \mathbf{z} \rangle = \langle \mathbf{x}, \mathbf{z} \rangle + \langle \mathbf{y}, \mathbf{z} \rangle$.
4. $\langle \mathbf{x}, \mathbf{x} \rangle \geq 0$ with equality if and only if $\mathbf{x} = \mathbf{0}$.

The *norm* associated with $\langle \cdot, \cdot \rangle$ is defined for $\mathbf{v} \in V$ as $\|\mathbf{v}\| = \sqrt{\langle \mathbf{v}, \mathbf{v} \rangle}$.

If a vector space V has an inner product, then we call V an *inner product space*.

Example A.14
The standard inner product and norm on $\mathbb{C}^n$ are given by

$$\langle (z_1, \ldots, z_n)^t, (w_1, \ldots, w_n)^t \rangle_2 = z_1\overline{w_1} + \ldots + z_n\overline{w_n}$$

and

$$\left\| (z_1, \ldots, z_n)^t \right\|_2 = \sqrt{z_1\overline{z_1} + \ldots + z_n\overline{z_n}}.$$

For example, in $\mathbb{C}^3$ we have

$$\langle (1, i, 0)^t, (-i, 5, 5 + i) \rangle_2 = 1 \cdot i + i \cdot 5 + 0 \cdot (5 - i) = 6i.$$

and

$$\left\|(1, i, 0)^t\right\|_2 = \sqrt{1 \cdot 1 + i \cdot (-i) + 0 \cdot 0} = \sqrt{2}.$$

Let $\mathbf{z}, \mathbf{w} \in \mathbb{C}^n$. Regard $\mathbf{z}$ and $\mathbf{w}$ as column vectors. Note that $\langle \mathbf{z}, \mathbf{w} \rangle_2 = \mathbf{w}^* \mathbf{z}$ where $\mathbf{w}^*$ is the conjugate transpose of $\mathbf{w}$.

Definition A.15 Let V be an inner product space and $\mathbf{v}_1, \ldots, \mathbf{v}_n \in V$. We say that $\{\mathbf{v}_1, \ldots, \mathbf{v}_n\}$ is an *orthogonal* set of vectors if $\left\langle \mathbf{v}_i, \mathbf{v}_j \right\rangle = 0$ whenever $i \neq j$.

If in addition to being orthogonal, we have that $\langle \mathbf{v}_i, \mathbf{v}_i \rangle = 1$ for $i = 1, 2, \ldots, n$, then we say that the set of vectors is *orthonormal*. That is, the vectors all have norm 1 and are mutually orthogonal.

Proposition A.16
Let V be an inner product space and $\mathbf{v}_1, \ldots, \mathbf{v}_n \in V$. If $\mathbf{v}_1, \ldots, \mathbf{v}_n$ form an orthogonal set of nonzero vectors, then $\mathbf{v}_1, \ldots, \mathbf{v}_n$ are linearly independent.

Proof
See [56, p. 327]. ◬

Example A.17
In $\mathbb{C}^4$, the vectors $(1, i, -1, -i)^t$, $(1, -1, 1, -1)^t$, $(1, -i, -1, i)^t$, and $(1, 1, 1, 1)$ are orthogonal. By Proposition A.16 they are linearly independent. Since $\mathbb{C}^4$ has dimension 4, by Proposition A.12 the vectors form a basis for $\mathbb{C}^4$.

Proposition A.18 (Gram-Schmidt)
Let V be an inner product space. Then there exists an othonormal basis for V.

Proof
See [56, p. 328]. ◬

Example A.19
Each vector in Example A.17 has norm 2. We can get an orthonormal basis for $\mathbb{C}^4$ by taking the vectors from Example A.17 and dividing them each by 2 to get $(1/2, i/2, -1/2, -i/2)^t$, $(1/2, -1/2, 1/2, -1/2)^t$, $(1/2, -i/2, -1/2, i/2)^t$, and $(1/2, 1/2, 1/2, 1/2)$.

Proposition A.20 (Cauchy-Schwarz inequality)
Let V be an inner product space with inner product $\langle \cdot, \cdot \rangle$ and associated norm $\|\cdot\|$. If $\mathbf{x}, \mathbf{y} \in V$, then

$$\left|\langle \mathbf{x}, \mathbf{y} \rangle\right| \leq \|\mathbf{x}\| \, \|\mathbf{y}\| .$$

Proof
See [56, p. 320]. ◬

Proposition A.21
Let V be an inner product space. If $\mathbf{v}, \mathbf{w} \in V$ and $\langle \mathbf{v}, \mathbf{w} \rangle = 0$, then $\|\mathbf{v} + \mathbf{w}\|^2 = \|\mathbf{v}\|^2 + \|\mathbf{w}\|^2$.

Proof
Exercise for the reader. Use the definition of $\|\cdot\|$.

Definition A.22 Let V be a vector space with subspaces U and W. We say that V is the *direct sum* of U and W, and write $V = U \oplus W$, if every element $\mathbf{v} \in V$ can be expressed uniquely in the form $\mathbf{v} = \mathbf{u} + \mathbf{w}$ for some $\mathbf{u} \in U$ and $\mathbf{v} \in V$.

If n is a positive integer, we write nV for $V \oplus V \oplus \cdots \oplus V$, where the direct sum is repeated n times.

Example A.23
Consider the subspaces $U = \{(z_1, 0)^t \mid z_1 \in \mathbb{C}\}$ and $W = \{(0, z_2)^t \mid z_2 \in \mathbb{C}\}$ of $\mathbb{C}^2$. Then, $\mathbb{C}^2 = U \oplus W$.

Definition A.24 Let V be an inner product space with subspaces U and W. We say that V is the *orthogonal direct sum* of U and W if $V = U \oplus W$ and $\langle \mathbf{u}, \mathbf{w} \rangle = 0$ for all $\mathbf{u} \in U$ and $\mathbf{w} \in W$.

Example A.25
Let U and W be as in Example A.23. Then $\mathbb{C}^2$ is the orthogonal direct sum of U and W.

Definition A.26 Suppose that W is a subspace of an inner product space V. The *orthogonal complement* of W, denoted by $W^\perp$, is defined to be

$$W^\perp = \{\mathbf{v} \in V \mid \langle \mathbf{v}, \mathbf{w} \rangle = 0 \text{ for all } \mathbf{w} \in W\}.$$

Proposition A.27
If V is an inner product space and W is a subspace of V, then $W^\perp$ is a subspace of V, and V is equal to the orthogonal direct sum $V = W \oplus W^\perp$.

Proof
See [56, pp. 332–33].

Example A.28
Consider the vector space $\mathbb{C}^4$. Let

$$W = \text{span}\{(1, 1, 1, 1)^t\}.$$

Then

$$W^\perp = \text{span}\{(1, i, -1, -i)^t, (1, -i, -1, i)^t, (1, -1, 1, -1)^t\}$$

and $\mathbb{C}^4 = W \oplus W^\perp$.

Definition A.29 Let A be an $n \times n$ matrix with real entries. We say that A is *symmetric* if $A^t = A$, where A^t is the transpose of A. That is, $A_{i,j} = A_{j,i}$ whenever $1 \le i, j \le n$.

Example A.30
The matrix

$$A = \begin{pmatrix} 1 & 3 & 2 \\ 3 & 0 & -1 \\ 2 & -1 & 1/2 \end{pmatrix}$$

is symmetric. The matrix

$$B = \begin{pmatrix} 1 & 2 & 3 \\ 2 & 0 & 6 \\ -1 & 2 & 10 \end{pmatrix}$$

is not symmetric.

The following lemma is a useful fact about real symmetric matrices. It states that a real symmetric matrix is equal to its adjoint matrix.

Lemma A.31
Let A be an $n \times n$ real symmetric matrix, and let $\mathbf{x}, \mathbf{y} \in \mathbb{C}^n$. Then $\langle A\mathbf{x}, \mathbf{y} \rangle_2 = \langle \mathbf{x}, A\mathbf{y} \rangle_2$.

Proof
Let A^* and $\mathbf{y}^*$ denote the conjugate transposes of A and $\mathbf{y}$, respectively. Because A is a symmetric matrix with real entries, we have that $A^* = A$. Hence,

$$\langle A\mathbf{x}, \mathbf{y} \rangle_2 = \mathbf{y}^* A\mathbf{x} = \mathbf{y}^* A^* \mathbf{x} = (A\mathbf{y})^* \mathbf{x} = \langle \mathbf{x}, A\mathbf{y} \rangle_2 .$$

Definition A.32 Let A be an $n \times n$ matrix with complex entries. We say that A is *unitary* if $A^*A = I_n$, where $A^* = \overline{A^t}$ is the conjugate transpose of A and I_n is the $n \times n$ identity matrix.

Example A.33
The matrix

$$A = \begin{pmatrix} \frac{1}{\sqrt{2}} & \frac{i}{\sqrt{2}} \\ \frac{1}{\sqrt{2}} & \frac{-i}{\sqrt{2}} \end{pmatrix}$$

is unitary because

$$A^*A = \begin{pmatrix} \frac{1}{\sqrt{2}} & \frac{1}{\sqrt{2}} \\ \frac{-i}{\sqrt{2}} & \frac{i}{\sqrt{2}} \end{pmatrix} \begin{pmatrix} \frac{1}{\sqrt{2}} & \frac{i}{\sqrt{2}} \\ \frac{1}{\sqrt{2}} & \frac{-i}{\sqrt{2}} \end{pmatrix} = \begin{pmatrix} 1 & 0 \\ 0 & 1 \end{pmatrix}$$

Remark A.34
The reader should verify the following fact. An $n \times n$ matrix A is unitary if and only if the columns of A form an orthonormal basis for $\mathbb{C}^n$.

3. THE MATRIX OF A LINEAR TRANSFORMATION

The ideas in this section are used heavily in Chapter 6.

Definition A.35 Given a vector space V we write $\beta = [\mathbf{v}_1, \mathbf{v}_2, \dots, \mathbf{v}_n]$ to denote an *ordered basis* for V. That is, β is a basis with a particular ordering.

With an ordered basis one can define the coordinates of a vector with respect to that basis.

Definition A.36 Let V be a vector space with ordered basis $\beta = [\mathbf{v}_1, \mathbf{v}_2, \dots, \mathbf{v}_n]$. Given $\mathbf{x} = a_1\mathbf{v}_1 + a_2\mathbf{v}_2 + \cdots + a_n\mathbf{v}_n \in V$ we write

$$[\mathbf{x}]_\beta = (a_1, a_2, \dots, a_n)^t$$

for the *coordinates* of $\mathbf{x}$ with respect to the basis β.

Example A.37
Consider the vector space $\mathbb{C}^2$, the basis

$$\beta = [(1, 2)^t, (-1, 1)^t],$$

and the vector $\mathbf{v} = (5, 4)^t$. Then $\mathbf{v} = 3 \cdot (1, 2)^t - 2 \cdot (-1, 1)^t$. Hence $[\mathbf{v}]_\beta = (3, -2)^t$.

Let $\beta' = [(1, 0)^t, (0, 1)^t]$ be the standard ordered basis for $\mathbb{C}^2$. Then $\mathbf{v} = 5 \cdot (1, 0)^t + 4 \cdot (0, 1)^t$. Hence, $[\mathbf{v}]_{\beta'} = (5, 4)^t$.

Definition A.38 Let $L : V \to W$ be a linear transformation between two vector spaces V and W. Suppose that $\beta = [\mathbf{v}_1, \mathbf{v}_2, \dots, \mathbf{v}_n]$ is an ordered basis for V and γ is an ordered basis for W. The matrix

$$[L]_\beta^\gamma = \left([L(\mathbf{v}_1)]_\gamma \;\middle|\; [L(\mathbf{v}_2)]_\gamma \;\middle|\; \cdots \;\middle|\; [L(\mathbf{v}_n)]_\gamma \right)$$

is called the *matrix for L with respect to β and γ*. If $V = W$ and $\beta = \gamma$, then we write $[L]_\beta$ for $[L]_\beta^\gamma$.

The following proposition sheds light on the definition.

Proposition A.39
Let V and W be vector spaces with ordered bases β and γ, respectively, and let $L : V \to W$ be a linear transformation. Then

$$[L(\mathbf{v})]_\gamma = [L]_\beta^\gamma [\mathbf{v}]_\beta$$

for every $\mathbf{v} \in V$.

Proof
See [56, p. 84]. ⊛

Example A.40
Consider the vector space $\mathbb{C}^2$ and the linear transformation $L : \mathbb{C}^2 \to \mathbb{C}^2$ given by

$$L((z_1, z_2)^t) = (z_1 + z_2, 2z_1 - z_2)^t.$$

Consider the ordered basis

$$\beta = \left[(1,1)^t, (-1,1)^t\right].$$

Then

$$\begin{aligned} L((1,1)^t) &= (2,1)^t &&= \tfrac{3}{2}\cdot(1,1)^t - \tfrac{1}{2}\cdot(-1,1)^t, \text{ and} \\ L((-1,1)^t &= (0,-3)^t &&= -\tfrac{3}{2}\cdot(1,1)^t - \tfrac{3}{2}\cdot(-1,1)^t. \end{aligned}$$

Therefore

$$[L]_\beta = \left(\left[L(1,1)^t\right]_\beta \;\middle|\; \left[L(-1,1)^t\right]_\beta \right) = \begin{pmatrix} \frac{3}{2} & -\frac{3}{2} \\ -\frac{1}{2} & -\frac{3}{2} \end{pmatrix}.$$

Let $\mathbf{v} = (1,2)^t$. Then

$$(1,2)^t = \frac{3}{2}\cdot(1,1)^t + \frac{1}{2}\cdot(-1,1)^t.$$

Hence, $[\mathbf{v}]_\beta = (3/2, 1/2)^t$. Notice that

$$L(\mathbf{v}) = (3,0)^t = \frac{3}{2}\cdot(1,1)^t + \frac{-3}{2}\cdot(-1,1)^t.$$

Hence $[L(\mathbf{v})]_\beta = (3/2, -3/2)^t$. The reader should verify that $[L(\mathbf{v})]_\beta = [L]_\beta[\mathbf{v}]_\beta$.

Remark A.41
Think of the matrix $[L]_\beta$ as a function that computes L but takes coordinate vectors in terms of β as its input and gives coordinate vectors in terms of β as its output.

Definition A.42 Let V be a vector space with ordered bases β and β'. Let $I : V \to V$ be the identity transformation. That is, $I\mathbf{v} = \mathbf{v}$ for all $\mathbf{v} \in V$. The matrix $[I]_\beta^{\beta'}$ is called the *change-of-basis matrix from* β *to* β'.

The change-of-basis matrix does exactly what its name says: it changes vector coordinates in terms of one basis into vector coordinates in terms of another basis.

Example A.43
Consider the vector space $\mathbb{C}^2$. Let β be as in Example A.40 and

$$\beta' = \{(1,0)^t, (0,1)^t\}$$

be the standard basis for $\mathbb{C}^2$. Let $\mathbf{v} = (1,2)^t$. Then $[\mathbf{v}]_\beta = (3/2, 1/2)^t$ and $[\mathbf{v}]_{\beta'} = (1,2)^t$. Note that

$$[I]_\beta^{\beta'} = \begin{pmatrix} 1 & -1 \\ 1 & 1 \end{pmatrix}.$$

The reader should verify that $[I]_\beta^{\beta'}[\mathbf{v}]_\beta = [\mathbf{v}]_{\beta'}$. That is, the matrix $[I]_\beta^{\beta'}$ turns β coordinates into β' coordinates. The reader should verify this for vectors other than $\mathbf{v}$. We see shortly that this fact is true in general.

One has the following important properties of the foregoing objects.

Proposition A.44
Let V be a vector space, and let $L : V \to V$ be a linear transformation. Suppose that β and β' are ordered bases for V. Furthermore, let $Q = [I]_\beta^{\beta'}$. Then

1. Q is invertible and $Q^{-1} = [I]_{\beta'}^{\beta}$.
2. For any $\mathbf{v} \in V$, we have that $[\mathbf{v}]_{\beta'} = Q[\mathbf{v}]_\beta$.
3. $[L]_\beta = Q^{-1}\,[L]_{\beta'}\,Q$.

Proof
See [56, pp. 103–4]. ◬

Example A.45
Consider the vector space $\mathbb{C}^2$. Let L, β, β', and $[I]_\beta^{\beta'}$ be as in Examples A.40 and A.43. Let $Q = [I]_\beta^{\beta'}$. The reader should verify that

$$[L]_{\beta'} = \begin{pmatrix} 1 & 1 \\ 2 & -1 \end{pmatrix}$$

and $[L]_\beta = Q^{-1}[L]_{\beta'}Q$.

Remark A.46
Think of the expression $[L]_\beta = Q^{-1}[L]_{\beta'}Q$ from Proposition A.44 in the following way: Q turns β coordinate vectors into β' coordinate vectors. Then $[L]_{\beta'}$ computes L, but its input and output are coordinate vectors in terms of β'. Last, Q^{-1} turns β' coordinate vectors back into β coordinate vectors. Hence the composition of these three steps is $[L]_\beta$.

Another useful fact is the following.

Proposition A.47
Let V be a vector space and β an ordered basis for V. Let $L_1 : V \to V$ and $L_2 : V \to V$ be linear transformations. Then $[L_1 \circ L_2]_\beta = [L_1]_\beta [L_2]_\beta$.

Proof
See [56, p. 82].

4. EIGENVALUES OF LINEAR TRANSFORMATIONS

Definition A.48 Let A be an $n \times n$ complex matrix. The *characteristic polynomial* of A is the polynomial

$$p_A(x) = \det(A - xI),$$

where I is the $n \times n$ identity matrix.

Definition A.49 Let A be an $n \times n$ complex matrix. The *eigenvalues* of A are the complex roots of the equation $p_A(x) = 0$ (counted with multiplicity). If λ is a root of p_A, then the *algebraic multiplicity* of λ is the largest positive integer k for which $(x - \lambda)^k$ is a factor of $p_A(x)$.

Example A.50
Let $A = \begin{pmatrix} 0 & 1 & 1 \\ 0 & 0 & 1 \\ 0 & 0 & 5 \end{pmatrix}$. Then

$$p_A(x) = \det \begin{pmatrix} -x & 1 & 1 \\ 0 & -x & 1 \\ 0 & 0 & 5 - x \end{pmatrix} = -x^2(x - 5).$$

So the eigenvalues of A are $\lambda = 0, 5$ with algebraic multiplicities 2 and 1, respectively.

Suppose that $p_A(\lambda) = 0$ for some $\lambda \in \mathbb{C}$. Then $\det(A - xI) = 0$, which implies that $A - xI$ is a noninvertible matrix. Hence, $A - xI$ is not one-to-one. Thus, there exists a nonzero element $\mathbf{v} \in \mathbb{C}^n$ such that $\mathbf{v} \in \ker(A - \lambda I)$. So, $A\mathbf{v} = \lambda\mathbf{v}$. This leads us to the next definition.

Definition A.51 Let A be a $n \times n$ complex matrix and λ be an eigenvalue of A. We say that a nonzero vector $\mathbf{v} \in \mathbb{C}^n$ is an *eigenvector* (or *eigenfunction*) of A with respect to λ if $A\mathbf{v} = \lambda\mathbf{v}$. The *eigenspace* of A corresponding to λ is the subspace of $\mathbb{C}^n$ given by

$$E_\lambda(A) = \{\mathbf{v} \in \mathbb{C}^n | A\mathbf{v} = \lambda\mathbf{v}\}.$$

Note that $\mathbf{0} \in E_\lambda(A)$. The zero vector is not an eigenvector of A but is included in $E_\lambda(A)$ to make it a subspace of $\mathbb{C}^n$.

The *geometric multiplicity* of an eigenvalue λ is the dimension of $E_\lambda(A)$ as a vector space over $\mathbb{C}$.

Example A.52
In Example A.50, we saw that the eigenvalues of

$$A = \begin{pmatrix} 0 & 1 & 1 \\ 0 & 0 & 1 \\ 0 & 0 & 5 \end{pmatrix}$$

are $\lambda = 0, 5$ with algebraic multiplicities 2 and 1, respectively.

Let us compute the geometric multiplicity of 0. To find a basis for $E_0(A)$ we must solve the equation $A\mathbf{x} = 0\mathbf{x}$ for $\mathbf{x}$. This reduces to solving the system

$$\begin{array}{rcrcl} x_2 & + & x_3 & = & 0 \\ & & x_3 & = & 0 \\ & & 5x_3 & = & 0 \end{array}.$$

Hence, a basis for $E_0(A)$ is $\{(1, 0, 0)^t\}$. Thus, the geometric multiplicity of 0 is $\dim(E_0(A)) = 1$. Note that the algebraic multiplicity of 0 is 2.

Similarly, a basis for $E_5(A)$ is $\{(6, 5, 25)^t\}$. Thus, the geometric multiplicity of 5 and the algebraic multiplicity of 5 are both equal to 1.

In general, if λ is an eigenvalue of a matrix A, then the geometric multiplicity of λ is less than or equal to the algebraic multiplicity of λ. However, for symmetric matrices, we get equality.

Theorem A.53 (Spectral theorem for symmetric real matrices)
Suppose that A is an $n \times n$ symmetric matrix with real entries. Then

1. All the eigenvalues of A are real.
2. There is an orthonormal basis of $\mathbb{C}^n$ consisting of n real eigenvectors of A.
3. If λ is an eigenvalue of A, the algebraic multiplicity of λ equals the geometric multiplicity of λ.

Proof
See [56, pp. 257 and 362].

Note that the eigenvectors in the theorem are orthogonal with respect to the standard inner product on $\mathbb{C}^n$.

Example A.54
Consider the symmetric matrix

$$A = \begin{pmatrix} 0 & 1 & 1 \\ 1 & 0 & 1 \\ 1 & 1 & 0 \end{pmatrix}.$$

One easily computes that $p_A(x) = -(x-2)(x+1)^2$. So the eigenvalues of A are $\lambda = -1, 2$ with algebraic multiplicities 2 and 1, respectively. A basis for $E_{-1}(A)$ is $\{(-2, 1, 1)^t, (0, 1, -1)^t\}$, and a basis for $E_2(A)$ is $\{(1, 1, 1)^t\}$. Notice that the dimensions of the eigenspaces match up with the multiplicities of the eigenvalues. Also, note that the eigenvectors that are given are orthogonal to one another. By dividing them by their lengths, one gets an orthonormal basis of eigenvectors for $\mathbb{C}^3$.

Definition A.55 An *orthogonal matrix* O is a matrix with real entries whose columns form an orthonormal set of vectors; that is, where $O^t = O^{-1}$.

Corollary A.56
Let A be an $n \times n$ symmetric matrix with eigenvalues $\lambda_1, \ldots, \lambda_n$. Then there exists an $n \times n$ orthogonal matrix O such that

$$O^t A O = \begin{pmatrix} \lambda_1 & 0 & \cdots & 0 \\ 0 & \lambda_2 & \cdots & 0 \\ \vdots & \vdots & \ddots & \vdots \\ 0 & 0 & \cdots & \lambda_n \end{pmatrix}.$$

Proof
By Theorem A.53, there exists an orthonormal set of real vectors $\mathbf{v}_1, \ldots, \mathbf{v}_n$ such that $A\mathbf{v}_i = \lambda_i \mathbf{v}_i$ for $i = 1, \ldots, n$. Construct the matrix O whose columns are the vectors $\mathbf{v}_i$. That is, let

$$O = \left(\mathbf{v}_1 \middle| \mathbf{v}_2 \middle| \cdots \middle| \mathbf{v}_n \right).$$

Then

$$\begin{aligned} O^t A O &= \begin{pmatrix} \mathbf{v}_1^t \\ \vdots \\ \mathbf{v}_n^t \end{pmatrix} \left(A\mathbf{v}_1 \middle| A\mathbf{v}_2 \middle| \cdots \middle| A\mathbf{v}_n \right) \\ &= \begin{pmatrix} \lambda_1 \langle \mathbf{v}_1, \mathbf{v}_1 \rangle_2 & \lambda_2 \langle \mathbf{v}_1, \mathbf{v}_2 \rangle_2 & \cdots & \lambda_n \langle \mathbf{v}_1, \mathbf{v}_n \rangle_2 \\ \vdots & \vdots & \ddots & \vdots \\ \lambda_1 \langle \mathbf{v}_n, \mathbf{v}_1 \rangle_2 & \lambda_2 \langle \mathbf{v}_n, \mathbf{v}_2 \rangle_2 & \cdots & \lambda_n \langle \mathbf{v}_n, \mathbf{v}_n \rangle_2 \end{pmatrix} \\ &= \begin{pmatrix} \lambda_1 & 0 & \cdots & 0 \\ 0 & \lambda_2 & \cdots & 0 \\ \vdots & \vdots & \ddots & \vdots \\ 0 & 0 & \cdots & \lambda_n \end{pmatrix}. \end{aligned}$$

Note that $O^t O = I$ by a similar proof. Thus, O is an orthogonal matrix. ⊛

Example A.57
Let A be as in Example A.54. Construct an orthogonal matrix O as follows: Take the orthogonal set of eigenvectors given in Example A.54 and divide them by their norms. Then put them in as the columns of our matrix O to get

$$O = \begin{pmatrix} \frac{1}{\sqrt{3}} & \frac{-2}{\sqrt{6}} & 0 \\ \frac{1}{\sqrt{3}} & \frac{1}{\sqrt{6}} & \frac{1}{\sqrt{2}} \\ \frac{1}{\sqrt{3}} & \frac{1}{\sqrt{6}} & \frac{-1}{\sqrt{2}} \end{pmatrix}.$$

The reader should verify that

$$O^t AO = \begin{pmatrix} 2 & 0 & 0 \\ 0 & -1 & 0 \\ 0 & 0 & -1 \end{pmatrix}.$$

Definition A.58 Let A be an $n \times n$ matrix. The *trace* of A, denoted by $\operatorname{tr}(A)$, is defined to be the sum of the diagonal entries of A. That is, $\operatorname{tr}(A) = A_{1,1} + \cdots + A_{n,n}$.

Example A.59
With A as in Example A.54, we have $\operatorname{tr}(A) = 0$. Note that the sum of the eigenvalues of A (counted with multiplicity) is $2 - 1 - 1 = 0$.

Lemma A.60
Let A be an $n \times n$ matrix with eigenvalues $\lambda_1, \ldots, \lambda_n$. Then $\operatorname{tr}(A) = \sum_{i=1}^{n} \lambda_i$.

Proof
See [117, p. 265]. ◈

Lemma A.61
Suppose that A and B are $n \times n$ matrices and there is an invertible matrix Q such that $A = Q^{-1}BQ$; that is, A and B are similar matrices. Then

1. The characteristic polynomials of A and B are the same.
2. A and B have the same eigenvalues.
3. The trace of A equals the trace of B.

Proof
(1) Suppose that A, B, and Q are $n \times n$ matrices, where Q is invertible and $A = Q^{-1}BQ$. Then

$$\begin{aligned} p_B(x) &= \det(B - xI) \\ &= \det(Q^{-1}) \det(B - xI) \det(Q) \\ &= \det(Q^{-1} BQ - xQ^{-1}Q) \\ &= \det(A - xI) \\ &= p_A(x). \end{aligned}$$

(2) This follows from (1), because the eigenvalues of a matrix are the roots of its characteristic polynomial.

(3) This follows from and (2) and Lemma A.60. ◬

Lemma A.62
Let A be an $n \times n$ symmetric matrix with eigenvalues $\lambda_1, \ldots, \lambda_n$ (counted with multiplicity). Then $\mathrm{tr}(A^k) = \sum_{i=1}^{n} \lambda_i^k$ for any positive integer k.

Proof
By Corollary A.56, there is an orthogonal matrix O such that $A = ODO^t$ where

$$D = \begin{pmatrix} \lambda_1 & 0 & \cdots & 0 \\ 0 & \lambda_2 & \cdots & 0 \\ \vdots & \vdots & \ddots & \vdots \\ 0 & 0 & \cdots & \lambda_n \end{pmatrix},$$

and $O^t = O^{-1}$. Hence, $A^k = OD^kO^t$. By Lemma A.61, $\mathrm{tr}(A^k) = \mathrm{tr}(D^k) = \sum_{i=1}^{n} \lambda_i^k$. ◬

5. EIGENVALUES OF CIRCULANT MATRICES

In this section, we present formulas for the eigenvalues of a circulant matrix.

Definition A.63 A matrix C is called *circulant* if it can be written in the form

$$C = \begin{pmatrix} c_0 & c_1 & c_2 & \cdots & c_{n-1} \\ c_{n-1} & c_0 & c_1 & \cdots & c_{n-2} \\ c_{n-2} & c_{n-1} & c_0 & \cdots & c_{n-3} \\ \vdots & \vdots & \vdots & \ddots & \vdots \\ c_1 & c_2 & c_3 & \cdots & c_0 \end{pmatrix}. \tag{40}$$

Proposition A.64
The eigenvalues of the circulant matrix C given in Equation 40 are

$$\chi_a = \sum_{j=0}^{n-1} c_j\, \xi^{aj},$$

where $\xi = \exp(2\pi i/n)$ and $a = 0, 1, 2, \ldots, n-1$.

Proof
Let $\psi_a = (1, \xi^a, \xi^{2a}, \ldots, \xi^{(n-1)a})^t$. We leave it to the reader to show that $C\psi_a = \chi_a \psi_a$. Therefore, ψ_a is an eigenvector of C with eigenvalue χ_a.

If $0 \leq b < a \leq n-1$, then by the geometric sum formula we have that

$$\langle \psi_a, \psi_b \rangle = \sum_{j=0}^{n-1} \xi^{(a-b)j} = \frac{\xi^{(a-b)n} - 1}{\xi^{(a-b)} - 1} = 0.$$

Thus, by Lemma A.16 the vectors $\{\psi_0, \dots, \psi_{n-1}\}$ form a basis for $\mathbb{C}^n$. Hence, they form a basis of eigenvectors for C.

Example A.65
Consider the circulant matrix

$$A = \begin{pmatrix} 2 & 0 & 1 \\ 1 & 2 & 0 \\ 0 & 1 & 2 \end{pmatrix}.$$

Let $\xi = \exp(2\pi i/3)$. The eigenvalues of A are

$$\lambda_0 = 2\xi^0 + 0\xi^0 + 1\xi^0 = 3, \text{ and}$$
$$\lambda_1 = 2\xi^0 + 0\xi^1 + 1\xi^2 = 2 + \cos(4\pi/3) + i\sin(4\pi/3) = 3/2 - i\sqrt{3}/2, \text{ and}$$
$$\lambda_2 = 2\xi^0 + 0\xi^2 + 1\xi^4 = 2 + \cos(8\pi/3) + i\sin(8\pi/3) = 3/2 + i\sqrt{3}/2.$$

Appendix B

Asymptotic Analysis of Functions

1. BIG OH

Throughout this section, let $\mathbb{N} = \{1, 2, 3, \ldots\}$ denote the set of natural numbers, and let $\mathbb{R}^+$ denote the set of positive real numbers.

Definition B.1 Let $f, g : \mathbb{N} \to \mathbb{R}^+$. We say that $f(n) = O(g(n))$ if there exist constants $N, c > 0$, where $f(n) \leq cg(n)$ for all $n > N$. Otherwise, we say that $f(n) \neq O(g(n))$.

That is, $f(n) = O(g(n))$ means that at some point f is bounded above by a constant times g. So the big oh notation measures growth rates of functions. The following lemma from [46, p. 440] is sometimes useful.

Lemma B.2
Let $f, g : \mathbb{N} \to \mathbb{R}^+$. If

$$\lim_{n\to\infty} \frac{f(n)}{g(n)} = c > 0,$$

then $f(n) = O(g(n))$ and $g(n) = O(f(n))$. If

$$\lim_{n\to\infty} \frac{f(n)}{g(n)} = \infty,$$

then $g(n) = O(f(n))$, but $f(n) \neq O(g(n))$.

Proof
Suppose that $\lim_{n\to\infty} f(n)/g(n) = c > 0$. Then there exists a constant $N > 0$ such that $f(n)/g(n) < (c + 1)$ for all $n > N$. Thus, $f(n) = O(g(n))$. Using a similar argument on the equation $\lim_{n\to\infty} g(n)/f(n) = 1/c > 0$, yields that $g(n) = O(f(n))$.

Suppose that $\lim_{n\to\infty} f(n)/g(n) = \infty$. Then, $\lim_{n\to\infty} g(n)/f(n) = 0$. Thus, there exists a constant $N > 0$ such that $g(n)/f(n) < 1$ for all $n > N$.

So, $g(n) = O(f(n))$. If, on the other hand, we had that $f(n) = O(g(n))$, there would exist constants $M, d > 0$ such that $f(n)/g(n) < d$ for all $n > M$. But this clearly contradicts that fact that $\lim_{n\to\infty} f(n)/g(n) = \infty$. Hence, $f(n) \neq O(g(n))$. ⊛

Example B.3
Since $\lim_{n\to\infty} n^2/\log(n) = \infty$, we have $\log(n) = O(n^2)$ but $n^2 \neq O(\log(n))$.

We end with two other commonly used notations.

Definition B.4 Let $f, g : \mathbb{N} \to \mathbb{R}^+$. We say that $f(n) = \Omega(g(n))$ if there exist constants $N, c > 0$, where $cg(n) \leq f(n)$ for all $n > N$. We say that $f(n) = o(g(n))$ if for each $c > 0$ there exists $N_c > 0$ such that $f(n) \leq cg(n)$ for all $n > N_c$.

2. LIMIT INFERIOR OF A FUNCTION

Definition B.5 Let X be a subset of the real numbers $\mathbb{R}$. If X is bounded from below, then the greatest lower bound of X is called the *infimum* of X and is denoted by $\inf(X)$.

Example B.6
We have $\inf((2, 10]) = 2$, $\inf([5, \infty)) = 5$,

$$\inf(\{1/n \mid n = 1, 2, 3, \dots\}) = 0.$$

Definition B.7 Let (a_n) be a sequence of real numbers that is bounded from below. The *limit inferior* of (a_n) is

$$\liminf_{n\to\infty} a_n = \lim_{n\to\infty} \inf(\{a_{n+1}, a_{n+2}, a_{n+3}, \dots\}).$$

Remark B.8
Suppose (a_n) is a sequence of real numbers that is bounded from below and $L = \liminf_{n\to\infty} a_n$. Then, for every $\epsilon > 0$, there exists an $N > 0$ such that $a_n > L - \epsilon$ for all $n > N$. That is, for every $\epsilon > 0$, $L - \epsilon$ is an eventual lower bound for the sequence (a_n). Another way to say this is that every number b less than L is an eventual lower bound for the sequence (a_n) and only a finite number of terms from the sequence fall below b.

Example B.9
Let

$$a_n = \begin{cases} 1 & \text{if } n \text{ is odd} \\ -1 & \text{if } n \text{ is even.} \end{cases}$$

Then, $\liminf_{n\to\infty} a_n = -1$.

Example B.10
Let

$$a_n = \begin{cases} 2 & \text{if } n \text{ is odd} \\ -\frac{1}{n} & \text{if } n \text{ is even.} \end{cases}$$

Then, $\liminf_{n\to\infty} a_n = 0$.

Remark B.11
If (a_n) is a sequence of real numbers and $\lim_{n\to\infty} a_n$ exists, then it can be shown that $\liminf_{n\to\infty} a_n = \lim_{n\to\infty} a_n$. Consult any advanced calculus or real analysis book for a proof.

References

1. M. Abért and L. Babai, *Finite groups of uniform logarithmic diameter*, Israel J. Math. **158** (2007), 193–203.
2. M. Ajtai, J. Komlos, and E. Szemeredi, *An* $o(n \log n)$ *sorting network*, Proc. 15th Annual ACM Symposium on Theory and Computing (New York), Association for Computing Machinery, 1983.
3. S. Akers and B. Krishnamurthy, *A group-theoretic model for symmetric interconnection networks*, IEEE Trans. Comput. **38** (1989), no. 4, 555–66.
4. N. Alon, *Eigenvalues and expanders*, Combinatorica **6** (1986), no. 2.
5. ———, *Eigenvalues, geometric expanders, sorting in rounds, and Ramsey theory*, Combinatorica **6** (1986), no. 3, 207–19.
6. N. Alon, A. Lubotzky, and A. Wigderson, *Semi-direct product in groups and zig-zag product in graphs: Connections and applications (extended abstract)*, 42nd IEEE Symposium on Foundations of Computer Science (Las Vegas, NV, 2001), IEEE Computer Soc., 2001, pp. 630–37.
7. N. Alon and V. D. Milman, λ_1, *isoperimetric inequalities for graphs, and superconcentrators*, J. Combin. Theory, Series B **38** (1985), 73–88.
8. N. Alon and Y. Roichman, *Random Cayley graphs and expanders*, Rand. Struct. Alg. **5** (1994), no. 2, 271–84.
9. N. Alon, O. Schwartz, and A. Shapira, *An elementary construction of constant-degree expanders*, Combin. Probab. Comput. **17** (2008), no. 3, 319–27.
10. A. Amit and N. Linial, *Random graph coverings. I. General theory and graph connectivity*, Combinatorica **22** (2002), no. 1, 1–18.
11. ———, *Random lifts of graphs: Edge expansion*, Combin. Probab. Comput. **15** (2006), no. 3, 317–32.
12. A. Amit, N. Linial, and J. Matoušek, *Random lifts of graphs: Independence and chromatic number*, Rand. Struct. Alg. **20** (2002), no. 1, 1–22.
13. F. Annexstein and M. Baumslag, *On the diameter and bisector size of Cayley graphs*, Math. Systems Theory **26** (1993), no. 3, 271–91.
14. L. Babai, G. Hetyei, W. M. Kantor, A. Lubotzky, and Á. Seress, *On the diameter of finite groups*, 31st Annual Symposium on Foundations of Computer Science, Vol. I, II (St. Louis, MO, 1990), IEEE Computer Soc. Press, 1990, pp. 857–65.
15. L. Babai, W. M. Kantor, and A. Lubotzky, *Small-diameter Cayley graphs for finite simple groups*, European J. Combin. **10** (1989), no. 6, 507–22.
16. R. Bacher and P. de la Harpe, *Exact values of Kazhdan constants for some finite groups*, J. Algebra **163** (1994), 495–515.

17. L. Bassalygo and M. Pinsker, *Complexity of an optimum nonblocking switching network without reconnections*, Prob. Pered. Inform. **9** (1973), no. 1, 84–87.
18. P. Bateman and H. Diamond, *Analytic number theory, an introductory course*, World Scientific, 2004.
19. B. Bekka, P. de la Harpe, and A. Valette, *Kazhdan's property (T)*, New Mathematical Monographs, vol. 11, Cambridge University Press, 2008.
20. M. Bellare, O. Goldreich, and S. Goldwasser, *Randomness in interactive proofs*, Comput. Complexity **3** (1993), no. 4, 319–54.
21. A. Benjamin, B. Chen, and K. Tucker, *Sums of evenly spaced binomial coefficients*, Mathematics Magazine, December 2010, pp. 370–73.
22. F. Bien, *Constructions of telephone networks by group representations*, Notices A.M.S. **36** (1989), no. 1, 5–22.
23. N. Biggs, *Algebraic graph theory*, 2nd ed., Cambridge University Press, 1993.
24. Y. Bilu and N. Linial, *Lifts, discrepancy and nearly optimal spectral gap*, Combinatorica **26** (2006), no. 5, 495–519.
25. B. Bollobás, *The isoperimetric number of random regular graphs*, Eur. J. Combin. **9** (1988), no. 3, 241–44.
26. ———, *Random graphs*, 2nd ed., Cambridge University Press, 2001.
27. R. C. Bose and R. J. Nelson, *A sorting problem*, J. Assoc. Comput. Mach. **9** (1962), 282–96.
28. J. Bourgain and A. Gamburd, *Uniform expansion bounds for Cayley graphs of* $\mathrm{SL}_2(\mathbb{F}_p)$, Ann. of Math. **167** (2008), no. 2, 625–42.
29. Jean Bourgain and Alex Gamburd, *Expansion and random walks in* $\mathrm{SL}_d(\mathbb{Z}/p^n\mathbb{Z})$. *I*, J. Eur. Math. Soc. (JEMS) **10** (2008), no. 4, 987–1011.
30. A. Broder and E. Shamir, *On the second eigenvalue of random regular graphs*, 18th Annual Symposium on Foundations of Computer Science (1987), 286–94.
31. R. Brooks, P. Perry, and P. Petersen, *On Cheeger's inequality*, Comm. Math Helv. **68** (1993), 599–621.
32. M. Burger, *Kazhdan constants for* $\mathrm{SL}(3, \mathbf{Z})$, J. Reine Angew. Math. **413** (1991), 36–67.
33. P. Buser, *On the bipartition of graphs*, Discrete Appl. Math. **9** (1984), no. 1, 105–9.
34. D. Cantor, *On nonblocking switching networks*, Networks **1** (1971).
35. T. Ceccherini-Silberstein, F. Scarabotti, and Filippo Tolli, *Weighted expanders and the anisotropic Alon-Boppana theorem*, Eur. J. Combin. **25** (2004), no. 5, 735–44.
36. D. Charles, K. Lauter, and E. Goren, *Cryptographic hash functions from expander graphs*, J. Cryptology **22** (2009), no. 1, 93–113.
37. P. Chiu, *Cubic Ramanujan graphs*, Combinatrica **12** (1992), 275–85.
38. F. R. K. Chung, *On concentrators, superconcentrators, generalizers, and nonblocking networks*, Bell System Tech. J. **58** (1979), no. 8, 1765–77.
39. ———, *Diameters and eigenvalues*, J. Am. Math. Soc. **2** (1989), no. 2, 187–96.
40. ———, *Spectral graph theory*, Regional Conference Series in Mathematics, no. 92, American Mathematical Society, 1997.
41. S. Cioabă, *Closed walks and eigenvalues of abelian Cayley graphs*, Comptes Rend. Math. **342** (2006), no. 9, 635–38.
42. ———, *Eigenvalues of graphs and a simple proof of a theorem of Greenberg*, Lin. Alg. Appl. **416** (2006), no. 2–3, 776–82.
43. ———, *On the extreme eigenvalues of regular graphs*, J. Combin. Theory Ser. B **96** (2006), no. 3, 367–73.

44. S. Cioabă and M. Ram Murty, *Expander graphs and gaps between primes*, Forum Math. **20** (2008), no. 4, 745–56.
45. G. Davidoff, P. Sarnak, and A. Valette, *Elementary number theory, group theory, and Ramanujan graphs*, London Mathematical Society, Student Texts, no. 55, Cambridge University Press, 2003.
46. M. Davis, R. Sigal, and E. Weyuker, *Computability, complexity, and languages: Fundamentals of theoretical computer science*, 2nd ed., Computer Science and Scientific Computing, Academic Press, 1994.
47. M. Dedeo, D. Lanphier, and M. Minei, *The spectrum of platonic graphs over finite fields*, Discrete Math. **307** (2007), 1074–81.
48. J. Derbidge, *Kazhdan constants of cyclic groups*, master's thesis, California State University, Los Angeles, 2010.
49. P. Diaconis, *Group representations in probability and statistics*, Institute of Mathematical Statistics Lecture Notes—Monograph Series, 11, Institute of Mathematical Statistics, 1988.
50. J. Dodziuk, *Difference equations, isoperimetric inequality and transience of certain random walks*, Trans. Amer. Math. Soc. **284** (1984), no. 2, 787–94.
51. K. Dsouza, *A combinatorial trace method to assay graph eigenvalues*, master's thesis, California State University, 2010.
52. K. Dsouza and M. Krebs, *A combinatorial trace method: Counting closed paths to assay graph eigenvalues*, to appear in Rocky Mountain J. Math.
53. P. Erdős, *Graph theory and probability*, Can. J. Math. **11** (1959), 34–38.
54. P. Erdős, R. L. Graham, and E. Szemeredi, *On sparse graphs with dense long paths*, Computers and mathematics with applications, Pergamon, 1976, pp. 365–69.
55. R. Feng and J. Kwak, *Zeta functions of graph bundles*, J. Korean Math. Soc. **43** (2006), no. 6, 1269–87.
56. S. Friedberg, A. Insel, and L. Spence, *Linear algebra*, 3rd ed., Prentice Hall, 1997.
57. J. Friedman, *On the second eigenvalue and random walks in random d-regular graphs*, Combinatorica **11** (1991), 331–62.
58. ———, *Relative expanders or weakly relatively Ramanujan graphs*, Duke Math. J. **118** (2003), no. 1, 19–35.
59. ———, *A proof of Alon's second eigenvalue conjecture and related problems*, Mem. Am. Math. Soc. **195** (2008), no. 910.
60. J. Friedman, R. Murty, and J. Tillich, *Spectral estimates for abelian Cayley graphs*, J. Combin. Theory, Series B **96** (2006), 111–21.
61. J. Friedman and J. Tillich, *Generalized Alon-Boppana theorems and error-correcting codes*, SIAM J. Discrete Math. **19** (2005), no. 3, 700–718 (electronic).
62. W. Fulton and J. Harris, *Representation theory: A first course*, Graduate Texts in Mathematics, no. 129, Springer, 1991.
63. O. Gabber and Z. Galil, *Explicit constructions of linear size superconcentrators*, 20th Annual Symposium on Foundations of Computer Science (San Juan, Puerto Rico, 1979), IEEE, 1979, pp. 364–70.
64. ———, *Explicit constructions of linear-sized superconcentrators*, J. Comp. Sys. Sci. **22** (1981), no. 3, 407–20.
65. T. Gelander and A. Żuk, *Dependence of Kazhdan constants on generating subsets*, Israel J. Math. **129** (2002), 93–98.
66. C. Godsil and G. Royle, *Algebraic graph theory*, Graduate Texts in Mathematics, no. 207, Springer, 2001.

67. Y. Greenberg, *Spectra of graphs and their covering trees,* Ph.D. dissertation, Hebrew University of Jerusalem, 1995 [in Hebrew].
68. M. Gromov, *Groups of polynomial growth and expanding maps,* Inst. Hautes Études Sci. Publ. Math. (1981), no. 53, 53–73.
69. S. Hoory, *A lower bound on the spectral radius of the universal cover of a graph,* J. Combin. Theory Ser. B **93** (2005), no. 1, 33–43.
70. S. Hoory, N. Linial, and A. Wigderson, *Expander graphs and their applications,* Bull. Amer. Math. Soc. **43** (2006), no. 4, 439–561.
71. M. Horton, H. M. Stark, and A. Terras, *Zeta functions of weighted graphs and covering graphs, Analysis on graphs and its applications,* Proc. Sympos. Pure Math., vol. 77, American Mathematics Society, pp. 29–50.
72. K. Ireland and M. Rosen, *A classical introduction to modern number theory,* 2nd ed., Graduate Texts in Mathematics, no. 84, Springer, 1990.
73. Dmitry Jakobson, Stephen D. Miller, Igor Rivin, and Zeév Rudnick, *Eigenvalue spacings for regular graphs,* Emerging applications of number theory (Minneapolis, MN, 1996), IMA Vol. Math. Appl., vol. 109, Springer, 1999, pp. 317–27.
74. S. Jimbo and A. Maruoka, *Expanders obtained from affine transformations,* Combinatorica **7** (1987), no. 4, 343–55.
75. M. Kassabov, *Kazhdan constants for* $\mathrm{SL}_n(\mathbb{Z})$, Internat. J. Algebra Comput. **15** (2005), no. 5–6, 971–95.
76. ———, *Symmetric groups and expander graphs,* Invent. Math. **170** (2007), no. 2, 327–54.
77. M. Kassabov, A. Lubotzky, and N. Nikolov, *Finite simple groups as expanders,* Proc. Natl. Acad. Sci. USA **103** (2006), no. 16, 6116–19.
78. D. A. Každan, *On the connection of the dual space of a group with the structure of its closed subgroups,* Funkcional. Anal. i Priložen. **1** (1967), 71–74.
79. M. Klawe, *Limitations on explicit constructions of expanding graphs,* SIAM J. Comput. **13** (1984), no. 1, 156–66.
80. M. Kotani and T. Sunada, *Zeta functions of finite graphs,* J. Math. Sci. (Tokyo) **7** (2000), no. 1, 7–25.
81. M. Krebs and A. Shaheen, *On the spectra of Johnson graphs,* Elec. J. Lin. Alg. **17** (2008), 154–67.
82. Z. Landau and A. Russell, *Random Cayley graphs are expanders:a simple proof of the Alon-Roichman theorem,* Elec. J. Combin. **11** (2004), no. R62.
83. D. Lanphier and J. Rosenhouse, *Cheeger constants of Platonic graphs,* Discrete Mathematics, Volume 277, pp. 101–13.
84. W. Ledermann, *Introduction to group characters,* Cambridge University Press, 1977.
85. N. Linial and D. Puder, *Words maps and spectra of random graph lifts,* http://www.citebase.org/abstract?id=oai:arXiv.org:0806.1993, 2008.
86. P. Loh and L. Schulman, *Improved expansion of random Cayley graphs,* Discrete Math. Theor. Comp. Sci. **6** (2004), 523–28.
87. A. Lubotzky, *Discrete groups, expanding graphs, and invariant measures,* Progress in Mathematics, vol. 125, Birkhauser Verlag, 1994.
88. ———, *Cayley graphs: Eigenvalues, expanders and random walks,* Surveys in combinatorics, 1995 (Stirling), London Math. Soc. Lecture Note Ser., vol. 218, Cambridge University Press, 1995, pp. 155–89.
89. A. Lubotzky, R. Phillips, and P. Sarnak, *Ramanujan graphs,* Combinatorica **8** (1988), no. 3, 261–77.

90. A. Lubotzky and B. Weiss, *Groups and expanders*, Expanding Graphs, DIMACS Series in Discrete Mathematics and Theoretical Computer Science, vol. 10, American Mathematical Society, 1993, pp. 95–109.
91. E. Lui, *The phi = beta conjecture and eigenvalues of random graph lifts*, http://www.citebase.org/abstract?id=oai:arXiv.org:0909.1231, 2009.
92. G. A. Margulis, *Explicit constructions of expanders*, Prob. Pered. Inform. **9** (1973), no. 4, 71–80.
93. ———, *Explicit constructions of concentrators*, Prob. Inform. Trans. **9** (1975), 325–32.
94. ———, *Explicit group-theoretical constructions of combinatorial schemes and their application to the design of expanders and concentrators*, Prob. Inform. Trans. **24** (1988), no. 1, 39–46.
95. G. M. Masson and B. W. Jordan Jr., *Generalized multi-stage connection networks*, Networks **2** (1972), 191–209.
96. R. Meshulam and A. Wigderson, *Expanders in group algebras*, Combinatorica **24** (2004), no. 4, 659–80.
97. S. Miller and T. Novikoff, *The distribution of the largest nontrivial eigenvalues in families of random regular graphs*, Experiment. Math. **17** (2008), no. 2, 231–44.
98. H. Mizuno and I. Sato, *Weighted Bartholdi zeta functions of some graphs*, Far East J. Math. Sci. **27** (2007), no. 2, 301–21.
99. B. Mohar, *Isoperimetric numbers of graphs*, J. Comb. Theory, Series B **47** (1989), 274–91.
100. M. Morgenstern, *Existence and explicit constructions of $q+1$ regular Ramanujan graphs for every prime power q*, J. Comb. Theory, Ser. B **62** (1994), 44–62.
101. M. Ram Murty, *Ramanujan graphs*, J. Ramanujan Math. Soc. **18** (2003), no. 1, 33–52.
102. M. Nathanson, *Elementary methods in number theory*, Graduate Texts in Mathematics, no. 195, Springer, 2000.
103. A. Nilli, *On the second eigenvalue of a graph*, Discrete Math. (1991), no. 91, 207–10.
104. I. Niven, H. Zuckerman, and H. Montgomery, *An introduction to the theory of numbers*, 5th ed., Wiley, 1991.
105. J. Ofman, *A universal automaton*, Trans. Moscow Math. Soc. **14** (1965), 200–215.
106. D. V. Osin, *Kazhdan constants of hyperbolic groups*, Funktsional. Anal. i Prilozhen. **36** (2002), no. 4, 46–54.
107. I. Pak and A. Żuk, *On Kazhdan constants and mixing of random walks*, Int. Math. Res. Not. (2002), no. 36, 1891–905.
108. M. Pinsker, *On the complexity of a concentrator*, 7th International Teletraffic Conference, Stockholm, June 1973, pp. 318/1–4.
109. N. Pippenger, *Superconcentrators*, SIAM J. Comput. **6** (1977), no. 2, 298–304.
110. ———, *Generalized connectors*, SIAM J. Comput. 7 (1978), 510–14.
111. ———, *Sorting and selecting in rounds*, SIAM J. Comput. **16** (1987), no. 6, 1032–38.
112. A. Pizer, *Ramanujan graphs and Hecke operators*, Bull. Amer. Math. Soc. (N.S.) **23** (1990), no. 1, 127–37.
113. F. Preparata and J. Vuillemin, *The cube-connected cycles: A versatile network for parallel computation*, Comm. ACM **24** (1981), no. 5, 300–9.
114. R. Read and R. Wilson, *An atlas of graphs*, Oxford Science Publications, Clarendon Press Oxford University Press, 1998.
115. R. Reeds, *Zeta functions on kronecker products of graphs*, Rose-Hulman Und. Math. J. 7 (2006), no. 1.

116. O. Reingold, S. Vadhan, and A. Wigderson, *Entropy waves, the zig-zag graph product, and new constant-degree expanders*, Ann. Math. **155** (2002), no. 1, 157–87.
117. D. Robinson, *A course in linear algebra with applications*, 2nd ed., World Scientific Publishing, 2006.
118. Y. Roichman, *Expansion properties of Cayley graphs of the alternating group*, J. Comb. Theory, Series A **79** (1997), no. 2, 281–97.
119. J. Rosenhouse, *Isoperimetric numbers of Cayley graphs arising from generalized dihedral groups*, J. Comb. Math. Comb. Comput. **42** (2002), 127–38.
120. E. Rozenman, A. Shalev, and A. Wigderson, *Iterative construction of Cayley expander graphs*, Theory Comput. **2** (2006), 91–120.
121. B. Sagan, *The symmetric group: Representations, combinatorial algorithms, and symmetric functions*, 2nd ed., Graduate Texts in Mathematics, no. 203, Springer, 2001.
122. P. Sarnak, *What is an expander?*, Notices Amer. Math. Soc. **51** (2004), no. 7, 762–63.
123. I. Sato, *Weighted Bartholdi zeta functions of graph coverings*, Discrete Math. **308** (2008), no. 12, 2600–6.
124. H. Schellwat, *Highly expanding graphs obtained from dihedral groups*, Expanding Graphs, DIMACS Series in Discrete Mathematics and Theoretical Computer Science, vol. 10, American Mathematical Society, 1993, pp. 117–23.
125. Á. Seress, *Permutation group algorithms*, Cambridge Tracts in Mathematics, vol. 152, Cambridge University Press, 2003.
126. J. Serre, *Linear representations of finite groups*, Graduate Texts in Mathematics, no. 42, Springer, 1977.
127. Y. Shalom, *Bounded generation and Kazhdan's property (T)*, Inst. Hautes Études Sci. Publ. Math. (1999), no. 90, 145–68.
128. ———, *Explicit Kazhdan constants for representations of semisimple and arithmetic groups*, Ann. Inst. Fourier (Grenoble) **50** (2000), no. 3, 833–63.
129. A. Terras, *Fourier analysis on finite groups and applications*, London Mathematical Society Student Texts, vol. 43, Cambridge University Press, 1999.
130. A. Terras and H. Stark, *Zeta functions of finite graphs and coverings*, Adv. Math. **121** (1996), 124–65.
131. ———, *Zeta functions of finite graphs and coverings, part ii*, Adv. Math. **154** (2000), 132–95.
132. ———, *Zeta functions of finite graphs and coverings, part iii*, Adv. Math. **208** (2007), 467–89.
133. D. Xiao, *The evolution of expander graphs*, Harvard honor's thesis, http://www.hcs.harvard.edu/thesis/repo/32/, April 2003.
134. A. Żuk, *Property (T) and Kazhdan constants for discrete groups*, Geom. Funct. Anal. **13** (2003), no. 3, 643–70.

Index

www.ingramcontent.com/pod-product-compliance
Ingram Content Group UK Ltd.
Pitfield, Milton Keynes, MK11 3LW, UK
UKHW010114180926
475524UK00002B/14